I0055814

TOPOLOGY WITH APPLICATIONS

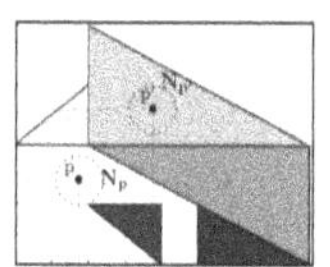

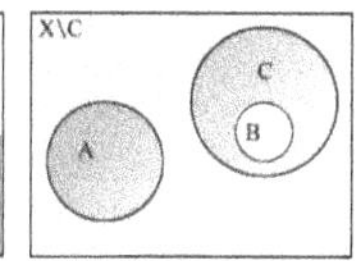

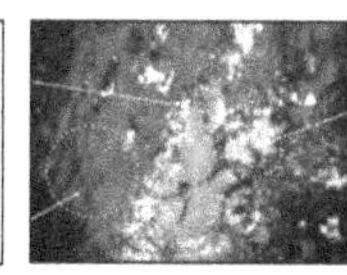

Topological Spaces via Near and Far

TOPOLOGY WITH APPLICATIONS

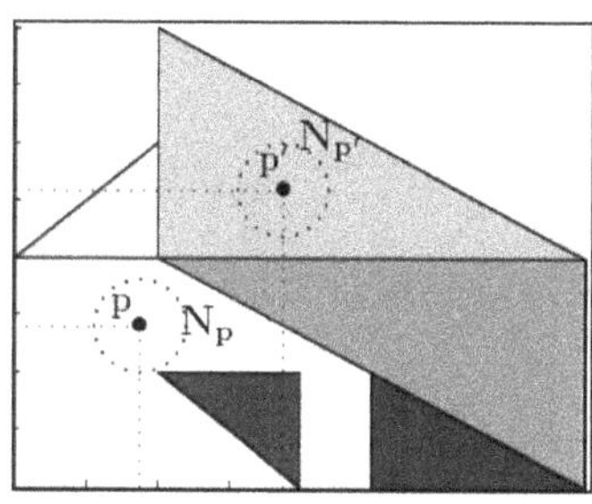

Topological Spaces via Near and Far

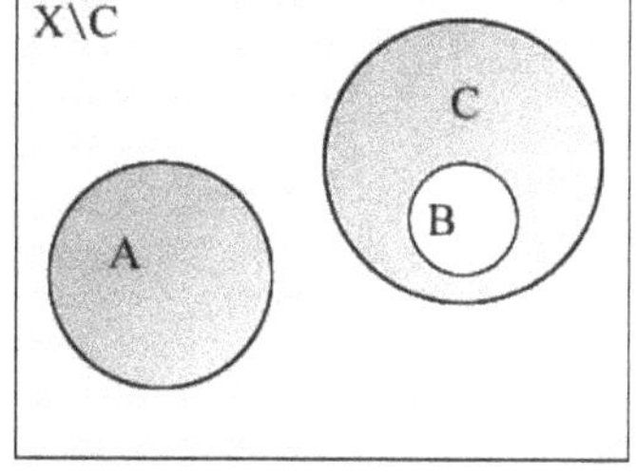

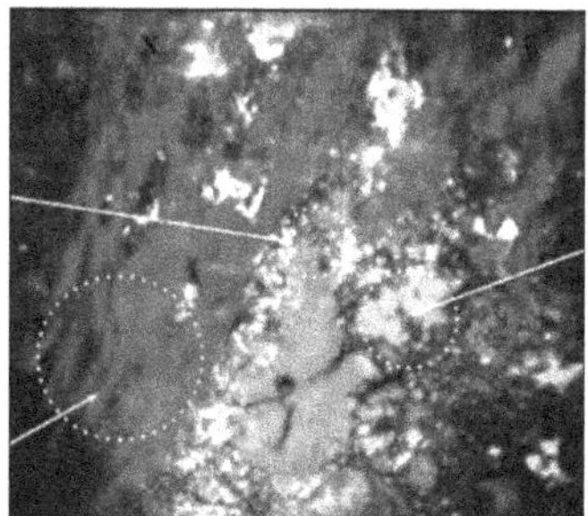

Somashekhar A. Naimpally
Lakehead University, Canada

James F. Peters
University of Manitoba, Canada

World Scientific

NEW JERSEY • LONDON • SINGAPORE • BEIJING • SHANGHAI • HONG KONG • TAIPEI • CHENNAI

Published by

World Scientific Publishing Co. Pte. Ltd.
5 Toh Tuck Link, Singapore 596224
USA office: 27 Warren Street, Suite 401-402, Hackensack, NJ 07601
UK office: 57 Shelton Street, Covent Garden, London WC2H 9HE

British Library Cataloguing-in-Publication Data
A catalogue record for this book is available from the British Library.

TOPOLOGY WITH APPLICATIONS
Topological Spaces via Near and Far

ISBN 978-981-4407-65-6

Printed in Singapore by Mainland Press Pte Ltd.

We dedicate this book to
Sudha and Sheela and Saroja

Foreword

Topological ideas originated in the 19th century, mainly in the works of Riemann and Poincaré, and finally led to the classical definitions of topological space either via the closure operator by Kuratowski, or via open sets. The latter one is the most popular now. However, such progress was preceded by many other attempts and approaches which appear to be useful in different cases and worthwhile to be developed.

One such approach is based on the notion of near sets introduced by F. Riesz in 1908. A set X with a nearness relation between its subsets is called a proximity space and every such structure induces a topology on X defined via the closure operator: we say that a point x lies in the closure of a subset U if the subset $\{x\}$ is near U. It appears that the same topology on X may correspond in this way to different proximities.

Moreover many topological results may be inherited from statements concerning proximity spaces. It has to be recalled that in the same manner the proximity structure is induced by the uniform relation introduced by A. Weil.

The proximity was rediscovered by V.A. Efremovič in the middle of 1930s and later A.D. Wallace arrived at the same concept. Initially, the proximity theory has developed by its own means with strong contributions by Efremovič and his school and the first author of this book, S.A. Naimpally.

Recently, the proximity theory found valuable applications which strongly justify their developments. This book was preceded by a short excellent survey[1] by the same authors of the main ideas of near sets, which

[1]Peters, Jim; Naimpally, Som. Applications of near sets. Notices Amer. Math. Soc. 59 (2012), no. 4, 536–542.

have important applications in image processing, image analysis, face recognition, and some other engineering and natural science problems.

In a sense, the way in which this theory came from its mathematical origin to applications is quite opposite to the development of fuzzy sets arising from applications and leading to a branch of modern set theory.

The targets of this book are three-fold. First, it is to expose the main constructions and ideas of set-theoretical topology through the magnifying glass of proximity and uniform spaces. Such an introduction to topology strongly benefits from the simple as well as rigorous notions of near and far at various levels. The motivation originates in advanced calculus and goes to metric spaces, topology, proximity, and uniformity. Diverse results are unified, as for example, in the proximal avatar of the result by A.D. Taimanov on extensions of continuous functions from dense subsets.

Second, this book gives an introduction to descriptive proximity rising from extensions of the usual spatial nearness relations in terms of descriptively near sets. This leads to many important results that are fundamental building blocks in understanding topology and its implications.

Third, this book demonstrates the utility of the notions of near and far in many diverse applications, ranging from cell biology and micropalaeontology to camouflage and forgery detection. The applications are made incisive by the accompanying use of near and far in bringing to light the subtleties typically hidden in physical structures such as the distinctively different shapes of microfossils and typical camouflage of animals in natural surroundings. Of particular interest in this book is the range of applications of the proposed approach to topology in study of climate change, mineral and fossil fuel exploration.

We strongly recommend this book as a concise and very original introduction of set-theoretical topology, nearness theory and their modern applications.

October 2012

Professor Iskander A. Taimanov, D. Sc.,
Member of Russian Academy of Sciences,
Chair of Geometry and Topology,
Novosibirsk State University, Russia

Preface

The main purpose in writing this book is to demonstrate the beneficial use of *near* and *far*, discovered by F. Riesz over a century ago, from the undergraduate to the research level in general topology and its applications. Use of *near* and *far* is intuitive yet rigourous at the same time, which is rare in mathematics. The *near* and *far* paradigm is based on many years of teaching and research by the authors. The book introduces topology and its many applications viewed within a framework that begins with metric spaces and deals with the usual topics such as continuity, open and closed sets, metric nearness, compactness and completeness and glides into topology, proximity and uniformity. Most topics are first studied in metric spaces and later in a topological space. The motivation for this approach is a straightforward, intuitive and yet rigourous rendition of topological concepts. The approach also unifies several scattered results in many areas. Many exercises come from the current literature and some occur in simplified form in metric spaces. The end result is a solid, workable framework for the study of topology with a variety of applications in science and engineering that include camouflage, cell biology, digital image analysis, forgery detection, general relativity, microscopy, micropalaeontology, pattern recognition, population dynamics, psychology and visual merchandising.

We gratefully acknowledge the insightful Foreword by I.A. Taimanov. This Foreword is especially important, since it lucidly brings together the principal highlights of this book and it serves as a commemoration of the seminal work on topology by A.D. Taimanov, who proved one of the most fundamental results in topology concerning extensions of continuous functions from dense subspaces.

Contents

Chapter 1

Basic Framework

> The specific attraction and in large part the significance of topology lies in the fact that its most important questions and theorems have an immediate intuitive content and thus teach us in a direct way about space, which appears as the place in which continuous processes occur.
>
> –Paul Alexandroff, 1932

1.1 Preliminaries

The framework for topology begins with an introduction to metric spaces. The motivation for this approach is that metric spaces are immediate generalizations of real and complex numbers and lie between them and topological spaces. Thus, we start with an intermediate step. Metric spaces have more structure than topological spaces and they provide stepping stones to a variety of important topics in topology. Every metric space is defined in terms of a metric that is a distance function. A metric space is a set with a structure determined by the properties of a metric. Perhaps the most appealing induced structure is the distance between a point and a set.

In analysis, the setting is the space of real numbers $\mathbb{R}$, the Euclidean space $\mathbb{R}^m$ or the space of complex numbers $\mathbb{C}$. The basic concept in analysis is *limit*, which depends on the distance $|x-y| \geq 0$ between two points x and y, where the distance arises from the norm $\| x - y \|_p$. In general, the L_p norm $\| \cdot \|_p = \left(\sum_{i=1}^{m} \cdot_i^{p}\right)^{\frac{1}{p}}$.

Moreover, all other concepts and results depend directly or indirectly upon limit, *e.g.*, continuity, uniform continuity, compactness, and connectedness. If X is a subset of any one of the sets $\mathbb{R}, \mathbb{R}^m, \mathbb{C}$, the distance between two points of X is non-negative, symmetric and satisfies the triangle inequality. Precisely, distance satisfies the following conditions for all $x, y, z \in X$.

d.1 $|x - y| \geq 0$,
d.2 $|x - y| = 0 \iff x = y$,
d.3 $|x - y| = |y - x|$ (symmetry),
d.4 $|x - z| \leq |x - y| + |y - z|$ (triangle inequality).

1.2 Metric Space

In 1906, M. Fréchet observed that given any non-empty set X, a distance function $d : X \times X \to \mathbb{R}$ can be defined and called it a metric. Thus was born the concept of a metric space. The pair (X, d) is called a **metric space**, where X is a non-empty set and d is a real-valued function on $X \times X$ called a *metric*, satisfying the following conditions for all $x, y, z \in X$ that are analogous to conditions d.1 to d.4.

M.1 $d(x, y) \geq 0$,
M.2 $d(x, y) = 0$, if and only if, $x = y$,
M.3 $d(x, y) = d(y, x)$ (symmetry),
M.4 $d(x, z) \leq d(x, y) + d(y, z)$ (triangle inequality).

Obviously, real numbers $\mathbb{R}$, Euclidean space $\mathbb{R}^m$ and complex numbers $\mathbb{C}$ with the usual distance are well-known examples of metric spaces. A trivial example results from defining the distance $d_{01}(x, y)$ to be zero if $x = y$ and 1, if $x \neq y$.
The metric d_{01} is called the discrete or 0-1 metric, which is useful as a counterexample.

Every normed vector space results in a metric space. The norm of a vector is a measure of its length. For example, consider the vector $\boldsymbol{x} = (x_1, \ldots, x_m)$ in $\mathbb{R}^m$. The norm $\| \boldsymbol{x} \|_2$ of $\boldsymbol{x}$ is defined by

$$\| \boldsymbol{x} \|_2 = \sqrt{x_1^2 + \cdots + x_m^2}.$$

Consider, then, vectors $\boldsymbol{x}, \boldsymbol{y}$ in $\mathbb{R}^m$ and define the Euclidean metric

$$d_2 : \mathbb{R}^m \times \mathbb{R}^m \to \mathbb{R},$$

where

$$d_2(\boldsymbol{x}, \boldsymbol{y}) = \| \boldsymbol{x} - \boldsymbol{y} \|_2 = \sqrt{\sum_{i=1}^{m} (x_i - y_i)^2}.$$

1.3 Gap Functional and Closure of a Set

The study of metric spaces depends on the fundamental concept of the limit point of a set that can be described in terms of nearness of a point to a set, first suggested by F. Riesz in 1908. First, we consider a more general nearness of two sets. Let d be a metric on a nonempty set X.

The nearness of sets is formalized with the **gap functional** $D(A, B)$, where,

$$D(A, B) = \begin{cases} inf\{d(a, b) : a \in A, b \in B\}, & \text{if } A \text{ and } B \text{ are not empty,} \\ \infty, & \text{if } A \text{ or } B \text{ is empty.} \end{cases}$$

When, for example, A is the singleton set $\{x\}$, we write x for $\{x\}$. For

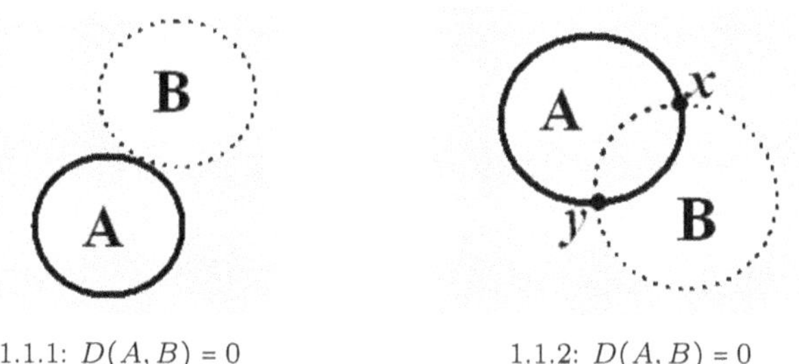

1.1.1: $D(A, B) = 0$ 1.1.2: $D(A, B) = 0$

Fig. 1.1 Sample near sets

$D(x, B) = 0$, the point x is said to be near B and called the **closure point** of B.

> **Near Sets**. Two sets are near each other, if and only if, the gap between them is zero, *i.e.*,
> (N1) there is at least one point near both sets, *or*
> (N2) the sets are asymptotically near each other.

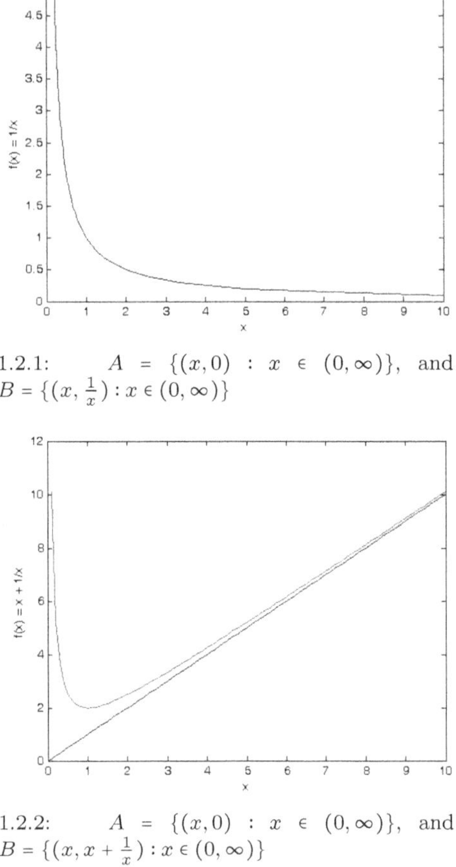

1.2.1: $A = \{(x, 0) : x \in (0, \infty)\}$, and $B = \{(x, \frac{1}{x}) : x \in (0, \infty)\}$

1.2.2: $A = \{(x, 0) : x \in (0, \infty)\}$, and $B = \{(x, x + \frac{1}{x}) : x \in (0, \infty)\}$

Fig. 1.2 Asymptotically near sets

The near set condition (N1) is illustrated in Fig. 1.1.1 and Fig. 1.1.2. In Fig. 1.1.1, the sets A and B (represented by the discs shown) have a common point. A and B in Fig. 1.1.2 are near each other, since $D(A, B) = 0$ for a number of points such as $\boldsymbol{x}, \boldsymbol{y}$. Recall that the asymptote of a curve is a line such that the distance between the curve and the line approaches zero as they tend toward infinity. This is the situation in Fig. 1.2.1, where set $A = \{(x, 0) : x \in (0, \infty)\}$ is asymptotically near $B = \left\{\left(x, \frac{1}{x}\right) : x \in (0, \infty)\right\}$ so that A and B satisfy near set condition (N2). Similarly, in Fig. 1.2.2, $A = \{(x, 0) : x \in (0, \infty)\}$ and $B = \left\{\left(x, x + \frac{1}{x}\right) : x \in (0, \infty)\right\}$ are asymptotically near sets.

In effect, sets A and B are **near sets** (denoted by $A\ \delta\ B$), if and only if, $A\ \delta\ B$, *i.e.*, the gap distance is zero. Notice that the gap distance $D(A, B)$ is always greater than or equal to zero (see Problem 1.21.7). In sum,

> A is near B, if and only if, $D(A, B) = 0$.

The **closure of a subset** $B \subset X$ (denoted $\mathrm{cl}\, B$) is the set of all points of X that are near B, *i.e.*,

$$\mathrm{cl}\, B = \{x \in X : D(x, B) = 0\}.$$

Here are some examples.

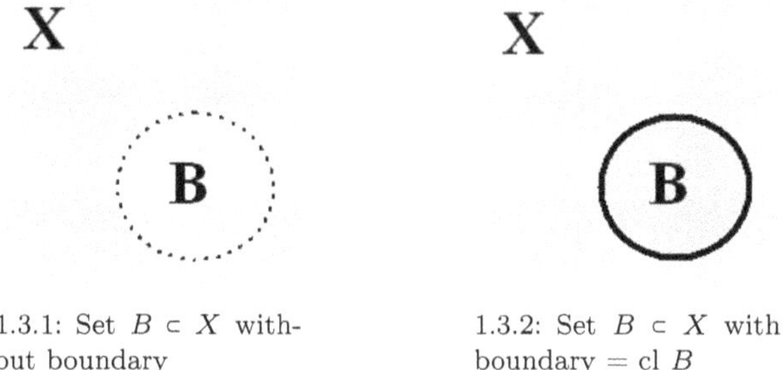

1.3.1: Set $B \subset X$ without boundary

1.3.2: Set $B \subset X$ with boundary $= \mathrm{cl}\ B$

Fig. 1.3 Sample closure of a subset

(1) A subset $B \subset X$ is shown in Fig. 1.3. The subset $B \subset X$ with only its interior points (*i.e.*, without its boundary) is represented by the circle with dotted edge in Fig. 1.3.1. The closure of subset $B \subset X$ includes the boundary of B represented by a solid edge of the circle in Fig. 1.3.2.
(2) For an open interval (a, b) in $\mathbb{R}$, $\mathrm{cl}\ (a, b) = [a, b]$.
(3) For an open interval (a, ∞) in $\mathbb{R}$, $\mathrm{cl}\ (a, \infty) = [a, \infty)$.

> $\{x\}\ \delta\ B$ denotes the fact that point x is near set B.

A consideration of nearness of a point to a set leads to

Theorem 1.1. *Suppose X is a non-empty set. For each point $x \in X$ and subsets $B, C \subset X$, we have*
K1 $x\ \delta\ B \Rightarrow B \neq \emptyset$,
K2 $\{x\} \cap B \neq \emptyset \Rightarrow x\ \delta\ B$,
K3 $x\ \delta\ (B \cup C) \Leftrightarrow x\ \delta\ B$ *or* $x\ \delta\ C$,
K4 $x\ \delta\ B$ *and* $b\ \delta\ C$ *for each* $b \in B \Rightarrow x\ \delta\ C$.

Proof.
K3:
K3 follows from the fact that $D(x, B \cup C) = min\{D(x,B), D(x,C)\}$. So x is near $B \cup C \iff x$ is near B or x is near C.
K4:
Suppose $x\ \delta\ B$ and $\varepsilon > 0$ is arbitrary. Since x is near B, there is a $b \in B$ such that $d(x,b) < \frac{\varepsilon}{2}$. Since b is near C, there is a $c \in C$ such that $d(b,c) < \frac{\varepsilon}{2}$. Hence, $d(x,c) \leq d(x,b) + d(b,c) < \frac{\varepsilon}{2} + \frac{\varepsilon}{2} = \varepsilon$, showing thereby that $D(x,C) = 0$, *i.e.*, $x\ \delta\ C$. □

In every metric space (X,d), there is a **Kuratowski operator** induced by the metric d, where x is a closure point of a set $B \subset X$, if and only if, x is near B or $D(x,B) = 0$. Let cl denote a closure operator on a set B. The properties (K1-K4) in Theorem 1.1 can be rewritten in the following way.
cl.1 $\mathrm{cl}\emptyset = \emptyset$,
cl.2 $B \subset \mathrm{cl}\,B$,
cl.3 $\mathrm{cl}(B \cup C) = \mathrm{cl}\,B \cup \mathrm{cl}\,C$,
cl.4 $\mathrm{cl}(\mathrm{cl}\,B) = \mathrm{cl}\,B$.

A point p is a **limit point** of a set $B \subset X$, if and only if, p is near $B - \{p\}$ or $D(p, B - \{p\}) = 0$. The set of limit points of a set B is denoted by B' and is called the *derived* set of B. The closure operator on a set B is denoted either by $\mathrm{cl}\,B$ or by $\bar{B}$. It is easy to see that the $\mathrm{cl}\ B = B \cup B'$.

1.4 Limit of a Sequence

A sequence in a set X is a function $f : \mathbb{N} \to X$, where $\mathbb{N}$ is the set of natural numbers. It is common to denote a sequence either by $\{x_n \mid n \in \mathbb{N}\}$ or, more concisely, by $\{x_n\}$, where $x_n = f(n)$ with $n \in \mathbb{N}$. A sequence $\{x_n\}$ in a metric space (X,d) is said to **converge to a limit** $\mathcal{L}$, if and only if, for every infinite subset $M \subset \mathbb{N}$, $\mathcal{L}$ is near $f(M)$, *i.e.*, $D(f(M), \mathcal{L}) = 0$.

$x_n \to \mathcal{L}$, if and only if, for every infinite subset $M \subset \mathbb{N}, \mathcal{L}$ is near $f(M)$.

Often the word *eventually* is used for 'there is an $m \in \mathbb{N}$ such that for all $n > m$'. Convergence of a sequence x_n to a limit $\mathcal{L}$ is equivalent to the usual definition in calculus and analysis texts, namely,

$$\text{for all } \varepsilon > 0, \text{ eventually, } D(x_n, \mathcal{L}) < \varepsilon.$$

For example, consider the convergence of $x_n = f(n)$ in Fig. 1.4, where

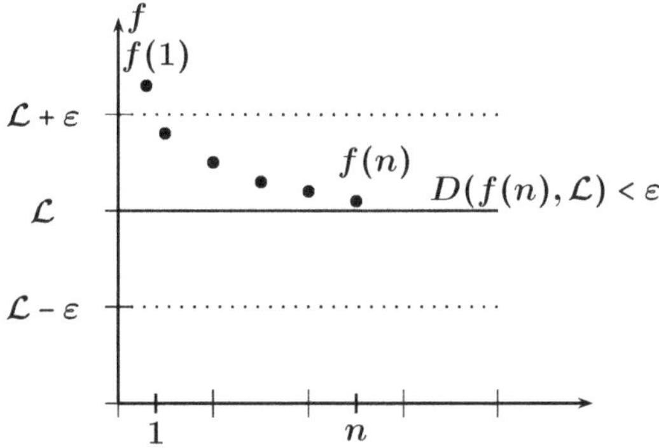

Fig. 1.4 Geometric view of convergence: *eventually* $D(f(n), \mathcal{L}) < \varepsilon$

eventually $D(f(n), \mathcal{L}) < \varepsilon$. In other words, x_n converges to $\mathcal{L}$, if and only if, the sequence $\{D(x_n, \mathcal{L})\}$ of real numbers converges to 0. It is interesting to note that the definition of the limit of a sequence in a metric space relies on the space $\mathbb{R}$ for its definition.

Theorem 1.2. *Put $n \in \mathbb{N}$ and let $x_n = f(n)$ denote a sequence in a metric space (X, d). The following assertions are equivalent.*

1.2(a) For every positive number ε, eventually $D(x_n, \mathcal{L}) < \varepsilon$.

1.2(b) For every infinite subset $M \subset \mathbb{N}$, $f(M)$ is near $\mathcal{L}$, i.e., $D(f(M), \mathcal{L}) = 0$.

Proof.

1.2(a) $\Rightarrow$ 1.2(b) Suppose M is an infinite subset of $\mathbb{N}$. For every $\varepsilon > 0$, *eventually* there is an $n \in M$ such that $D(x_n, \mathcal{L}) < \varepsilon$. Hence, $f(M)$ is near $\mathcal{L}$, *i.e.*, $D(f(M), \mathcal{L}) = 0$

1.2(b) $\Rightarrow$ 1.2(a) Assume 1.2(a) is not true, then there is an $\varepsilon > 0$ such that for each $n \in \mathbb{N}$, there is a $p \in \mathbb{N}, p > n$ and $D(x_p, \mathcal{L}) \geq \varepsilon$. The set of all such p's form an infinite set M satisfying $D(x_p, \mathcal{L}) \geq \varepsilon$, *i.e.*, $f(M)$ is not near $\mathcal{L}$, showing that 1.2(b) is false. □

It is a straightforward task to show that the limit of a sequence is *unique*, *i.e.*, no sequence can converge to more than one point (see Problem 1.21.24).

1.5 Continuity

Let (X, d_X), (Y, d_Y) be a pair of metric spaces and let $f : X \to Y$ be a function such that for each x, there is a unique $f(x) \in Y$. Intuitively, the function f is continuous at $p \in X$ provided that p is near x implies $f(p)$ is near $f(x)$. The above statement is not precise and can be made precise by substituting a non-empty set E for x. In other words, f is continuous at p is equivalent to the observation that f preserves the nearness of p to subsets E of X.

> The function f is **continuous at** $p \in X$, if and only if, for each subset E of X, p is near E implies $f(p)$ is near $f(E)$.

The usual ε, δ ('epsilon, delta') definition of continuity in calculus takes on the following avatar in metric spaces. A function f is continuous at $p \in X$, if and only if, for every $\varepsilon > 0$, there is a $\eta > 0$ such that, for every $x \in X$,

$$d_X(p, x) < \eta \text{ implies } d_Y(f(p), f(x)) < \varepsilon.$$

Theorem 1.3 helps take the mystery out of the usual definition of continuity at a point.

Theorem 1.3. *Let (X, d_X), (Y, d_Y) be a pair of metric spaces. The following assertions are equivalent for a function $f : X \to Y$.*

1.3(c.1) *The function* f *is* **continuous at** $p \in X$, i.e., *if for each subset* $E \subset X$, p *is near* E *implies* $f(p)$ *is near* $f(E)$, i.e.,

$$D_X(p, E) = 0 \text{ implies } D_Y(f(p), f(E)) = 0.$$

1.3(c.2) *For every* $\varepsilon > 0$, *there is a* $\eta > 0$ *such that, for every* $x \in X$,

$$d_X(p, x) < \eta \text{ implies } d_Y(f(p), f(x)) < \varepsilon.$$

Proof.

1.3(c.1) $\Rightarrow$ 1.3(c.2)

For $p \in X$ and any $\varepsilon > 0$, put

$$E = \{x \in X : d_Y(f(p), f(x)) \geq \varepsilon)\}.$$

Since $D_Y(f(p), f(E)) \geq \varepsilon$ and f is continuous at p, then $D_X(p, E) \geq \eta$ for some $\eta > 0$. Hence, given $\varepsilon > 0$, there is an $\eta > 0$ such that $d_Y(f(p), f(x)) \geq \varepsilon$ implies the distance $d_X(p, x) \geq \eta$, which is equivalent to 1.3(c.2).

1.3(c.2) $\Rightarrow$ 1.3(c.1)

Let $D_X(p, E) = 0$ and let $\varepsilon > 0$ be arbitrary. Then there is an $\eta > 0$ such that $d_X(p, x) < \eta$ implies $d_Y(f(p), f(x)) < \varepsilon$. Since $D_X(p, E) = 0$, there is $q \in E$ such that $d_X(p, q)$ and, hence, $d_Y(f(p), f(q)) < \varepsilon$. This shows that $D_Y(f(p), f(E)) = 0$. □

A function $f : X \to Y$ that is continuous at each point of X is said to be a function continuous from X to Y or, simply, f is continuous. Here are several examples.

(1) Constant functions are continuous.
(2) The identity function $f : X \to X$, given $f(x) = x$, is continuous.
(3) The composition of two continuous functions is continuous.
(4) If $f : X \to \mathbb{R}$ is a continuous function, then $|f|$ (*i.e.*, $|f|(x) = |f(x)|$) is continuous.

Theorem 1.4. *Let* E *be a non-empty subset of a metric space* (X, d). *The function* $f : X \to \mathbb{R}$ *defined by* $f(x) = D(x, E)$ *is continuous.*

Proof.

Let p be any point in X and assume $D(p, E) = r \geq 0$. If $\varepsilon > 0$, then there exists a $y \in E$ such that $d(p, y) < r + \frac{\varepsilon}{2}$. If $\eta = \frac{\varepsilon}{2}$, then for all $x \in X$, observe that $d(p, x) < \eta$ implies

$$d(x, y) \leq d(x, p) + d(p, y) < \eta + r + \frac{\varepsilon}{2} = r + \varepsilon.$$

This implies that for $d(p, x) < \eta$, $D(x, E) < r + \varepsilon$. In effect, the function f is continuous at p. □

1.6 Open and Closed Sets

Consider a metric space (X, d). In analogy with $\mathbb{R}^m$, the set

$$S_d(p, \varepsilon) = \{x \in X : d(p, x) < \varepsilon\},$$

with $\varepsilon > 0$, is called an **open ball** with centre p and radius ε. When it is clear from the context that an open ball is defined in terms of a metric d, we simply write $S(p, \varepsilon)$. Similarly, the set

$$B_d(p, \varepsilon) = \{x \in X : d(p, x) \le \varepsilon\},$$

with $\varepsilon > 0$, is called a **closed ball** with centre p and radius ε. A subset $E \subset X$ in the metric space (X, d) is termed an **open set**, if and only if, for each point $x \in E$, there is an $\varepsilon > 0$ such that the open ball $S(x, \varepsilon) \subset E$. This observation is equivalent to each of the following statements.

Open.1 A set E is open, if and only if, each point of E is far from E^c (complement of E),

Open.2 An open set E is a union of open balls.

The **interior** of a set $A \subset X$ is defined as the union of all open sets contained in A. The interior of A is denoted by int A or $\mathring{A}$.

A subset F is a **closed set**, if and only if, its complement is open. This is equivalent to asserting

> In a metric space (X, d), a subset $F \subset X$ is **closed**, if and only if, F contains all points near it.

Notice that *open* and *closed* are complementary concepts and not negatives of each other. Observe that

Obs.1 Mathematician's riddle:
How does a door differ from a set? Unlike a door, a non-empty set can be open, or closed, or both, or neither.

Obs.2 A set X and the empty set $\varnothing$ are closed as well as open, *i.e.*, **clopen**.

Obs.3 In the space of real numbers, $(0, 1]$ is neither open nor closed.

Obs.4 In the space of real numbers with $a, b \in \mathbb{R}$, an open interval $(a, b), a < b$ is an open set. **Note**: Lemma 1.1 presents the analogue of an open interval in a metric space.

Lemma 1.1. *An open ball $S(p, \varepsilon)$ in a metric space (X, d) is open.*

Proof.
If $x \in X$ is a point in $S(p,\varepsilon)$, then $d(p,x) = \eta < \varepsilon$. We show that the open ball $S(x,\varepsilon-\eta) \subset S(p,\varepsilon)$. If $y \in S(x,\varepsilon-\eta)$, then

$$d(p,y) \le d(p,x) + d(x,y) < (\varepsilon-\eta)+\eta = \varepsilon, i.e., \ y \in S(p,\varepsilon). \qquad \square$$

The family of open sets in a metric space satisfies three simple conditions. These conditions will be used later as a motivation for defining an abstract topological space.

Theorem 1.5. *In a metric space* (X,d),
T.*1* *X and the empty set $\varnothing$ are open;*
T.*2* *The union of open sets is open;*
T.*3* *The finite intersection of open sets is open.*

Proof.
T.1 See Problem 1.21.33.
T.2 Suppose that $\{E_\alpha : \alpha \in \Lambda\}$ is a family of open subsets in X and x belongs to their union $\bigcup\{E_\alpha : \alpha \in \Lambda\}$. Then x must be in E_λ for some $\lambda \in \Lambda$ and E_λ is open. So, there is an open ball $S(x,\epsilon) \subset E_\lambda$ that is contained in the union $\bigcup\{E_\alpha : \alpha \in \Lambda\}$. Hence, $\bigcup\{E_\alpha : \alpha \in \Lambda\}$ is open.
T.3 Suppose that $\{E_n : 1 \le n \le m\}$ is a finite family of open subsets in X and x belongs to their intersection $\bigcap\{E_n : 1 \le n \le m\}$. For each n, there is an $\varepsilon_n > 0$ and an open ball $S(x,\epsilon_n)$ contained in E_n. If $\varepsilon = min\{\epsilon_n : 1 \le n \le m\}$, then the open ball $S(x,\epsilon)$ is contained in the intersection $\bigcap\{E_n : 1 \le n \le m\}$. Hence, $\bigcap\{E_n : 1 \le n \le m\}$ is open. $\square$

For many purposes in topology and analysis, a family of open sets suffices and a metric is not needed. The family $\mathcal{F} = \mathcal{F}_d$ of open sets of a metric space (X,d) is called a **metric topology** (induced by d). It will be seen later that the following observation holds.

> For any set, one can obtain families of subsets satisfying the three conditions in Theorem 1.5. Each such family is called a **topology**. In effect, the term *topology* is a topic in mathematics as well as the name for each family of open sets.

In the space of real numbers $\mathbb{R}$ with the usual metric, every open subset is a union of open intervals. The family of open intervals is called a **base** for the usual metric topology. To tackle any problem in a metric space, a suitable base suffices. Recall that in real analysis, frequently used bases are the family of open intervals with rational end-points or of diameter less than $\frac{1}{n}$, where n is a natural number. In a metric space (X, ρ), the family of open balls $\{S(x, \varepsilon) : x \in X, \varepsilon > 0\}$ is a base, since every open set in X is a union of open balls.

Suppose $f : \mathbb{N} \to X$ is a sequence in a metric space (X, d). If (n_k) is a sequence of increasing natural numbers (*i.e.*, $n_{k+1} > n_k$ for all $k \in \mathbb{N}$), then the sequence $g : \mathbb{N} \to X$ defined by $g(k) = f(n_k)$ is called a **subsequence** of f. It is easy to show that if f converges to limit $\mathcal{L}$, then every subsequence of f converges to $\mathcal{L}$. In fact, a sequence converges to a limit, if and only if, all subsequences of the sequence converge to the same limit.

A point $c \in X$ is called a **cluster point** of a sequence f, if and only if, for every $\varepsilon > 0$ and each $m \in \mathbb{N}$, there is an $n > m$ such that $d(f(n), c) < \varepsilon$ or $f(n)$ belongs to $S(c, \varepsilon)$, the open ball with centre c and radius ε. It is common to use the word *frequently* instead of the expression

$$\text{for each } m \in \mathbb{N}, \text{ there is an } n > m.$$

So, a point $c \in X$ is a cluster point of the sequence f, if and only if, for every $\varepsilon > 0$, *frequently* $f(n)$ belongs to $S_d(c, \varepsilon)$. Obviously, c is a cluster point of the sequence f, if and only if, there is an infinite subset $M \subset \mathbb{N}$ such that c is near $f(M)$, *i.e.*, $D(c, f(M)) = 0$. The following result holds.

Lemma 1.2. *The point $c \in X$ is a cluster point of a sequence $f : \mathbb{N} \to X$, if and only if, there is a subsequence of f that converges to c.*

Unlike the uniqueness of a limit point of a sequence, it is possible for a sequence to have more than one cluster point. For example,

$$f : \mathbb{N} \to \mathbb{R}, \text{ defined by } f(n) = \begin{cases} 1, & \text{if } n \text{ is odd,} \\ 0, & \text{if } n \text{ is even} \end{cases}$$

has two cluster points 0 and 1, but neither cluster point is a limit of the sequence f.

1.7 Metric and Fine Proximities

After considering the relation 'a point is near a set' in a metric space, it is natural to consider the relation 'one set is near another set'. A nearness relation between sets is called a **proximity**. In a metric space (X, d), nearness between sets $A, B \subset X$ is defined using the gap functional $D(A, B)$, *i.e.*,

Metric Proximity between Sets:

A *is near* $B \iff D(A,B) = 0$, where

$$D(A,B) = \begin{cases} inf\{d(a,b) : a \in A, b \in B\}, & \text{if } A, B \neq \emptyset, \\ \infty, & \text{if } A = \emptyset \text{ or } B = \emptyset. \end{cases}$$

For simplicity, A *is near* B is expressed by $A\ \delta\ B$ or by $A\ \delta_d\ B$ (metric proximity), if a reference to a metric is needed. The negation A *is far from* B is expressed by $A\ \underline{\delta}\ B$ or by $A\ \underline{\delta_d}\ B$. The properties that are satisfied by metric proximity are enunciated in Theorem 1.6.

Theorem 1.6. *For all subsets $A, B, C \subset X$ and points $x, y \in X$ in a metric space (X, d), metric proximity satisfies*
P.*1* $A\ \delta\ B \implies B\ \delta\ A$ *(symmetry),*
P.*2* $A\ \delta\ B \implies A \neq \emptyset$ *and* $B \neq \emptyset$,
P.*3* $A \cap B \neq \emptyset \implies A\ \delta\ B$,
P.*4* $A\ \delta\ (B \cup C) \implies A\ \delta\ B$ *or* $A\ \delta\ C$ *(union axiom),*
P.*5* $A\ \underline{\delta}\ B \implies$ *there exists* $E \subset X$ *such that* $A\ \underline{\delta}\ E$ *and* $E^c\ \underline{\delta}\ B$,
P.*6* $\{x\}\ \delta\ \{y\} \implies x = y$.

Proof.
For proofs of properties P.1, P.2, P.3 and P.6, see Problems 1.21.41-1.21.44.
P.4: Notice that $D(A, B \cup C) = min\{D(A,B),\ D(A,C)\}$. So

$$\begin{aligned} A\ \delta\ (B \cup C) &\iff D(A, B \cup C) = 0 \\ &\iff D(A,B) = 0 \text{ or } D(A,C) = 0 \\ &\iff A\ \delta\ B \text{ or } A\ \delta\ C. \end{aligned}$$

P.5: $A\ \underline{\delta}\ B \iff \exists\ \varepsilon > 0$ such that $D(A,B) = \varepsilon$. Put

$$E = \{x \in X : D(x, B) \leq \frac{\varepsilon}{2}\}.$$

Hence, there is a subset $E \subset X$ such that $A\ \underline{\delta}\ E$ and $E^c\ \underline{\delta}\ B$. □

From P.4 and induction, it follows that a subset A is near $\bigcup\{B_k : 1 \leq k \leq m, k \in \mathbb{N}\}$, if and only if, A is near some B_k.

Lemma 1.3. *For all subsets $A, B, C \subset X$ and point $b \in B$ in a metric space (X, d),*

(a) If δ is a metric proximity, it satisfies a condition analogous to the Kuratowski property K4 in Theorem 1.1, namely,

$$A\ \delta\ B \text{ and } b\ \delta\ C \text{ for each } b \in B \implies A\ \delta\ C.$$

(b) If δ is a metric proximity, then $A\ \delta\ B \iff \mathrm{cl}\,A\ \delta\ \mathrm{cl}\,B$.
(c) Observe that

$$A\ \underline{\delta}\ B \text{ implies } (i)\ \mathrm{cl}\,A \subset B^c, \text{ and } \\ (ii)\ A \subset \mathrm{int}(B^c).$$

Proof.
(a) Suppose $A\ \delta\ B$ and $b\ \delta\ C$ for each $b \in B$ but $A\ \underline{\delta}\ C$. There is an $E \subset X$ such that $A\ \underline{\delta}\ E$ and $E^c\ \underline{\delta}\ C$. Since $A\ \delta\ B$, $B \not\subset E$, *i.e.*, B intersects E^c. Let $b \in B \cap E^c$. Then $b\ \delta\ C$ and this contradicts $E^c\ \underline{\delta}\ C$.
(b) $\Rightarrow$ By the union axiom P.4,

$$A\ \delta\ B, A \subset C, B \subset D \implies C\ \delta\ D.$$

Hence, $A\ \delta\ B \implies \mathrm{cl}\,A\ \delta\ \mathrm{cl}\,B$.
$\Leftarrow$ See Problem 1.21.45
(c) $A\ \underline{\delta}\ B$ implies $\mathrm{cl}\,A\ \underline{\delta}\ B$ implies $\mathrm{cl}\,A \subset B^c$. □

Suppose A, B are subsets of X. Then B is said to be a **neighbourhood** of A, if and only if, A is a subset of the interior of B, *i.e.*, $A \subset \mathrm{int}\,B$. If a neighbourhood is open, then it is termed an **open neighbourhood**. Similarly, if a neighbourhood is closed, it is termed a **closed neighbourhood**. Let $B^c = X - B$. Clearly,

$$\text{if } A\ \underline{\delta}\ B^c, \text{ then } A \subset \mathrm{int}(B^{cc}), \textit{i.e.}, \ A \subset \mathrm{int}\,B.$$

However, the converse is not true. For example, in $\mathbb{R}$ with the usual metric proximity δ,

$$A = (0,1), B = [0,1], A \subset \mathrm{int}\,B, \text{but } A\ \delta\ B^c.$$

The relation $A\ \underline{\delta}\ B^c$ (*i.e.*, *A far from* B^c) is denoted by $A \ll B$ and B is called a **proximal neighbourhood** of A. Thus, a proximal neighbourhood of a set A is also a neighbourhood but a neighbourhood need not be a proximal neighbourhood, unless A is a singleton. The properties P.1-P.6 in Theorem 1.6 can be rewritten using $\ll_\delta$ (if it is obvious from the context, the subscript δ is omitted).

Proximal Neighbourhood Properties:
PN.1 $A \ll B \iff B^c \ll A^c$,
PN.2 $\varnothing \ll X$,
PN.3 $A \ll B \implies A \subset B$,
PN.4 $A \subset B \ll C \subset D \implies A \ll D$,
PN.5 $A \ll B_k$ for $1 \leq k \leq n \iff A \ll \bigcap\{B_k : 1 \leq k \leq n\}$,
PN.6 $A \ll B \implies$ there is an E so that $A \ll E \ll B$.

Observe that metric proximity on a metric space (X, d) is **compatible** with the closure induced by the metric, *i.e.*, for a point $x \in X$ and subset $E \subset X$,

$$x \text{ is in the closure of } E \iff \{x\}\ \delta\ E.$$

Later, it will be seen that there are, in general, many compatible proximities on a metric space. For now, a proximity of paramount importance called **fine proximity** (denoted by δ_0) on X is introduced, next.

$$A\ \delta_0\ B \iff \operatorname{cl} A \cap \operatorname{cl} B \neq \varnothing.$$

It is easy to see that if δ is a metric proximity, then

$$A\ \delta_0\ B \iff \operatorname{cl} A \cap \operatorname{cl} B \neq \varnothing \implies \operatorname{cl} A\ \delta\ \operatorname{cl} B \implies A\ \delta\ B.$$

But the converse is not necessarily true. For example, in $\mathbb{R}$ (reals), put $A = \mathbb{N}$ (set of natural numbers) and let $B = \{n - \frac{1}{n} : n \in \mathbb{N}\}$. Then A is near B in the metric proximity but not so in the fine proximity, *i.e.*, $A\ \delta\ B$ but $A\ \underline{\delta}_0\ B$. It is a straightforward task to show that δ_0 satisfies conditions P.1, P.2, P.3, P.4 and P.6 (replacing δ with δ_0). Here is a sample proof of P.5 rewritten with δ_0 instead of δ.

P.5 $A\ \underline{\delta}_0\ B \implies$ there exists $E \subset X$ such that $A\ \underline{\delta}_0\ E$ and $E^c\ \underline{\delta}_0\ B$.

Proof. Given a metric space (X, d), suppose $A\ \underline{\delta}_0\ B$, *i.e.*, $\operatorname{cl} A$ and $\operatorname{cl} B$ are disjoint. Then the function $f : X \to [0, 1]$ defined by

$$f(x) = \frac{D(x, \operatorname{cl} A)}{D(x, \operatorname{cl} A) + D(x, \operatorname{cl} B)}$$

is continuous and satisfies $f(A) = 0$ and $f(B) = 1$. Put $E = f^{-1}[\frac{1}{2}, 1]$. Then, clearly $A\ \underline{\delta}_0\ E$ and $E^c\ \underline{\delta}_0\ B$. □

Let $(X, d), (Y, d')$ denote metric spaces and let $f : X \to Y$ denote a function that maps X to Y. If f is continuous, then it preserves nearness of points to sets. Its analogue in proximity spaces arises, if f preserves the nearness of pairs of sets.

Proximally Continuous Function:
Given metric spaces $(X,d),(Y,d')$, a function $f : X \to Y$ is **proximally continuous**, if and only if, f preserves the nearness of pairs of sets. That is, for all subsets $A, B \subset X$, a proximally continuous function f satisfies

$$A \ \delta_d \ B \implies f(A) \ \delta_{d'} \ f(B).$$

Equivalent conditions for the proximal continuity of f for all subsets $A, B \subset X$ and $E, F \subset Y$ are

PC.1 $D(A,B) = 0 \implies D'(f(A), f(B)) = 0$, or
PC.2 $D'(E,F) > 0 \implies D(f^{-1}(E), f^{-1}(F)) > 0$, or
PC.3 $E \ \underline{\delta}_{d'} \ F \implies f^{-1}(E) \ \underline{\delta}_d \ f^{-1}(F)$.

Analogous definitions of proximal continuity are obvious, when X and Y have other proximities. Also obvious is the fact that every proximally continuous function is continuous. But the converse is not true, which the following example shows.

Consider the function $f : \mathbb{R} \to \mathbb{R}$ from the space of real numbers to itself defined by $f(x) = x^2$ for $x \in \mathbb{R}$. The set $\mathbb{N}$ of natural numbers is near the set $E = \{n - \frac{1}{n} : n \in \mathbb{N}\}$ but the image sets $f(\mathbb{N})$ and $f(E)$ are far from each other, since the gap functional on them equals $inf\{|n^2 - (n - \frac{1}{n})^2|\} = 2$. However, the following important result is true (its utility will be shown later).

Lemma 1.4. *Let $(X,d),(Y,d')$ be metric spaces. Assume X has fine proximity δ_0 and Y has any compatible proximity λ and let $f : X \to Y$ be a function from X to Y. Then f is continuous, if and only if, f is proximally continuous.*

Proof. We need prove only that the continuity of f implies that f is proximally continuous. Suppose A and B are subsets of X that are near in the fine proximity δ_0, *i.e.*, $\mathrm{cl}\,A$, $\mathrm{cl}\,B$ intersect. Hence, $f(\mathrm{cl}\,A)$, $f(\mathrm{cl}\,B)$ intersect. Since f is continuous, $f(\mathrm{cl}\,A) \subset \mathrm{cl}\,f(A)$ and $f(\mathrm{cl}\,B) \subset \mathrm{cl}\,f(B)$. This implies that $\mathrm{cl}\,f(A)$, $\mathrm{cl}\,f(B)$ intersect and, hence, $f(A) \ \lambda \ f(B)$. □

1.8 Metric Nearness

In the setting of a metric space (X, d), we have studied (a) topology: a point is near a set and (b) proximity: nearness of pairs of sets. It is natural to consider when families of sets are near. This was done by V.M. Ivanova and A.A. Ivanov for finite families and by Herrlich for sets of arbitrary cardinality. This generalisation is significant, since it unifies many structures and simplifies problems that are studied in topology, something that will become apparent, later. We now give a brief introduction to these concepts in the setting of metric spaces to provide motivation for their study in abstract spaces.

Prior to this, we used gap functionals to define metric topology and metric proximity. We now consider an alternate, equivalent approach to nearness that supports generalisation. For a subset A in a metric space (X, d) and $\varepsilon > 0$, consider $S(A, \varepsilon)$ (ε-enlargement of A) such that

$$S(A, \varepsilon) = \bigcup\{S(a, \varepsilon) : a \in A\},$$

i.e., $S(A, \varepsilon)$ is the union of open balls of radius ε with centres at points of A. Then consider another approach to **metric proximity**. A is near B (denoted $A\delta B$), if and only if, for every $\varepsilon > 0$, ε-enlargements of A and B intersect, *i.e.*,

$$S(A, \varepsilon) \cap S(B, \varepsilon) \neq \varnothing.$$

With this alternative approach to metric proximity in mind and the analogy with the metric proximity axioms, the axioms for a Herrlich form of near structure are natural. Before we proceed, we introduce some notation.

$$\begin{aligned}
\mathcal{A}, \mathcal{B} &\text{ denote families of subsets of } X,\\
\eta\mathcal{A}, \text{ or } \mathcal{A} \in \eta, &\text{ denotes } \mathcal{A} \text{ is near},\\
\underline{\eta}\mathcal{A} &\text{ denotes } \mathcal{A} \text{ is not near},\\
A\ \eta\ B &= \eta\{A, B\},\\
\mathcal{A} \vee \mathcal{B} &= \{A \cup B : A \in \mathcal{A}, B \in \mathcal{B}\},\\
\mathrm{cl}_\eta E &= \{x \in X : \{x, E\} \in \eta\},\\
\mathrm{cl}_\eta \mathcal{A} &= \{\mathrm{cl}_\eta A : A \in \mathcal{A}\},\\
\mathcal{P}X &\text{ denotes the power set of } X,\\
\mathcal{P}^{n+1}X &= \mathcal{P}(\mathcal{P}^n X) \text{ for } n \in \mathbb{N}.
\end{aligned}$$

In a metric space (X, d), a **Herrlich near structure** or **Herrlich nearness** on X is defined in the following way. $\mathcal{A}$ is near or $\eta\mathcal{A}$, if and only if, for each $\varepsilon > 0$, ε-enlargements of members of $\mathcal{A}$ intersect, *i.e.*, $\bigcap\{S(A,\varepsilon) : A \in \mathcal{A}\} \neq \varnothing$. It is apparent that Herrlich nearness satisfies (N.1)-(N.5):

(N.1) $\bigcap\{A : A \in \mathcal{A}\} \neq \varnothing \Rightarrow \eta\mathcal{A}$,
(N.2) $\underline{\eta}\mathcal{A}$ and $\underline{\eta}\mathcal{B} \Rightarrow \underline{\eta}(\mathcal{A} \vee \mathcal{B})$,
(N.3) $\eta\mathcal{A}$ and, for each $B \in \mathcal{B}$, there is an $A \in \mathcal{A} : A \subset B \Rightarrow \eta\mathcal{B}$,
(N.4) $\varnothing \in \mathcal{A} \Rightarrow \underline{\eta}\mathcal{A}$,
(N.5) $\eta(\mathrm{cl}_\eta\mathcal{A}) \Rightarrow \eta\mathcal{A}$ (Herrlich axiom).

For $\eta \subset \mathcal{P}^2X$, η is called a Čech near structure or Čech nearness, if and only if, conditions (N.1)-(N.4) are satisfied. In fact, metric nearness satisfies a condition that is stronger than (N.5). It is the vestigial form of the triangle inequality (similar to T.3 of Theorem 5). Metric nearness is equivalent to metric uniformity. If $(X,d),(Y,d')$ are metric spaces, then a function $f : X \to Y$ is called a **near map**, if and only if, it preserves nearness of families and this is equivalent to the uniform continuity of f. Conditions (N.1)-(N.5) provide motivation for studying nearness in abstract spaces in the sequel to this chapter.

1.9 Compactness

In classical analysis, several important results are proved for closed, bounded subsets in the space of real numbers $\mathbb{R}$ or in the space of complex numbers $\mathbb{C}$. For example,

(1) Every sequence in a closed, bounded set has a convergent subsequence.
(2) Every continuous function takes a closed, bounded set onto a closed, bounded set.
(3) Every continuous function on a closed, bounded set is uniformly continuous.
(4) Every open cover of a closed, bounded set E (*i.e.*, a family of open sets whose union contains the set E) has a finite subcover (Heine-Borel-Lebesgue Theorem), *i.e.*, X is compact.
(5) Every infinite, bounded subset of a non-empty set E has a limit point in E (Bolzano-Weierstrass).

After a lot of experimentation in metric spaces and general topological spaces, principally by P. Alexandroff and P. Urysohn, the Heine-Borel-Lebesgue theorem was found to be very helpful. And, nowadays, this theorem is used as the definition of **compactness**. It is considered one of the most important concepts in analysis and topology.

Open Covers and Compact Subsets:
A family $\mathcal{F}$ of open sets of a metric space (X, d) is called an **open cover** of a subset $E \subset X$, if and only if, each point $x \in E$ belongs to some set $G \subset \mathcal{F}$. A subfamily of $\mathcal{F}$ that is a cover of E is called a **subcover** of E. A subset E in a metric space (X, d) is **compact**, if and only if, every open cover of E has a *finite* subcover. If $E = X$, then (X, d) is termed a **compact metric space**.

A sample open cover of E is shown in Fig. 1.5. In this Figure, the open sets in $\mathcal{F}$ are represented by regions bounded by dotted boundaries. Part of each open set overlaps with a part of E to form an open cover.

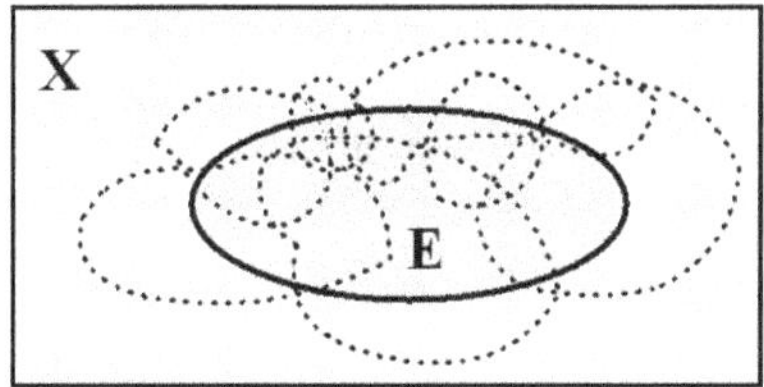

Fig. 1.5 Open cover of $E \subset X$

If $\mathcal{F}$ is an open cover of E and $X \notin \mathcal{G}$, then the family $\mathcal{H}$ of complements of members of $\mathcal{F}$, *i.e.*, $\mathcal{H} = \{G^c : G \in \mathcal{F}\}$, is a family of closed sets such that each point of E does not belong to at least one member of $\mathcal{H}$. Consequently, the definition of compactness can be given in terms of closed sets. That is, if the intersection of a family $\mathcal{H}$ of closed subsets of E does not contain any point of E, then there is a finite subfamily of $\mathcal{H}$ whose intersection does not contain any point of E.

Recall that a family $\mathcal{F}$ has the **finite intersection property**, if and only if, the intersection of the members of each finite subfamily of $\mathcal{F}$ is non-empty. Thus, a subset E of a metric space X is compact, if and only if, the members of every family $\mathcal{H}$ of closed subsets of E with the finite intersection property have non-empty intersection. Later, it will be shown that this characterisation of compact sets is useful in generalising sequences to filters in general topology.

Here are some observations about compactness.

Ct.1 Closed and bounded sets of $\mathbb{R}^m$ are compact. However, $\mathbb{R}$ is not compact, since the open cover $\{(-\infty, n) : n \in \mathbb{N}\}$ of $\mathbb{R}$ has no finite subcover. See Problem 1.21.51.

Ct.2 Compactness is a generalisation of **finiteness**. For example, the following results that are trivial for finite sets remain true if *finite* is replaced by *compact*. Note that every function on a finite metric space is continuous.

finite.1 A finite set is bounded.

finite.2 A function on a finite set is bounded.

finite.3 A function on a finite set is uniformly continuous.

Ct.3 If a sequence $f : \mathbb{N} \to X$ converges to a limit $\mathcal{L}$, then the subset $E = f(N) \cup \mathcal{L}$ is compact and closed.

Proof. If $\mathcal{F}$ is an open cover of E, then there is one member $G \subset \mathcal{F}$ that contains $\mathcal{L}$ and so, eventually, $f(n) \in G$. Hence, there are only finitely many $f(n)$'s outside G. In effect, there is a finite subcover of $\mathcal{F}$. Since (*) $\mathcal{L}$ is the only limit of the sequence f, E is closed. □

Ct.4 Every closed subset of a compact metric space is compact.

Proof. Let E be a closed subset of a compact metric space (X, d) and let $\mathcal{F}$ be an open cover of E. Then $\{E^c\} \cup \mathcal{F}$ is an open cover of the compact set X and, hence, has a finite subcover $\mathcal{H}$. Then $\mathcal{F}$ has $\mathcal{H} - \{E^c\}$ as a finite subcover of E. □

Ct.5 A compact subset E of a metric space (X, d) is closed and bounded but the converse is not true.

Proof. See Problem 1.21.53. □

A compact subset behaves like a point or a finite set. For instance, in terms of metric proximity, two disjoint, closed subsets, one of which is finite are *far*.

Theorem 1.7. *Two disjoint, closed subsets A, B of a metric space (X, d), one of which is compact, are far in metric proximity.*

Proof. Suppose $A \subset X$ is compact. Then each point $a \in A$ being far from B has an open neighbourhood N_a far from B. Observe that $\{N_a : a \in A\}$ is an open cover of A and, since A is compact, it has a finite subcover $\{N_{a_k} : 1 \le k \le m\}$ whose union is far from B. Hence, A is far from B and, therefore, the gap $D_\rho(A, B) > 0$. □

Two disjoint, non-compact, closed subsets $A, B \subset X$ need not be far in the metric proximity. For example, in $\mathbb{R}$, put $A = \mathbb{N}$ and $B = \{(n - \frac{1}{n}) : n \in \mathbb{N}\}$. Then A and B are disjoint, non-compact, closed subsets and also *not far* in the metric proximity. The result in Theorem 1.7 also holds for the fine proximity or any compatible proximity, satisfying

P.5 $A \underline{\delta} B$ implies that there is a subset $E \subset X$ such that $A \underline{\delta} E$ and $E^c \underline{\delta} B$.

1.10 Lindelöf Spaces and Characterisations of Compactness

This section introduces characterisations of compactness in metric spaces that are analogous to those in real and complex analysis. Such characterisations of compactness stem from the ramifications of an important result from E.L. Lindelöf concerning an m-dimensional, Euclidean space $\mathbb{R}^m$. Recall that a family $\mathcal{F}$ of sets is a **base** for a topology $\mathfrak{T}$, if and only if, $\mathcal{F}$ is a subfamily of $\mathfrak{T}$ and for each point x of the space and for each neighbourhood U of x, there is a member $V \in \mathcal{F}$ such that $x \in V \subset U$.

In particular, the space $\mathbb{R}^m$ has a countable, open base. That is, $\mathbb{R}^m$ has a family $\mathcal{B}$ of open sets with the following properties.

$\mathcal{B}$.1 The family $\mathcal{B}$ is countable, *i.e.*, in 1-1 correspondence with the natural numbers.

$\mathcal{B}$.2 For each point x in an open subset $G_x \subset \mathbb{R}^m$, there is a $B_x \in \mathcal{B}$ such that $x \in B_x \subset G_x$.

For example, $\mathbb{R}$ has a countable base of open intervals with rational endpoints and their product m-times gives a countable open base for $\mathbb{R}^m$. This led Lindelöf to the discovery that every open cover of $\mathbb{R}^m$ has a countable subcover. The proof is straightforward.

Proof.
If $\mathcal{F}$ is any open cover of $\mathbb{R}^m$, then, for each $x \in \mathbb{R}^m$, there is an open superset G_x from $\mathcal{F}$. Then there is a $B_x \in \mathcal{B}$ such that $x \in B_x \subset G_x$. Choose only one such G_x for a B_x. Since $\mathcal{B}$ is countable, the family $\{G_x\}$ of supersets of B_xs is a countable subcover of $\mathcal{F}$. □

This property of $\mathbb{R}^m$ observed by Lindelöf gives rise to what are known as Lindelöf spaces.

> A space that has the property that every open cover has a countable subcover is called Lindelöf.

In particular, every metric space X is a Lindelöf space, if and only if, every open cover of X has a countable subcover. Recall that a set is **countably infinite**, if and only if, there is a 1-1 correspondence between the natural numbers $\mathbb{N}$ and the elements of the set. A set that is either finite or countably infinite is termed **countable**. A metric space is called **countably compact**, if and only if, every countable open cover has a finite subcover. It is obvious that every Lindelöf, countably compact space is compact.

An interesting result for metric spaces is the fact that countable compactness equals compactness. This is so because the countable compactness of a space implies the existence of a countable base. And, this in turn, implies that such a space is Lindelöf.

Suppose that the metric space X is countably compact. First, we show that for each $\varepsilon > 0$, there is a finite cover of X consisting of open balls of radius ε. If this is not true, then we can find inductively an infinite set of points $\{x_k : k \in \mathbb{N}\}$ such that the family of open balls $\mathcal{A} = \{S(x_k, \varepsilon) : k \in \mathbb{N}\}$ is not a cover of X and each x_k does not belong to $\{S(x_m, \varepsilon) : m > k\}$. If there are points in X that are not in the union of members of $\mathcal{A}$, we cover them by the union H of open balls with radius $\frac{\varepsilon}{2}$. Then the countable, open cover $\mathcal{A} \cup \{H\}$ of X has no finite subcover, a contradiction.

So, for each each $n \in \mathbb{N}$, a countably compact metric space X has a *finite*, open cover of open balls $\mathcal{B}_n = \{S(x_{mn}, \frac{1}{n}) :$ finitely many $x_{mn} \in X\}$. We show that $\{\mathcal{B}_n : n \in \mathbb{N}\}$ is a countable base for X. This follows from the fact that if a point x is in an open set G, then there is $\varepsilon > 0$ such that $S(x, \varepsilon)$ is a subset of G. Choose a natural number n such that $2\frac{1}{n} < \varepsilon$. Observe, then, that there is an open ball $S(x_{mn}, \frac{1}{n})$ containing x and $S(x_{mn}, \frac{1}{n})$ is contained in $S(x, \varepsilon)$, a subset of G. These observations pave the way for characterisations of compactness in metric spaces that are analogues of those in real and complex analysis.

Theorem 1.8. *In a metric space (X, ρ), the following assertions are equivalent.*
(a) A subset $E \subset X$ is compact.
(b) A subset $E \subset X$ is countably compact.
(c) Every sequence in E has a subsequence converging to a point in E.
(d) Every sequence in E has a cluster point in E.
(e) Every infinite subset of E has a limit point in E (Bolzano-Weierstrass).

Proof.
(a) $\Longleftrightarrow$ (b) The fact that (a) is equivalent to (b) has already been shown.
(a) $\Longrightarrow$ (d) Let E be a compact subset of X and let (x_n) be a sequence in

E. If (x_n) has no cluster point, then it has a non-convergent subsequence (x_{n_k}) of distinct terms. Hence, the infinite open cover $\{X - x_{n_k} : k \in \mathbb{N}\}$ has no finite subcover and that contradicts the compactness of E.
(d) $\Longrightarrow$ (b) If $E \subset X$ is not countably compact, then there is a countable, open cover $\{G_n\}$ such that, for each $n \in \mathbb{N}$, G_{n+1} contains a point that is not in $\bigcup\{G_k : 1 \leq k \leq n\}$. Let x_1 be a point in G_1. Having selected $x_1, x_2, \ldots, x_n$, choose inductively $x_{n+1} \in G_{n+1}$ that is not in each member of $\{G_k : 1 \leq k \leq n\}$. Then the sequence (x_n) has no cluster point.
(c) $\Longleftrightarrow$ (d) $\Longleftrightarrow$ (e) follow from Problem 1.21.51. □

Theorem 1.9. *A continuous function from one metric space to another metric space preserves compactness.*

Proof. First version of proof.
Suppose f is a continuous function on a metric space X to a metric space Y and $E \subset X$ is a compact subset of X. If (y_n) is any sequence in $f(E)$, pick x_n in $f^{-1}(y_n)$. Since E is compact, (x_n) has a subsequence (x_{n_k}) converging to a point in E. Then $(f(x_{n_k}))$ is a subsequence of (y_n) converging to $f(x)$ in $f(E)$. Hence, $f(E)$ is compact. □

Proof. Second version of proof.
If $\mathcal{F}$ is an open cover of $f(E)$, then $f^{-1}(\mathcal{F}) = \{f^{-1}(G) : G \subset \mathcal{F}\}$ is an open cover of E. Since E is compact, there is a finite subcover $\{f^{-1}(G_k) : G_k \in \mathcal{F}, 1 \leq k \leq m\}$ of E. Clearly, $\{G_k \in \mathcal{F}, 1 \leq k \leq m\}$ is a finite subcover of $f(E)$. Hence, $f(E)$ is compact. □

From Theorem 1.9, a well-known result in calculus has an analogue, namely,

Implication of Theorem 1.9:
A real-valued, continuous function on a compact metric space is bounded and attains its bounds.

A metric space (X, ρ) is **locally compact**, if and only if, each point $x \in X$ has a compact neighbourhood G, *i.e.*, x is in the interior of G. The space of complex numbers $\mathbb{C}$ and Euclidean space $\mathbb{R}^m$ are not compact but are locally compact. For example, each point $x \in \mathbb{R}$ has a compact neighbourhood $[x - \varepsilon, x + \varepsilon]$, where $\varepsilon > 0$. Compactness is a valuable property, but is quite strong. Local compactness is weaker than compactness and is sufficient for many purposes, which is evident from real and complex analysis. A compact metric space is locally compact but the converse is not true, something that is evident, if one considers $\mathbb{R}$. Also, notice that $\mathbb{R}$ is locally compact but $\mathbb{Q}$ (space of rational numbers) is not locally compact.

1.11 Completeness and Total Boundedness

In real analysis, it is shown that the space of rational numbers $\mathbb{Q}$ with the usual metric is not *complete* and its completion is the space of real numbers. In this section, the concept of *completeness* in metric spaces is considered. Motivation for Cauchy sequences is via proximity and leads to the traditional definition. The construction of a completion of a metric space is postponed. Naturally, the study of completeness entails a consideration of a Cauchy sequence with discussion and results viewed via proximity.

Suppose $f : \mathbb{N} \to X$ is a sequence in a metric space (X, ρ) and that $f(n) \to p \in X$. Then, for any two infinite subsets $A, B \subset \mathbb{N}$, the point p is near both $f(A)$ and $f(B)$. It follows from Problem 1.21.50 that $f(A)\ \delta\ f(B)$, where δ is the metric proximity. That the converse is not true can be seen from an example.

In $X = (0, 1)$, the sequence defined by $f(n) = \frac{1}{n}$ has no limit in X but for any two infinite subsets A and B in $\mathbb{N}$, $f(A)\ \delta\ f(B)$. The importance of this property of convergent sequences was brought into prominence by A. Cauchy and bears his name. A sequence $f : \mathbb{N} \to X$ in a metric space (X, ρ) is called a **Cauchy sequence**, provided that for all infinite subsets $A, B \subset \mathbb{N}$, $f(A)\ \delta\ f(B)$. These remarks confirm the first result.

Cauchy Sequence:
Every convergent sequence in a metric space is Cauchy but a Cauchy sequence need not converge.

Using the above definition, the following result from calculus is given a new proof.

Lemma 1.5. *Every Cauchy sequence $f : \mathbb{N} \to \mathbb{R}^n$ is bounded.*

Proof.
If f is not bounded, then there is an $n_1 \in \mathbb{N}$ such that $|f(n_1)| > |f(1)| + 1$. Repeating the same argument, there is an $n_2 \in \mathbb{N}$ and $|f(n_2)| > |f(n_1)| + 1$. In this manner, we inductively find n_i such that $n_{i+1} > n_i$ for all $i \in \mathbb{N}$ and

$|f(n_{i+1})| > |f(n_i)| + 1$. Clearly,

$$A = \{n_{2i} : i \in \mathbb{N}\} \text{ and } B = \{n_{2i+1} : i \in \mathbb{N}\}$$

are infinite and $f(A)$ is far from $f(B)$, *i.e.*, f is not Cauchy. □

From calculus, the importance of the space of real numbers $\mathbb{R}$ is apparent and that $\mathbb{R}$ is complete, when compared to the space of rational numbers that is not complete. A metric space (X, ρ) is termed **complete**, provided that every Cauchy sequence in X converges.

Theorem 1.10. *If f is a Cauchy sequence in a metric space (X, ρ) and a subsequence $f \cdot g$ of f converges to $l \in X$, then f itself converges to l.*

Proof.
Suppose f does not converge to l. Then there is an infinite subset $A \subset \mathbb{N}$ such that $l \underline{\delta} f(A)$, *i.e.*, there is an $r > 0$ such that $f(A) \cap (l - r, l + r) = \emptyset$. Since $f \cdot g$ converges, there is an infinite subset $B \subset \mathbb{N}$ with $f \cdot g(B) \subset (l - \frac{r}{2}, l + \frac{r}{2})$. Clearly, $f(B)$ is also an infinite subset of $\mathbb{N}$, but $f(A)$ is far from $f \cdot g(B)$, and that contradicts the fact that f is Cauchy. □

Theorem 1.11. *$\mathbb{R}^m$ is complete.*

Proof.
Let $f : \mathbb{N} \to \mathbb{R}^m$ be a Cauchy sequence. Then f is bounded, *i.e.*, there is an $M > 0$ such that $f(\mathbb{N}) \subset M$, a bounded set in $\mathbb{R}^m$ and the closure of M is compact. So f has a convergent subsequence. From Theorem 1.10, f is also convergent. □

Theorem 1.12. *Every closed subset of a complete metric space is complete.*

Proof.
Suppose Y is a closed subset of a complete metric space (X, ρ) and f is a Cauchy sequence in Y. Then f is Cauchy in X and, hence, f converges to some $l \in X$. Since Y is closed, $l \in Y$ and so f converges in Y. □

Theorem 1.13. *Every compact metric space is complete and every discrete metric space is complete.*

Proof.
From Theorem 1.8, every sequence in a compact metric space has a subsequence converging to a limit. If the sequence is Cauchy, then, from Theorem 1.10, the sequence converges to the same limit. Hence, compactness implies completeness. In a discrete metric space, a Cauchy sequence is eventually constant. □

Lemma 1.6. (Efremovič Lemma)
Suppose that (X, ρ) is a metric space, δ, a metric proximity, and $(a_n), (b_n)$ are two sequences in X with $\rho(a_n, b_n) \geq r > 0$ for all $n \in \mathbb{N}$. Then

$$\text{there exists an infinite subset } P \subset \mathbb{N} \text{ such that } \{a_p : p \in P\} \ \underline{\delta} \ \{b_p : p \in P\}.$$

Proof. For each $n \in \mathbb{N}$, define

$$B_n = \{m \in \mathbb{N} : \rho(a_n, b_m) \leq \frac{r}{4}\},$$
$$C_n = \{m \in \mathbb{N} : \rho(b_n, a_m) \leq \frac{r}{4}\}.$$

If k, l are in B_n, then $\rho(b_k, b_l) \leq \rho(b_k, a_n) + \rho(a_n, b_l) \leq \frac{r}{4} + \frac{r}{4} \leq \frac{r}{2}$ and so $\rho(a_k, b_l) \geq \frac{r}{2}$. If not and $\rho(a_k, b_l) < \frac{r}{2}$, then, by the triangle inequality, obtain $\rho(a_k, b_k) \leq \rho(a_k, b_l) + \rho(b_l, b_k) < r$, a contradiction. If the set B_n is infinite for some $n \in \mathbb{N}$, put $P = B_n$. The case where C_n is infinite for some $n \in \mathbb{N}$ is similar.

So the case where B_n and C_n are both finite remains to be considered. Put $n_1 = 1$ and let n_2 be the first natural number larger than n_1 and each member of $B_{n_1} \cup C_{n_1}$. Since $n_2 \notin B_{n_1}$, $\rho(a_{n_1}, b_{n_2}) > \frac{r}{4}$ and, similarly, $\rho(a_{n_2}, b_{n_1}) > \frac{r}{4}$. Inductively, let n_{k+1} be the least natural number larger than n_k and each element in $B_{n_k} \cup C_{n_k}$. Then $P = \{n_k : k \in \mathbb{N}\}$ satisfies the condition $\{a_p : p \in P\} \ \underline{\delta} \ \{b_p : p \in P\}$. □

The following result calls attention to the equivalence of the nearness approach to defining a Cauchy sequence and the traditional epsilonitic definition.

Theorem 1.14. *For a metric space (X, ρ), the following are equivalent.*
(Cauchy.a) A sequence $f : \mathbb{N} \to X$ is Cauchy, i.e., *for all infinite subsets $A, B \subset \mathbb{N}$, $f(A) \ \delta \ f(B)$.*
(Cauchy.b) For every $\varepsilon > 0$, there is an $N \in \mathbb{N}$ such that, for all $m, n \in \mathbb{N}, m > N, n > N, \rho(f(m), f(n)) < \varepsilon$.

Proof.
(Cauchy.a) $\Longrightarrow$ (Cauchy.b) Suppose that this is not true. Then there is an $\varepsilon > 0$ such that, for all $N \in \mathbb{N}$, there are $m > N, n > N$ with $\rho(f(m), f(n)) \geq \varepsilon$. So, inductively choosing m, n, obtain two increasing sequences $(m_i), (n_i)$ of natural numbers such that $\rho(f(m_i), f(n_i)) \geq \varepsilon$ for all $i \in \mathbb{N}$. Then, by Efromovič Lemma 1.6, there is an infinite subset $P \subset \mathbb{N}$ such that

$\{f(m_p) : p \in P\}$ is far from $\{f(n_p) : p \in P\}$, *i.e.*, f is not (Cauchy.a). $\Rightarrow\Leftarrow$

(Cauchy.b) $\Longrightarrow$ (Cauchy.a) Suppose $\varepsilon > 0$ is assigned and A, B are infinite subsets of $\mathbb{N}$. Then there is an $N \in \mathbb{N}$ such that, for all $m, n \in \mathbb{N}, m > N, n > N, \rho(f(m), f(n)) < \varepsilon$. But since A and B are infinite, we can find $m \in A, n \in B$ satisfying $m > N, n > N$ and, hence, $f(A)\ \delta\ f(B)$, *i.e.*, f is (Cauchy.a). □

Recall the definition of uniform continuity:

> **Uniform Continuity**:
> Suppose $(X, \rho), (Y, \eta)$ are metric spaces. A function $f : X \to Y$ is termed **uniformly continuous**, if and only if, for every $\varepsilon > 0$, there is a $\kappa > 0$ such that, for all $x, y \in X, \rho(x, y) < \kappa$ implies $\eta(f(x), f(y)) < \varepsilon$.

It is important to observe that κ depends on ε and is independent of x and y. Assume $A, B \subset X$ and let δ and δ' denote metric proximities in X, Y, respectively. Then a function $f : X \to Y$ is termed **proximally continuous** provided that

$$A\ \delta\ B \text{ in } X \Longrightarrow f(A)\ \delta'\ f(B) \text{ in } Y.$$

Theorem 1.15. (Efremovič Theorem)
Let $(X, d), (Y, d')$ denote metric spaces. Then a function $f : X \to Y$ is uniformly continuous, if and only if, f is proximally continuous.

Proof.
($\Rightarrow$) Let f be uniformly continuous and let A be near B in X. For any $\varepsilon > 0$, there is a $\kappa > 0$ such that $d(a, b) < \kappa$ implies $d'(f(a), f(b)) < \varepsilon$. Since A is near B, there are $a \in A, b \in B$ with $d(a, b) < \kappa$. So, $d'(f(a), f(b)) < \varepsilon$, *i.e.*, $f(A)$ is near $f(B)$ in Y.
($\Leftarrow$) Suppose f is proximally continuous but not uniformly continuous. Then, for each $n \in \mathbb{N}$, we have $x_n, y_n \in X$ such that $d(x_n, y_n) < \frac{1}{n}$ and $d'(f(x_n), f(y_n)) \geq \varepsilon$. From the Efremovič Lemma 1.6, there is an infinite subset $P \subset \mathbb{N}$ such that $\{f(x_p) : p \in P\}$ is far from $\{f(y_p) : p \in P\}$. Since $\{x_p : p \in P\}$ is near $\{y_p : p \in P\}$, this implies f is not proximally continuous. $\Rightarrow\Leftarrow$ □

From Theorem 1.13, every compact metric space is complete. However, the converse is not true. So a natural question arises: *Is there a property of a metric space which, together with completeness, implies compactness*? Such a property does exist and termed **total boundedness** or **precompactness** (PC). A motivation for the discovery of PC is to look at the following:

(i) A metric space is compact, if and only if, every sequence has a convergent subsequence, and

(ii) a metric space is complete, if and only if, every Cauchy sequence converges. Hence, the additional condition that is required for a complete space to be compact is that *every sequence has a Cauchy subsequence.*

Theorem 1.16. *In a metric space* (X, d)*, the following are equivalent.*

(a) Every sequence in $E \subset X$ *has a Cauchy subsequence.*

(b) For every $\varepsilon > 0$*, the* ε*-cover* $\{S(x, \varepsilon) : x \in E\}$ *of* E *has a finite subcover, where* $S(x, \varepsilon) = \{y \in E : d(x, y) < \varepsilon\}$.

(c) For every $\varepsilon > 0$*, there is a* finite *maximal* ε*-discrete subset* $F \subset X$*,* i.e., $E \subset \bigcup\{S(x.\varepsilon) : x \in F\}$ *and, for any two distinct points* $x, y \in F$*,* $d(x, y) \geq \varepsilon$.

A subset $E \subset X$ *satisfying this property is termed totally bounded or precompact.*

Proof.
Let $\varepsilon > 0$ be arbitrary and let x_1 be a point in $E \subset X$. If $S(x_1, \varepsilon)$ covers E, we are done. Otherwise, let x_2 be a point in $E - S(x_1, \varepsilon)$. Inductively, let x_{n+1} be a point in $E - \bigcup\{S(x_k, \varepsilon) : 1 \leq k \leq n\}$, if it is not empty. It is easy to see that $\{x_k\}$ is a discrete set with the distance between any two points $\geq \varepsilon$. Hence, the set $\{x_k\}$ is infinite, if there is a sequence in E that has no Cauchy subsequence. This shows the equivalence of (a) and (b), while the equivalence of (b) and (c) is considered in Problem 1.21.60. □

Corollary 1.1.

(a) Every compact subset in a metric space is totally bounded.

(b) A metric space is compact, if and only if, it is complete and totally bounded.

1.12 Connectedness

Intuitively, a metric space is *connected* if it is in one piece. This is not easy to make precise. So, it is customary to define its negative, namely, a disconnected metric space. Informally, this means that a metric space is *disconnected* (not connected), if and only if, it can be split into two parts A, B such that every point in A is far from B and every point in B is far from A. For example, consider a metric space that consists of the regions of Italy and notice that Sicily is separated from the rest of Italy by the sea.

A metric space (X, ρ) is **disconnected**, if and only if, $X = A \cup B$, where A and B are non-empty, disjoint, closed (or both are open) subsets of X.

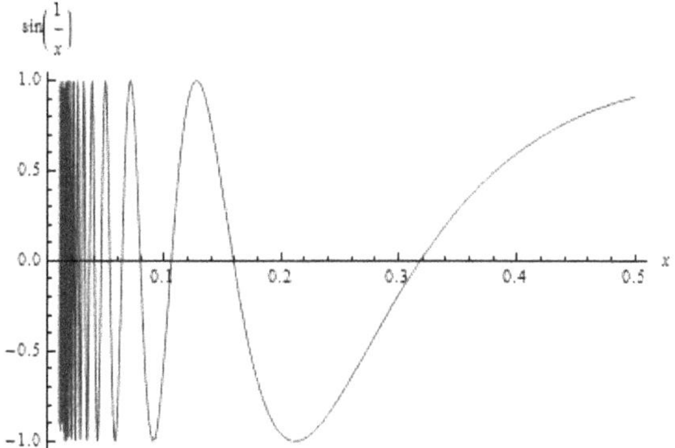

Fig. 1.6 Sine curve

A **connected space** is a space that is not disconnected. A subset $E \subset X$ is connected, if and only if, E cannot be expressed as a union of two non-empty subsets A and B such that $\text{cl}\, A \cap B = \varnothing$ and $A \cap \text{cl}\, B = \varnothing$. The **disconnectedness** of subset E in a metric space is expressed by writing that there are $A, B \subset E$ that are *mutually separated* in E.

For example, $\mathbb{R}^m$ is connected but $\mathbb{N}$ is disconnected. An interesting and useful example of a connected set in $\mathbb{R}^2$ is what is known as the **topologist's sine curve**, a subspace in $\mathbb{R}^2$ defined by

$$T = \{(0, y) : -1 \leq y \leq 1\} \cup \{(x, sin(\frac{1}{x})) : x > 0\}.$$

T consists of two parts, namely, a portion of the y-axis and a portion of $\mathbb{R}^2$ to the right of the y-axis (see Fig. 1.6). Intuitively, T seems to be disconnected but, in fact, from the definition of a connected space, T is connected.

Theorem 1.17 gives a non-intuitive, positive characterisation of connected spaces.

Theorem 1.17. *A metric space (X, ρ) is connected, if and only if, every continuous function from X to the discrete metric space $\{0, 1\}$ is constant.*

Proof. A surjective function $f : X \to \{0, 1\}$ is continuous, if and only if,

$f^{-1}(0)$ and $f^{-1}(1)$ are disjoint and closed. So X is connected, if and only if, f is constant. □

The intermediate value theorem (Bolzano's theorem) from calculus is a special case of Theorem 1.18. For the proof of this Theorem, recall that subsets A, B in a space X are **separated**, if and only if, $\mathrm{cl}\, A \cap B = \varnothing$ and $A \cap \mathrm{cl}\, B = \varnothing$.

Theorem 1.18. *Every continuous function from one metric space to another preserves connected sets.*

Proof.
Given metric spaces X, Y, suppose $E \subset X$ is a connected subset in X and f is a continuous function from X to Y. If $f(E)$ is not connected, is a union of non-empty subsets $A, B \subset Y$ such that $\mathrm{cl}\, A \cap B = A \cap \mathrm{cl}\, B = \varnothing$. Then

$$E = [E \cap f^{-1}(A)] \cup [E \cap f^{-1}(B)] \text{ and } [E \cap f^{-1}(A)] \text{ and } [E \cap f^{-1}(B)],$$

are separated in E, *i.e.*, E is not connected. ($\Rightarrow\Leftarrow$) □

If x is a point in an open subset G of the space $\mathbb{R}$, then there is a connected open interval $(x - \varepsilon, x + \varepsilon) \subset G$ for some $\varepsilon > 0$. Such spaces are termed *locally connected.* A metric space is **locally connected**, if and only if, each neighbourhood of any point contains a connected open neighbourhood of that point. The space $\mathbb{R}$ is locally connected but the topologist's sine curve (see Fig. 1.6) is not locally connected at $(0, 0)$.

A **component** C of a metric space X is a maximal connected subset of X. That is, C is connected and is not properly contained in another connected subset. Any two distinct components of a metric space are disjoint and closed. In a discrete metric space, singleton subsets are the only components. The same is true for the space $\mathbb{Q}$ of rational numbers.

In complex analysis, the usual domain of discussion is a subset in which any two points can be connected by an arc. A metric space X is termed **pathwise connected**, if and only if, for any two points $a, b \in X$, there is a continuous function $f : [0, 1] \to X$ with $f(0) = a, f(1) = b$. Notice that the topologist's sine curve is not path connected. Recall that a **homeomorhism** (also called a **topological transformation**) is a continuous, 1-1 mapping from a space X onto a space Y such that f^{-1} is continuous. A space is **arcwise connected**, if and only if, for any two points $a, b \in X$, there is a homeomorhism $f : [0, 1] \to X$ with $f(0) = a, f(1) = b$.

1.13 Chainable Metric Spaces

As in the case of compactness, there was a lot of experimentation to define connectedness. An early attempt was made by Cantor in 1883 to define connectedness in a metric space (X,d) via ε-chains. Given points p,q in X and $\varepsilon > 0$, an **ε-chain** of length n from p to q is a finite sequence $\{x_m\}, 0 \le m \le n$ such that

$$x_0 = p, x_n = q, \text{ and } d(x_{m-1}, x_m) < \varepsilon.$$

The metric space X is called **ε-chainable**, if and only if, each pair of points in X can be joined by an ε-chain. Finally, X is chainable, if and only if, for each $\varepsilon > 0$, each pair of points can be joined by an ε-chain.

The space of rational numbers with the usual metric is chainable but not connected. On the other hand, every connected metric space X is chainable. To show this, suppose X is not chainable. Then there is an $\varepsilon > 0$ and points p,q in X such that there is no ε-chain from p to q. Let E be the set of all points x of X such that there is a ε-chain from p to x. Then $S(E,\varepsilon)$, the union of all open balls with centre at points of E and radius ε, is open. But $S(E,\varepsilon)$ is also closed, since its complement

$$F = X - X(E,\varepsilon) = \bigcup\{S(x,\varepsilon) : x \in E^c\} \text{ is nonempty,}$$

since F contains q and is open. This contradicts the fact that X is connected.

The definition of chainability reminds us of total boundedness. To see the connection, let a chainable metric space X be called **uniformly ε-chainable**, if and only if, there is a natural number n such that each pair of points in X can be joined by an ε-chain of length at most n. And X is **uniformly chainable**, if and only if, it is uniformly ε-chainable for each $\varepsilon > 0$. Every totally bounded metric space X is uniformly chainable. For every $\varepsilon > 0$, there is an n such that

$$X = \bigcup\{S(x_m,\varepsilon) : x_m \in X, 0 \le m \le n\}.$$

Then each pair of points can be joined by a chain of length at most 2 + the length of the largest chain between x_m and $x_l, 0 \le m,l \le n$. A uniformly chainable space need not be totally bounded.

A locally compact space is one in which every point has a compact neighbourhood. A **uniformly locally compact** metric space is one in which there exists an $\varepsilon > 0$ such that all closed ε-balls are compact. Such a space is complete, since eventually the members of any Cauchy sequence can be included in one of the compact ε-balls.

It is now obvious that a chainable metric space that is compact is uniformly chainable and uniformly locally compact. Beer showed that the converse of the above statement is true.

Theorem 1.19. *A chainable metric space is compact, if and only if, the space is uniformly chainable and uniformly locally compact.*

1.14 UC Spaces

We recall the following results from complex analysis.

(a) Every continuous function on a compact set in $\mathbb{C}$ to $\mathbb{C}$ is uniformly continuous.

(b) Every function on an r-discrete set (in which any two points are at a distance greater than $r > 0$) to $\mathbb{C}$ is uniformly continuous.

(c) A subset E of complex numbers equals $A \cup B$, where A is compact and B is r-discrete, if and only if, every continuous function on E to $\mathbb{C}$ is uniformly continuous.

The above results from analysis led to a study of analogous results in a metric space. A metric space is called **UC**, if and only if, for each metric space Y, every continuous function on X to Y is uniformly continuous. Uniform continuity equals proximal continuity in metric spaces (*cf.* §1.11, Theorem 15). Continuity equals proximal continuity, if and only if, the metric proximity of the domain equals fine proximity, *i.e.*, two sets are near, if and only if, their closures intersect. Combining the above facts, we have the following result.

Theorem 1.20. *Let X be a metric space. The following are equivalent:*

(a) X is UC.

(b) Every pair of disjoint closed subsets of X are at a positive distance apart.

(c) Every continuous function on X to an arbitrary metric space Y is proximally continuous.

(d) Every continuous function on X to $\mathbb{R}$ is proximally continuous.

(We find it amusing that, while (d) is a trivial case of (c), it actually implies (c)!)

Proof. Only (d) implies (c) needs a proof. Suppose (c) is not true. Then there is a metric space Y, near sets A, B in X, and a continuous function $f : X \to Y$ such that $f(A)$ and $f(B)$ are far in Y. Then there is a function

$g : Y \to \mathbb{R}$ such that $gf(A)$ and $gf(B)$ are far in $\mathbb{R}$. So the function gf is continuous on X to $\mathbb{R}$, but gf is not proximally continuous. So (d) is not true. □

That a UC metric space X is complete follows easily. If X is not complete, then X has a Cauchy sequence (x_n) of distinct points which does not converge. Then the set $\{x_{(2n+1)} : n \in \mathbb{N}\}$ is near $\{x_{(2n)} : n \in \mathbb{N}\}$ in the metric proximity, but they are far in the fine proximity, a contradiction. So we have the result

Compact implies UC which, in turn, implies complete.

We conclude this section with an analogue of Theorem 20. Recall that p is a limit point of X, provided p is near $X - \{p\}$. The set of limit points of X, denoted by X', is called the **derived set** of X.

Theorem 1.21. *In a UC metric space* (X, d)
(a) the derived set X' *of* X *is compact, and*
(b) for every $\varepsilon > 0$, $Y = X - S(X', \varepsilon)$ *is uniformly discrete,* i.e., *there is an* $\eta > 0$ *such that, for any two distinct points* $x, y \in Y, d(x, y) > \eta$.

Proof. The proofs of (a) and (b) are similar.
(a) If X' is not compact, there exists a non-convergent sequence (x_n) of distinct points. Since each point of X' is a limit point of X, for each $n \in \mathbb{N}$, there is a point $y_n \in X$ such that $0 \le d(x_n, y_n) \le \frac{1}{n}$. Since (x_n) does not converge, the sequence (y_n) also does not converge. Then the sets $\{x_n : n \in \mathbb{N}\}$ and $\{y_n : n \in \mathbb{N}\}$ are near in the metric proximity but far in the fine proximity, a contradiction.
(b) If Y is uniformly discrete for some $\varepsilon > 0$, then, for each $n \in \mathbb{N}$, there are distinct points $x_n, y_n \in Y$ such that $0 \le d(x_n, y_n) \le \frac{1}{n}$. Then the sets $\{x_n : n \in \mathbb{N}\}$ and $\{y_n : n \in \mathbb{N}\}$ are near in the metric proximity. However, since Y has no limit points, they are far in the fine proximity, a contradiction. □

UC spaces have been studied extensively and are variously named after Lebesgue, Nagata, and Atsuji.

1.15 Function Spaces

The concept of an *arbitrary function* was almost unknown at the beginning of the nineteenth century. But the idea of *pointwise convergence* existed since the early days of calculus, especially in the study of power series and trigonometric series. The idea of putting a topological structure or convergence on a family of functions began with Riemann and a systematic study began towards the end of the nineteenth century. Uniform convergence was discovered in 1847-1848 by Stokes and Seidel independently. In a paper dated 1841 but published in 1894, Weierstrass used the notion of *uniform convergence.* Under the leadership of Weierstrass and Riemann, uniform convergence and related topics were studied during the last third of the nineteenth century. The topic of uniform convergence and its various ramifications were studied further by Arzelà, Dini, Volterra and others[1].

Let $(X, d), (Y, d')$ be metric spaces. Let $\mathcal{F}$ be the family of functions on X to Y. There are many situations in calculus that need convergences on $\mathcal{F}$. And there are many convergences. Classically, there have been three prominent convergences, namely, *pointwise convergence*, *uniform convergence* and *uniform convergence on compacta (compact sets).*

Let (f_n) be a sequence of functions in $\mathcal{F}$ with $f \in \mathcal{F}$. It is customary to use '$(f_n) \to f$' to mean '(f_n) converges to f'. The first convergence, which was quite natural, was pointwise convergence.

> (f_n) is said to **converge pointwise** to f, if and only if, for each $x \in X$, the sequence $((f_n(x))) \to f(x)$ in Y.

That is, for every $\varepsilon > 0$ and for each $x \in X$, there is a natural number m, depending on ε and x, such that, for all $n > m$, $d'(f_n(x), f(x)) < \varepsilon$. This convergence does not emerge from a metric on $\mathcal{F}$, but is topological in nature. So topology existed in the literature in analysis long before it was formalised in the twentieth century! It was soon found that pointwise convergence does not preserve continuity, *i.e.*,

if each f_n is continuous and (f_n) converges pointwise to f, *then* f need not be continuous.

Let $(f_n), f$ be functions on [0,1] to $\mathbb{R}$ defined by

$$\begin{aligned} f_n(x) &= x^n, \\ f(x) &= 0, \text{ for } 0 \leq x \leq 1, \text{ and} \\ f(1) &= 1. \end{aligned}$$

[1]For further information, there is Hobson's choice [92].

Then $(f_n) \to f$ pointwise. So although each f_n is continuous, the pointwise limit of f of the sequence (f_n) is not continuous. A natural problem arises in defining a convergence that preserves continuity and this led to the study of uniform convergence.

> (f_n) is said to **converge uniformly** to f, if and only if, for every $\varepsilon > 0$, there is a natural number m, *independent of* $x \in X$, such that, for all $n > m$ and for all $x \in X, d'(f_n(x), f(x)) < \varepsilon$.

Obviously, uniform convergence is stronger than pointwise convergence. The following example shows that it is strictly stronger.

Let $(f_n), f$ be functions on $\mathbb{R}$ to $\mathbb{R}$, defined by

$$f_n(x) = (x + \frac{1}{n})^2 \text{ and } f(x) = x^2.$$

It is easy to see that $(f_n) \to f$ pointwise for each fixed $x \in X$, $(x + \frac{1}{n})^2 \to x^2$. But for $x = n$,

$$|f_n(x) - f(x)| = \left|n + \frac{1}{n}\right|^2 > 2\,.$$

Hence, (f_n) does not converge *uniformly* to f. It is useful to know that unlike pointwise convergence on the family of functions $\mathcal{F}$, uniform convergence does arise from a metric.

Theorem 1.22. *Let $(X, d), (Y, d')$ be metric spaces and let $(f_n), f$ be a function on X to Y. If each f_n is (uniformly) continuous and $(f_n) \to f$ uniformly, then f is (respectively, uniformly) continuous.*

Proof. Let (x_n) be any sequence in X converging to $\mathcal{L}$ in X. That is, for any $\varepsilon > 0$, there is a natural number p such that, for all $n > p$, $d(x_n, \mathcal{L}) < \frac{\varepsilon}{3}$. Since $(f_n) \to f$ uniformly, eventually for all $x \in X$, $d'(f_n(x), f(x)) < \frac{\varepsilon}{3}$. For a particular n, f_n is continuous and so there is a neighbourhood U of p such that for all $x \in U$, $d'(f_n(x), f_n(p)) < \varepsilon/3$. Hence, for all $x \in U$,

$$d'(f(x), f(p)) < d'(f(x), f_n(x)) + d'(f_n(x), f_n(p)) + d'(f_n(p), f(x)) < \varepsilon.$$

This shows that f is continuous. We omit the case of uniform continuity, which is similar. □

The example in the previous paragraph shows that uniform convergence is not a necessary condition for a sequence of continuous functions to converge to a continuous function.

In classical analysis, both real numbers and complex numbers have the following property: a function that is continuous on each compact subset is continuous. This leads to the concept of *uniform convergence on compacta.*

> Let $(X,d),(Y,d')$ be metric spaces and let $(f_n), f$ be functions on X to Y. Then (f_n) is said to **converge uniformly on compacta** to f, if and only if, on each compact subset E of X, $(f_n) \to f$ uniformly. That is, for every $\varepsilon > 0$, there is a natural number m, *independent of* x in E, such that, for all $n > m$ and for all $x \in E$, $d'(f_n(x), f(x)) < \varepsilon$.

Since uniform convergence is quite strong, many mathematicians, including Arzela, Dini and others, have found necessary and sufficient conditions for a sequence of continuous functions to converge to a continuous function.

1.16 Completion

In real analysis, it is shown that the space of rational numbers $\mathbb{Q}$ is not complete. For example, the sequence whose n^{th} term is $(1+\frac{1}{n})^n$ is Cauchy but it does not have a limit in $\mathbb{Q}$. When the rational numbers are plotted on a straight line, one finds gaps that are filled up with irrational numbers, *i.e.*, the space $\mathbb{Q}$ is *completed* to obtain the space of real numbers $\mathbb{R}$. There are two well-known methods that are used to complete $\mathbb{Q}$.

($\mathbb{Q}$.1) Dedekind section that separates the rational numbers into two parts, namely, left and right, and

($\mathbb{Q}$.2) Weierstrass completion method that embeds rational numbers into the space of equivalent classes of Cauchy sequences.

Dedekind's section method cannot be extended to a metric space that need have a linear order like the rationals have. By contrast, Weierstrass's embedding method is used in constructing the completion of a metric space. There is also an isometrical embedding method used to embed a metric space in a suitable space of real-valued, bounded, continuous functions.

Given metric spaces $(X,d),(Y,d')$, let $f: X \to Y$ be a function from X to Y. The function f is termed an **isometry** from X onto Y, if and only if, f is bijective and, for all $p, q \in X$, $d(p,q) = d'(f(p), f(q))$. Let Z be a proper subset of space Y. If f is an isometry from space X onto Z, then X is said to be **isometrically embedded** into Y.

Metric Space Completion:
Completion of a metric space X is accomplished by isometrically embedding X onto a subset of a metric space known to be complete and then take its closure. Use the fact that $\mathbb{R}$ is a complete metric space. Then show that the space $C^*(X)$ of bounded, real-valued, continuous functions with the *sup* metric d' is complete. The final step is to embed X isometrically into $C^*(X)$.

Theorem 1.23. *The metric space* $(C^*(X), d')$ *is complete.*

Proof.
Let (X, d) be a metric space and let $C^*(X)$ denote the space of bounded, real-valued, continuous functions with the metric d' defined by

$$d'(f,g) = sup\{|f(x) - g(x)| : x \in X\}.$$

To see that $(C^*(X), d')$ is complete, consider a Cauchy sequence $(f_n) \in C^*(X)$. Then, for each $x \in X$. the sequence of real numbers $(f_n(x))$ is Cauchy. Since the space of real numbers is complete, the sequence $(f_n(x)) \to r \in \mathbb{R}$, a real number that we call $f(x)$. Since the convergence of (f_n) is uniform, the function f is continuous and obviously bounded, *i.e.*, $f \in C^*(X)$. Hence, $(C^*(X), d')$ is complete. □

Starting with the metric space (X, d), fix a point $a \in X$ and define a map μ from X to $C^*(X)$ defined by $\mu(p) = f_p$, where $f_p(x) = d(p,x) - d(a,x)$. Clearly, by the triangle inequality, $|d(p,x) - d(a,x)| \leq d(p,a)$, which shows that f_p is bounded and f_p is continuous, since f_p is the difference of two continuous functions. Hence, $f_p \in C^*(X)$. Given $p, q \in X$,

$$\begin{aligned} d'(f_p, f_q) &= sup\{|f_p(x) - f_q(x)| : x \in X\} \\ &= sup\{|d(p,x) - d(q,x)| : x \in X\} \leq d(p,q). \end{aligned}$$

Since $d(p,x) - d(q,x)$ takes the value $d(p,q)$ at the point q, $d(p,q) \leq d'(f_p, f_q)$ and, hence, $d(p,q) = d'(f_p, f_q)$, which shows that μ is an isometry from X into $C^*(X)$. Then take the closure of the isometric image to obtain the completion of the metric space (X, d). This line of reasoning proves what is known as the **Kuratowski Embedding Theorem**.

Theorem 1.24. *Every metric space (X, d) can be embedded isometrically into $C^*(X)$, the complete metric space of bounded, real-valued, continuous functions.*

1.17 Hausdorff Metric Topology

For a metric space (X, d), the set of all non-empty, closed subsets of X (denoted $CL(X)$) is called a **hyperspace**. The focus of this section is on what is known as the Hausdorff metric topology on the hyperspace $CL(X)$. The Hausdorff metric topology was discovered by D. Pompeiu and by F. Hausdorff. This topology has been used in a number of areas such as function space topologies, approximation theory, and optimisation.

In connection with the Hausdorff metric topology, recall the definitions of the gap distance and Hausdorff metric. Let A, B denote subsets of X. The **gap distance** between A and B (denoted $D(A, B)$) is defined to be

$$D(A, B) = inf\{d(a, b) : a \in A, b \in B\}.$$

For $\varepsilon > 0$, $S(A, \varepsilon) = \{x \in A : D(x, A) < \varepsilon\}$ (called the ε-enlargement of A).

The **Hausdorff metric** d_H on $CL(X)$ is defined in two equivalent ways.

$(d_H.1)$

$$d_H(A, B) = \begin{cases} inf\{\varepsilon > 0 : A \subset S(B, \varepsilon), B \subset S(A, \varepsilon)\}, & \text{if } \varepsilon \text{ exists,} \\ \infty, & \text{if no such } \varepsilon \text{ exists.} \end{cases}$$

$(d_H.2)$ $d_H(A, B) = sup\{|D(x, A) - D(x, B)| : x \in X\}$.

Proof. Equivalence of $d_H.1$ and $d_H.2$.

Let ρ, ρ' denote the formulae in $d_H.1$ and $d_H.2$, respectively. We show that ρ is equivalent to ρ'. Let a be any element of $A \subset X$. Then, from $d_H.2$, $|D(a, A) - D(a, B)| = D(a, B) \le \rho'$. This implies, for any $\varepsilon > 0$, that $D(a, B) < \rho' + \varepsilon$. Hence, $A \subset S(B, \rho' + \varepsilon)$ and, similarly, $B \subset S(A, \rho' + \varepsilon)$. In effect, $\rho \le \rho' + \varepsilon$. Since ε is arbitrary, $\rho \le \rho'$. For $x \in X$ and $\varepsilon > 0$, there is an $a \in A$ such that $d(x, a) < D(x, A) + \varepsilon$. Then there is a $b \in B$ such that $d(a, b) < D(a, B) + \varepsilon \le \rho + \varepsilon$. Hence,

$$D(x, B) \le d(x, b) \le d(x, a) + d(a, b) < D(x, A) + \rho + 2\varepsilon.$$

So, $D(x, B) - d(x, A) \le \rho + 2\varepsilon$. Interchanging A and B and noting that ε, we have $\rho' \le \rho$. Hence, $\rho' = \rho$. ☐

The **Hausdorff metric topology** is a topology induced by the metric d_H. In the Hausdorff metric topology, definition (d_H.2) shows that a sequence of closed sets (A_n) converges to a closed set A, if and only if, the sequences of functions $(D(x, A_n))$ converges *uniformly* to the function $D(x, A)$ on X. The Hausdorff metric topology on $CL(X)$ depends on the metric and not on the topology on X. Let X be the set of positive real numbers with metric $d(x,y) = |\frac{x}{x+1} - \frac{y}{y+1}|$ and the usual metric $d' = |x - y|$. The metrics d and d' induce the usual topology on X. However, in the Hausdorff metric topology, if A_n is the set of the first n natural numbers,

$$d_H(\mathbb{N}, A_n) = \frac{1}{(n+1)(n+2)} \to 0, \text{ but } d'_H(\mathbb{N}, A_n) = \infty.$$

Hence, although d, d' are compatible metrics on X, d_H, d'_H are not compatible on X. It is important to observe that the Hausdorff metric plays a significant role in fractal sets and measures.

1.18 First Countable, Second Countable and Separable Spaces

Let $(X, \rho), \mathbf{T}$ denote a metric space and metric topology, respectively. A *metric topology* with respect to X is the family of all open sets of X. Recall that a **base** for the topology $\mathbf{T}$ is a subfamily $\mathbf{B}$ of $\mathbf{T}$ such that each open set is a union of members of $\mathbf{B}$. In a similar manner, a **local base** at $x \in X$ is a family of neighbourhoods of x such that each neighbourhood of x contains a member of the family. For example, the family of open balls $\{S(x, \frac{1}{n}) : n \in \mathbb{N}\}$ is a local base at each $x \in X$. Also, notice that this family is countable.

> **First Countable & Second Countable Spaces:** When a space has a countable neighbourhood base at each point, the space is termed **first countable** (denoted $\mathcal{C}_I$). Thus, every metric space is first countable. If a space has a countable base for the topology $\mathbf{T}$, then the space is termed **second countable** (denoted $\mathcal{C}_{II}$). A simple example of a metric space that is not second countable is an uncountable set with the discrete metric whose value is 1 for distinct points.

An interesting property of the space of real numbers $\mathbb{R}$ is that it has a countable dense subset, namely, the set of rational numbers $\mathbb{Q}$. A metric space is termed **separable**, if and only if, the space has a countable dense subset. The real numbers provide an example of a separable space. A simple example of a metric space that is not separable is an uncountable set with the discrete metric. On the other hand, every second countable metric space is separable. For instance, consider a set that contains one point from each member of the countable base. The countable set so obtained is dense. A valuable property satisfied by a separable metric space is that it is second countable! Suppose (X, ρ) has a countable dense subset $\mathcal{D}$. Then the family of open balls $\{S(x, \varepsilon) : x \in \mathcal{D},\ \varepsilon \in \mathbb{Q}^+\}$ is countable, where $\mathbb{Q}^+$ is the set of positive rationals. If U is an open set in X and $p \in X$, then there is an open ball $S(p, \eta) \subset U$, where $\eta > 0$. Let $q \in \mathcal{D}$ be such that $d(p, q) < \frac{\eta}{3}$. Then the open ball $S(q, \frac{\eta}{3})$ contains p and $S(q, \frac{\eta}{3}) \subset S(p, \eta) \subset U$.

There is another property of metric spaces reminiscent of the Heine-Borel-Lebesgue definition of compactness, namely, every open cover of a space X has a finite subcover. Lindelöf introduced an analogous notion, *i.e.*, every open cover of X has a countable subcover. This property is satisfied by the space of real {complex} numbers but not by an uncountable, discrete metric space. In metric spaces, the three properties (separable, second countable and Lindelöf) are equivalent but not, in general, in topological spaces, something that will be considered in detail, later. For now, consider

Theorem 1.25. *In a metric space* (X, ρ), *the following are equivalent.*
(MS.1) X *is separable.*
(MS.2) X *is second countable.*
(MS.3) X *is Lindelöf.*

1.19 Dense Subspaces and Taimanov's Theorem

Extensions of continuous functions from dense subspaces of metric spaces are the focus of this section. Recall that a subset $\mathcal{D}$ of a space X is a **dense subset** in X, if and only if, the closure of $\mathcal{D}$ equals X, or, equivalently, $\mathcal{D}$ intersects every non-empty open set in X. Consider the following examples where there are no continuous extensions of continuous functions from the dense subspace (0,1) to the whole space [0,1].

(1) Put $f : (0, 1) \to \mathbb{R}$ defined by $f(x) = \frac{1}{x}$. Obviously, f is continuous but has no extension to [0,1], since f is unbounded (see Fig. 1.7.1 with a horizontal line at $y = 2$ and the x-axis not shown).

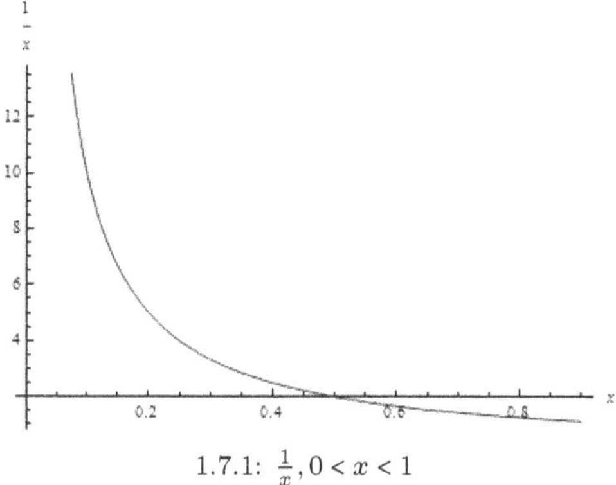

1.7.1: $\frac{1}{x}, 0 < x < 1$

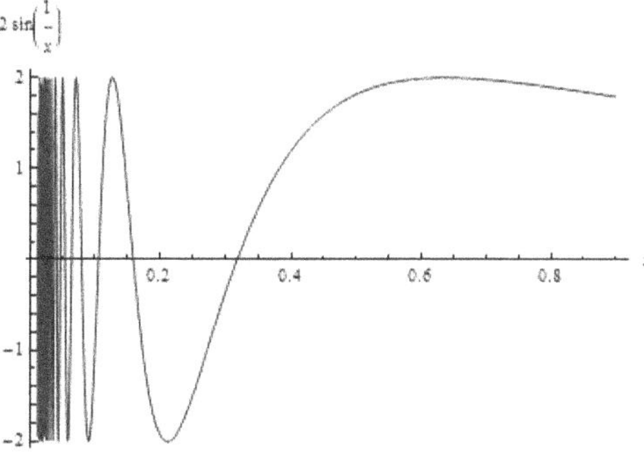

1.7.2: $2sin(\frac{1}{x}), 0 < x < 1$

Fig. 1.7 Sample functions in dense subspace (0,1)

(2) Put $g : (0,1) \to \mathbb{R}$ defined by $g(x) = 2sin(\frac{1}{x})$. This function is bounded but has no limit as $x \to 0$ and has no extension to [0,1] (see Fig. 1.7.2).

One of the most interesting and important problems in analysis is formulated as follows.

Problem Concerning Dense Subsets:
Let $(X, d), (Y, d')$ denote metric spaces and let $\mathcal{D}$ denote a dense subset of X. Given a continuous function $f : \mathcal{D} \to Y$, find necessary and/or sufficient conditions for the existence of a continuous extension $\hat{f}$ of f, *i.e.*, $\hat{f} : X \to Y$ is continuous and the restriction $\hat{f}|_{\mathcal{D}}$ of $\hat{f}$ to $\mathcal{D}$ equals f.

Notice that if an extension $\hat{f}$ exists, then it must be unique. This follows from the fact that if there were two continuous extensions $\hat{f}, \hat{g}$, then the set $\{x \in X : \hat{f}(x) = \hat{g}(x)\}$ contains $\mathcal{D}$ and would be closed in X.

This problem in its various avatars will be pursued later but, for now, consider the following special case. A sufficient condition that is well-known in the general case is that f be *uniformly continuous.* A.D. Taimanov first gave necessary and sufficient conditions for the existence of a continuous extension that led to various generalisations which were unified into a single result using proximity.

A first step towards the solution of the dense subsets problem is the assignment of a *necessarily unique* value $\hat{f}(x)$ for each $x \in X$ and this value should agree with $f(x)$, whenever $x \in \mathcal{D}$. For each $x \in X$, there is a sequence $\phi : \mathbb{N} \to \mathcal{D}$ that converges to x. This convergent sequence is Cauchy. Since we want $\hat{f}$ to be continuous, we would like $\hat{f} \circ \phi = f \circ \phi : \mathbb{N} \to Y$ to converge $\hat{f}(x) \in Y$. A necessary condition for this to occur is that $f \circ \phi$ is Cauchy in Y and, further, a sufficient condition is that Y be complete.

Now suppose that Y is complete and that f preserves Cauchy sequences. To have an unambiguous definition of $\hat{f}(x)$ for each $x \in X$, ensure that if any two sequences $\phi_1, \phi_2 \in \mathcal{D}$ converge to x, then $f \circ \phi_1, f \circ \phi_2$ both converge to the same limit in Y.

Finally, show that $\hat{f} : X \to Y$ is continuous, *i.e.*, if a sequence $\psi : \mathbb{N} \to X$ converges to $x \in X$, then the sequence $\hat{f} \circ \psi(n)$ converges to $\hat{f}(x) \in Y$. For each $n \in \mathbb{N}, \psi(n) \in X$, it is the case that there is a sequence $\gamma_v \in \mathcal{D}$ that converges to $\hat{f} \circ \psi(n)$ in Y. So, for each $n \in \mathbb{N}$, there is an $a_n \in \mathcal{D}$ with $|a_n - \psi(n)| < \frac{1}{n}$ and $|\hat{f}(x) - \hat{f} \circ \psi(n)| < \frac{1}{n}$. The sequence $\lambda : \mathbb{N} \to \mathcal{D}$ defined by $\lambda(n) = a_n$ converges to $x \in X$ because

$$|x - \lambda(n)| < |x - \psi(n) + \psi(n) - \lambda(n)| < |x - \psi(n)| + \frac{1}{n}$$

and both terms are eventually as small as possible. As a result, the sequence $f \circ \lambda$ converges to $\hat{f}(x)$. In a similar fashion, show that $|\hat{f}(x) - \hat{f} \circ \psi(n)|$ is eventually as small as possible. This line of reasoning proves Theorem 1.26.

Theorem 1.26. *Suppose $\mathcal{D}$ is a dense subset of a metric space (X, ρ) and that (Y, η) is a complete metric space. Then a necessary condition that a continuous function $f : \mathcal{D} \to Y$ has a continuous extension $\hat{f} : X \to Y$ is that f preserves Cauchy sequences. Further, if X is complete, then the condition is also sufficient.*

A uniformly (proximally) continuous function $f : \mathcal{D} \to Y$ preserves Cauchy sequences. To see this, let E, F be infinite subsets of $\mathbb{N}$ and let ϕ denote a Cauchy sequence in $\mathcal{D}$. Then, $\phi(E)$ is near $\phi(F)$ in $\mathcal{D}$ and, since f is proximally continuous, $f \circ \phi(E)$ is near $f \circ \phi(F)$ in Y. This shows that $f \circ \phi$ is Cauchy in Y. Next, show that $\hat{f}$ is uniformly continuous.

Let $\varepsilon > 0$ be assigned. Since f is uniformly continuous, there is a $\kappa > 0$ such that, for all $r, s \in \mathcal{D}$, $\rho(p,q) < \kappa$ implies $\eta(f(r), f(s)) < \frac{\varepsilon}{3}$. Now suppose that $x, y \in X$ with $\rho(x,y) < \frac{\kappa}{3}$. Since $\mathcal{D}$ is dense in X and $\hat{f}$ is continuous, there are $p, q \in \mathcal{D}$ such that

$$\rho(p,x) < \frac{\kappa}{3},\ \rho(q,y) < \frac{\kappa}{3},\ \eta(f(p), \hat{f}(x)) < \frac{\varepsilon}{3},\ \eta(f(q), \hat{f}(y)) < \frac{\varepsilon}{3}.$$

Combining these inequalities, it follows that $\eta(\hat{f}(x), \hat{f}(y)) < \varepsilon$, showing thereby that $\hat{f}$ is uniformly continuous. This proves a well-known result (see Theorem 1.27).

Theorem 1.27. *Suppose $\mathcal{D}$ is a dense subset of a metric space (X,ρ) and that (Y,η) is a complete metric space. Then a uniformly continuous function $f : \mathcal{D} \to Y$ has a uniformly continuous extension to X.*

Next, suppose that Y is compact and define a proximity δ on $\mathcal{D}$ such that

$$E\ \delta\ F \iff \mathrm{cl}_X E \cap \mathrm{cl}_X F \neq \varnothing.$$

That is, $E\ \delta\ F$ in $\mathcal{D}$, if and only if, $E\ \delta_0\ F$, where δ_0 is the fine proximity on X. Since Y is compact, the metric proximity δ' on it is the fine proximity. Suppose $f : \mathcal{D} \to Y$ has a continuous extension $\hat{f} : X \to Y$. Then $\hat{f}$ is proximally continuous and so is f. Hence, proximally continuous is a necessary condition for the existence of $\hat{f}$.

To show that *proximally continuity* is also a sufficient condition for the existence of $\hat{f}$, it is enough to show that if f is proximally continuous, then f preserves Cauchy sequences.

Proof. Suppose f is proximally continuous and ϕ is a Cauchy sequence in $\mathcal{D}$. If P, Q are any two infinite subsets of $\mathbb{N}$, then, since ϕ is Cauchy, $\phi(P)\ \delta\ \phi(Q)$ in X. The proximal continuity of f implies that $f \circ \phi(P)\ \delta'\ f \circ \phi(Q)$ in Y, showing thereby that f preserves Cauchy sequences. □

Theorem 1.28 is a special case of a result proved by A.D. Taimanov. For generalisations of this result, see Section 10.2.

Theorem 1.28. Taimanov
Suppose $\mathcal{D}$ is a dense subset of a metric space (X,d) and that (Y,d') is a compact metric space. A necessary and sufficient condition that f has a continuous extension from X to Y is that, for all disjoint, closed subsets E, F in Y, $\mathrm{cl}_X f^{-1}(E)$ and $\mathrm{cl}_X f^{-1}(F)$ are disjoint in X.

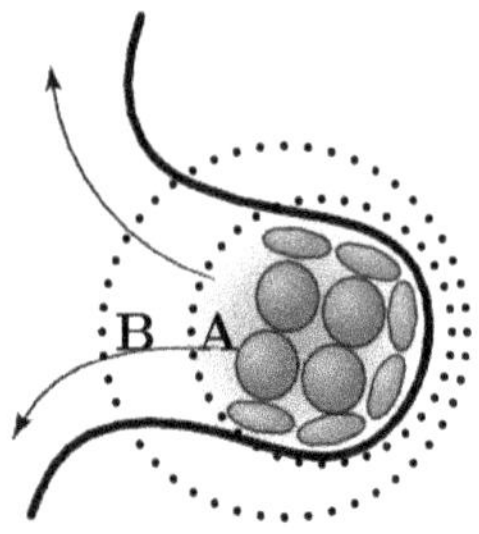

Fig. 1.8 Stem cell niche proximal neighbourhood

1.20 Application: Proximal Neighbourhoods in Cell Biology

Let B denote a set of tissue cells that form a **niche** (or microenvironment [238]), A, a set of stem cells. In a **simple niche**, stem cells reside in tissue, and over time produce progeny cells while self-renewing [166] (see, also, [197; 136]. In other words, B is a neighbourhood of A, since $A \subset \text{int}(B)$. A proximal neighbourhood for storage stem cells [166] is shown in Fig. 1.8. In addition, $A \underline{\delta} B^c$, *i.e.*, it can be shown that B is a proximal neighbourhood of A (see §1.7 for details). In fact, in keeping with the cell biology view of stem cells, there is a **fine proximity** relation δ between stem cells in A and partner cells $C \subset B$ (*i.e.*, δ satisfies axioms (P.1)-(P.4)), where migration of stem cells interact with surrounding tissue cells. Hence, $\text{cl}\, A \cap \text{cl}\, B \neq \emptyset$ (a relationship expressed by $A\ \delta_0\ B$). In other words, the proximity relation δ_0 provides a basis for a topological framework for representing, analysing, and classifying stem cell niches. The application of the proximity space approach in cell biology is discussed in detail in terms of NLO microscopy in Chapter 4.

1.21 Problems

A. Metric Space Conditions

(1.21.1) Show that $(\mathbb{R}^m, d_2)$ is a metric space.

(1.21.2) Using only conditions M.2 and M.4 for a metric, derive conditions M.1 and M.3.

(1.21.3) Give an example of a *pseudometric space.*

(1.21.4) Put $d(x, y) = |x - y|$ for $x, y \in \mathbb{R}$. Show that the triangle inequality holds for d.

(1.21.5) Let $p, q \in \mathbb{R}^2$ (points with coordinates $(p_1, p_2), (q_1, q_2)$, respectively, in the Euclidean plane. Show that $d(p,q) = min\{|p_1 - q_1|, |p_2 - q_2|\}$ is not a metric.

(1.21.6) Verify that condition M.4 in a metric space (X, d) is satisfied by what is known as the *reverse triangle inequality*[2]: $|d(x,z) - d(y,z)| \le d(x,y)$ for any $x, y, z \in X$.

B. Gap Functional, Closure and Nearness

(1.21.7) Prove that $D_d(A, B) \ge 0$ for every A and B. Hint: Use the fact that d is a metric.

(1.21.8) For the metric space $(\mathbb{R}, d)$, consider $B(0,8) = \{y \in X : d(0,y) < 8\}$, and give a $y \in \mathbb{R}$ (reals) such that $y\ \delta\ B(0,8)$, *i.e.*, a y near $B(0,8)$.

(1.21.9) A point x is a limit point of a set A if x belongs to $\text{cl}(A - \{x\})$. Give three examples of limit points for $A = (0,1]$ on the real line $\mathbb{R}$.

(1.21.10) Give an example of near sets that satisfy near set condition N.1. See, *e.g.*, Fig. 1.1.

(1.21.11) Give an example of near sets that satisfy the asymptotically near set condition N.2. See, *e.g.*, Fig. 1.2.

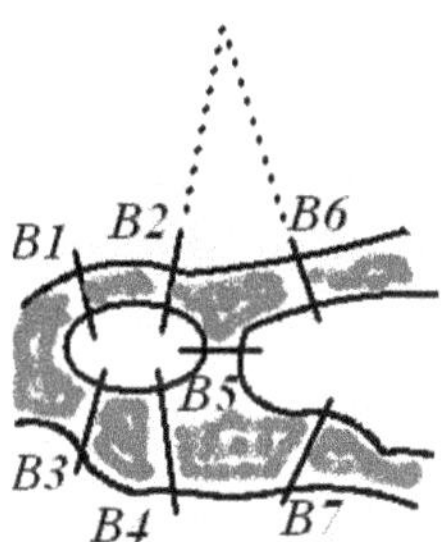

Fig. 1.9 Bridges (planar view)

(1.21.12) The city of Königsberg in Prussia is set on both sides of the Pregel River and includes two large islands connected to each other and the mainland by seven bridges. For simplicity, the bridges can be represented by lines in the Euclidean plane (see Fig. 1.9). For example, $A = \{a :$ point a lies on line $B2\}$ and the bridge

[2]The reverse triangle inequality appears in [217].

lines extend in either direction to infinity (see, *e.g.*, $B2, B6$ in the geometric view of the bridges). Identify all pairs of bridges that are near each other. For each pair of near sets, indicate which near set condition is satisfied.

(1.21.13) Let A, B denote sets. Show that

$$A \subset B \Rightarrow \operatorname{cl} A \subset \operatorname{cl} B.$$

(1.21.14) Compare the assertion

$$D(x, A) = 0 \iff x \in \operatorname{cl} A.$$

with the definition of closure. What can you conclude?

(1.21.15) d_2 (Euclidean distance) is a metric.

(1.21.16) Taxicab or Manhattan metric: $d_1(x, y) = |x_1 - x_2| + |y_1 - y_2|$.

(1.21.17) Max metric: $d_{max}(x, y) = max\{|x_1 - x_2|, |y_1 - y_2|\}$. Show that if min is substituted for max, then d_{min} is not a metric.

C. Convergence, Limit of a Sequence, Nearness

(1.21.18) Metrics d, d' on X are said to be **equivalent metrics** if they preserve convergent sequences, *i.e.*, for each sequence (x_n) and $\mathcal{L} \in X$,

$$d(x_n, \mathcal{L}) \to 0 \iff d'(x_n, \mathcal{L}) \to 0.$$

(1.21.19) Prove that if d is a metric on X, then $\frac{d}{d+1}$ is also an equivalent metric. Notice that since $0 \leq \frac{d}{d+1} < 1$, it follows that every metric space has an equivalent *bounded* metric. Another equivalent bounded metric is given by $d^* = min\{d, 1\}$.

(1.21.20) Given two sequences a_n, b_n in a metric space $\langle X, \rho \rangle$, state a condition that makes the following assertion true: $\{a_n\}\ \delta\ \{b_n\}$.

(1.21.21) Given

$$\{\frac{1}{2^{m_n}} \mid m, n \in \mathbb{N}\}, \text{where } 1 \leq m_1 < \cdots < m_n,$$

from C. Kuratowski [118], let m_i denote a Fibonacci number in the sequence

$$1, 1, 2, 3, \ldots, m_i, m_{i+1}, m_{i+2}, \ldots,$$

where $m_{i+2} = m_{i+1} + m_i$, *i.e.*, after 1, 1, each Fibonacci number is the sum of the previous two numbers. Does $\{\frac{1}{2^{m_n}}\}$ converge to a limit?

(1.21.22) Given

$$\{\frac{t_n}{3^n} \mid t, n \in \mathbb{N}\},\ \text{where } t_n \in \{0, 2\},$$

obtain, for example, $\frac{0}{3}, \frac{2}{9}, \frac{0}{27}, \ldots$. G. Cantor [36], p. 590, suggested that t take on both 0 and 2 values. Assume the choice of t_n is random for each n and consider the convergence of this Cantor sequence.

(1.21.23) Given

$$\{\frac{t_n}{3^n} \mid t, n \in \mathbb{N}\},\ \text{where } 0 \le t \le 1.$$

from C. Kuratowski [118], p. 11, obtain, for example, $\frac{\frac{1}{2}}{3}, \frac{\frac{1}{4}}{9}, \frac{\frac{1}{8}}{27}, \ldots,$ where $t = \frac{1}{2}$. Does

$$\left\{\frac{t}{3^n}\right\} \text{ converge to a limit}$$

for all values of t in [0,1]?

(1.21.24) Show that the limit of a sequence is *unique*, *i.e.*, no sequence can converge to more than one point. Hint: the proof is similar to convergence proofs in calculus and analysis.

D. Continuity

(1.21.25) Prove that constant functions are continuous.

(1.21.26) Prove that the identity function $f : X \to X$, given $f(x) = x$, is continuous.

(1.21.27) Prove that if $f : X \to \mathbb{R}$ is a continuous function, then $|f|$ (*i.e.*, $|f|(x) = |f(x)|$) is continuous.

(1.21.28) Give a definition of continuity of a function at a point in terms of neighbourhoods.

(1.21.29) Give a graphical interpretation of the situation described in Theorem 1.4.

(1.21.30) Show that

(a) A point p is a limit point of a set E, if and only if, there is a sequence (x_n) of distinct points in $E - \{p\}$ that converges to p.

(b) A point $c \in X$ is a cluster point of a sequence f, if and only if, there is a subsequence of f that converges to c.

(1.21.31) Let (X,d), (Y,d') be a pair of metric spaces and let $f : X \to Y$. Prove that if

$$x_0 \in \text{cl}\, E \text{ implies } f(x_0) \in \text{cl}\, f(E) \text{ for every } E \subset X,$$

then f is continuous at x_0.

(1.21.32) Let (X,d), (Y,d') be a pair of metric spaces and let $p \in X$. Show that the following are equivalent.

(a) For every $\varepsilon > 0$, there is an $\kappa > 0$ such that for each $x \in X$ with

$$d(x,p) < \kappa \text{ implies } d'(f(x)), f(p)) < \varepsilon.$$

(b) For each sequence $(x_n) \to p$ in X, $f(x_n) \to f(p)$ in Y.

(c) For each subset $E \subset X$, $D(p,E) = 0$ implies $D'(f(p)), f(E)) = 0$.

(d) For each open set $E \subset Y$, $f^{-1}(E)$ is open in X.

(e) for each closed set $F \subset Y$, $f^{-1}(F)$ is closed in X.

(f) For every subset $E \subset X$, $f(\text{cl}\, E) \subset \text{cl}\, f(E)$.

E. Open and Closed Sets

(1.21.33) In a metric space (X, ρ_X), prove that X and the empty set $\varnothing$ are open sets.

(1.21.34) Prove Lemma 1.2.

(1.21.35) Prove that a set E is closed, if and only if, $E = \text{cl}\, E$.

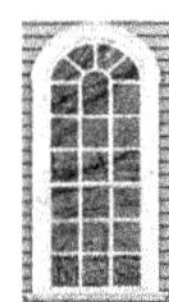

Fig. 1.10 Mullioned window

(1.21.36) A mullion is a structural element that divides adjacent panes of glass into a grid (see Fig. 1.10). How does a mullioned window with a missing pane of glass differ from a set? **Hint**: The window frame remains after the glass is removed.

(1.21.37) In a metric space, prove that every pair of distinct points is contained in disjoint open sets. Such a space is called a **Hausdorff space**.

(1.21.38) In a metric space, prove that every pair of disjoint closed sets are contained in disjoint open sets. Such a space is called a **normal space**.

(1.21.39) Show that every closed set in a metric space is a countable intersection of open sets. Such a set is called a G_δ set.

(1.21.40) **Complement of Problem 1.21.39**. Show that every open set in a metric space is a countable union of closed sets. Such a set is called an F_σ set.

F. Metric and Fine Proximities

(1.21.41) Prove property P.1 for Theorem 1.6.

(1.21.42) Prove property P.2 for Theorem 1.6.

(1.21.43) Prove property P.3 for Theorem 1.6.

(1.21.44) Prove property P.6 for Theorem 1.6.

(1.21.45) Prove $\Rightarrow$, Part (b) for Lemma 1.3.

(1.21.46) Prove P.1-P.5 with PN.1-PN.6.

(1.21.47) Let $\mathbb{R}$ denote the set of real numbers with the usual metric. Show that the following relations satisfy P.1-P.5 and are compatible proximities.

(a) **Proximity induced by one-point compactification.** $A\ \delta_1\ B \iff \mathrm{cl}\,A \cap\ \mathrm{cl}\,B \neq \varnothing$ or both A, B are unbounded.

(b) **Proximity induced by two-point compactification.** $A\ \delta_2\ B \iff \mathrm{cl}\,A \cap\ \mathrm{cl}\,B \neq \varnothing$ or both A, B are unbounded above or both are unbounded below.

(c) Show that $\delta_\rho, \delta_0, \delta_1, \delta_2$ are distinct.

(d) **Coarse proximity**. $A\ \delta_c\ B \iff \mathrm{cl}\,A \cap\ \mathrm{cl}\,B \neq \varnothing$ or both A, B are infinite. Show that δ_c does not satisfy P.5 but satisfies

$$A\ \delta_c\ B \text{ and } b\ \delta_c\ C, \text{for each } b \in B \text{ implies } A\ \delta_c\ C.$$

(1.21.48) Let $\mathbb{C}$ denote the set of complex numbers with the usual metric. Show that δ_1 defined by $A\ \delta_1\ B \iff \mathrm{cl}\,A \cap\ \mathrm{cl}\,B \neq \varnothing$ or both A, B are unbounded satisfies P.1-P.5 and is compatible. This proximity is induced by the one-point compactification of $\mathbb{C}$ known as **Riemann sphere**.

(1.21.49) **Theorem of Efremovič**. Let $(X,\rho), (Y,\xi)$ denote metric spaces and let $f : X \to Y$ be a function from X to Y. Then f is termed **uniformly continuous**, if and only if, for every $\varepsilon > 0$, there is an $\eta > 0$ such that for all $x, y \in X, \rho(x,y) < \eta$ implies $\xi(f(x), f(y)) < \varepsilon$. Show that f is uniformly continuous, if and only if, it proximally continuous.

(1.21.50) For a metric space $(X,\rho), p \in X, A, B \subset X$, show that if p is near the pair of subsets A, B, then A, B are near each other in both metric and fine proximities.

G. Compactness

(1.21.51) To complete the proof of Theorem 1.8, prove (c) $\Longleftrightarrow$ (d) $\Longleftrightarrow$ (e). Hint: see Problem 1.21.30.

(1.21.52) Prove that the open interval $(0,1)$ in $\mathbb{R}$ is not compact. Hint: consider subfamilies of the family of open sets $\{(\frac{1}{n},1) : n \in \mathbb{N}, n > 1\}$.

(1.21.53) Prove that the assertion *Every closed and bounded subset of a metric space is compact* is false.

(1.21.54) Show that the finite union of compact sets is compact.

(1.21.55) Show that in a compact metric space (X,d), metric proximity δ_d equals fine proximity δ_0.

(1.21.56) Show that the space of natural numbers $\mathbb{N}$, with the usual metric d, is not compact but that $\delta_d = \delta_0$.

(1.21.57) The **diameter** of a subset $E \subset X$ in a metric space (X,d) is denoted by diam$E$ and diam$E$ = $sup\{d(x,y) : x,y \in E\}$. Show that the following observations hold.

(diam.1) The diameter of a subset $E \subset X$ equals the diameter of its closure.

(diam.2) A set $E \subset X$ is bounded, if and only if, diam$E$ is finite.

(diam.3) A compact subset $E \subset X$ has a finite diameter and there are points $p,q \in E$ such that diam$E = d(p,q)$.

(1.21.58) Show that a continuous bijection on a compact space to another space is a homeomorphism.

(1.21.59) Suppose a space (X,d) is compact and f is a self-map such that $d(f(x),f(y)) \geq d(x,y)$. Prove that f is an isometry, *i.e.*, $d(f(x),f(y)) = d(x,y)$. Let f be a self-map that is 1-1, onto and $d(f(x),f(y)) \leq d(x,y)$. Show that f is isometry.

H. Completeness and Total Boundedness

(1.21.60) Prove the equivalence of (b) and (c) in Theorem 1.16.

(1.21.61) Show that a closed, bounded subset of $\mathbb{R}^m$ is compact.

(1.21.62) Show that every subsequence of a Cauchy sequence is Cauchy.

(1.21.63) **Hausdorff metric**. Let (X,d) denote a metric space and let $CL(X)$ denote the family of all non-empty, closed subsets of X.

For $E \subset CL(X)$ and $\varepsilon > 0$, put $S(E,\varepsilon) = \bigcup\{S(x,\varepsilon) : x \in E\}$. Show that $d_H : CL(X) \times CL(X) \to \mathbb{R}$ defined by

$$d_H(A,B) = \begin{cases} inf\{\varepsilon : A \subset S(B,\varepsilon) \text{ and } B \subset S(A,\varepsilon)\}, & \text{if } \varepsilon > 0, \\ \infty, & \text{otherwise} \end{cases}$$

is an infinite-valued metric on $CL(X)$. d_H is known as the Hausdorff metric. See G. Beer [23] for the literature up to 1993. For a more recent view of the Hausdorff metric, see J. Henrikson [85].

(1.21.64) Show that (X,d) is complete (respectively, totally bounded, compact), if and only if, $(CL(X), d_H)$ is complete (respectively, totally bounded, compact).

(1.21.65) **Strong uniform continuity** Given metric spaces $(X,d), (Y,d')$, suppose that $f : X \to Y$ is continuous and K is a compact subset of X. Prove that, for every $\varepsilon > 0$, there is a $\kappa > 0$, such that, for all points $p,q \in X$, $d(p,q) < \kappa$ and $\{p,q\} \cap K \neq \emptyset$ implies $d'(f(p), f(q)) < \varepsilon$. Show that this property of f on a subset of X implies uniform continuity on E which is strictly weaker [25], Prop. 1.2.

(1.21.66) **Cantor's Completeness Theorem** Recall that a family of sets $\mathfrak{N}$ is a **nest**, if and only if, whenever $A, B \in \mathfrak{N}$, then either $A \subset B$ or $B \subset A$ [106]. Show that a metric space (X,d) is complete, if and only if, every nest of non-empty, closed sets with diameters tending to 0, has a unique point of intersection.

(1.21.67) Given metric spaces $(X,d), (Y,d')$, let $\mathbb{B} = \mathbb{B}(X,Y)$ denote the family of all bounded, continuous functions from X to Y. For $f,g \in \mathbb{B}$, put

$$\mathcal{B}(f,g) = sup\{d'(f(x), g(x)) : x \in X\}.$$

Show that if Y is complete, then so is $(\mathbb{B}, \mathcal{B})$.

(1.21.68) **Banach Fixed Point Theorem**. Let (X,ρ) denote a complete metric space. Let f be a contractive self-map on X, *i.e.*, there is a λ such that $0 < \lambda < 1$ and $\rho(f(p), f(q)) \leq \lambda \cdot \rho(p,q)$, for all $p,q \in X$. Prove that the function f has a unique fixed point, *i.e.*, there is a point $z \in X$ such $f(z) = z$.

(1.21.69) Show that a uniformly continuous function preserves Cauchy sequences but that the converse is not true.

(1.21.70) Assume (X,d) is a totally bounded metric space and let $f : X \to X$ be a function defined on X. If there exist points $p,q \in X$ such that $d(f(p), f(q)) > d(p,q)$, then there exist points $s, r \in X$ such that $d(f(r), f(s)) < d(r,s)$. Notice that there are no restrictions on the function f, *i.e.*, f is *not* required to be continuous or injective or surjective. Show that Problem 1.21.59 follows from the above result (see, also, [161]).

(1.21.71) Given a metric space (X, ρ) and metric proximity δ. Let $\mathfrak{B}$ be the family of all finite unions of closed balls in (X, ρ) with non-negative radii. We say that a closed subset $E \subset X$ is **relatively totally bounded** with respect to $\underset{\bullet}{\delta}$ in an open set W, if and only if, there is a $B \subset \mathfrak{B}$ such that $E \ll B \ll W$ with respect to $\underset{\bullet}{\delta}$. Prove the following statements.

(a) A subset $F \subset X$ is totally bounded, if and only if, for every $\varepsilon > 0$, each closed subspace $E \subset F$ is relatively totally bounded with respect to δ in $S(E, \varepsilon)$.

(b) A subset $F \subset X$ is compact, if and only if, each closed subspace $E \subset F$ is relatively totally bounded with respect to $\underset{\bullet}{\delta}$ in every open set U containing E.

I. Connectedness

(1.21.72) Prove that the topologist's sine curve is not path connected (see Fig. 1.6).

(1.21.73) Show that the following assertions are equivalent in a metric space X.

(a) X is connected.

(b) Every proper subset of X has a non-empty boundary in X.

(c) No proper subset of X is both open and closed.

(d) Every real-valued, continuous function preserves connected sets.

(1.21.74) Show that if a subset E of a metric space is connected, then $\mathrm{cl}\, E$ is connected.

(1.21.75) Put $C([0,1])$ equal to the set of real-valued, continuous functions on the closed interval $[0,1]$ with the *sup* metric. Show that the metric space $(C([0,1]), sup)$ is connected.

(1.21.76) Let $f : \mathbb{R} \to \mathbb{R}$ be a function. Show that if f preserves compact sets as well as connected sets, then f is continuous.

(1.21.77) Let X and Y denote metric spaces and assume X is locally connected. Show that $f : X \to Y$ is continuous, if and only if, f preserves compact sets as well as connected sets.

(1.21.78) Show that every pathwise connected space is connected but the converse is not true.

(1.21.79) Show that the following subsets of $\mathbb{R}$ are not pairwise homeomorphic: (0,1), [0,1), and [0,1].

(1.21.80) A metric space X is **proximally connected**, if and only if, every real-valued, proximally continuous function from X to the discrete space $\{0, 1\}$ is constant. Compare this concept with the assertion that X *is connected.*

J. Miscellaneous Problems

(1.21.81) Show that a subset E of $\mathbb{R}$ is UC, if and only if, $E = A \cup B$, where A is compact and B is uniformly discrete.

(1.21.82) Suppose $f : (X, d) \to (Y, d')$ is continuous. Prove that $\rho(x, y) = d(x, y) + d'(f(x), f(y))$ is a compatible metric on X and makes $f : (X, \rho) \to (Y, d')$ uniformly continuous.

(1.21.83) Prove the following characterization of a UC metric space (X, d):

(a) For each open cover $\mathcal{G}$ of X, there is an $\varepsilon > 0$ (called Lebesgue number) such that each subset of X with diameter less than ε lies in some member of $\mathcal{G}$.

(b) The derived set X' (the set of all limit points of X) is compact and there is an $\varepsilon > 0$ such that any two points y, z at distances greater than ε from X' are at a distance greater than η for some $\eta > 0$.

(1.21.84) A sequence (x_n) in a metric space is called **pseudo-Cauchy**, if and only if, for each $\varepsilon > 0$, frequently $d(x_m, x_n) < \varepsilon$.

(a) Show that every Cauchy sequence is pseudo-Cauchy.

(b) Prove that a metric space is UC, if and only if, every pseudo-Cauchy sequence of distinct terms has a cluster point.

(1.21.85) Show that on the space of real numbers, a function that is continuous on each compact subset is continuous. Such a space called a k-space.

(1.21.86) Let $(X, d), (Y, d')$ be metric spaces and let $(f_n), f$ be functions on X to Y. Let $\mathcal{F}$ be the family of functions on X to Y. For f, g in $\mathcal{F}$, set

$$\rho(f, g) = sup\{d'(f(x), g(x)) : x \in X\}.$$

Show that ρ is an infinite-valued metric on $\mathcal{F}$ (it is called the *metric of uniform convergence*). Show that (f_n) converges uniformly to f, if and only if, $\rho(f_n, f)$ converges to zero. (To get the usual metric, we may replace d' by equivalent bounded metric.)

Chapter 2

What is Topology?

2.1 Topology

Topology is a branch of mathematics that is an abstraction of analysis and geometry. Real and complex analysis leads to a metric space (starting in Sect. 1.2), that is a set together with a distance function defined on pairs of points satisfying certain conditions. The next step is a topological space that is a set together with a structure leading to the study of continuous functions from one topological space to another. In school geometry, one begins with a study of maps that preserve congruence, followed by those preserving similarity. Next comes conical projection in which all conics are equivalent to a circle. Eventually one arrives at continuity (from ch. 1, it can be observed that this involves the preservation of nearness of points to sets).

The term *topology*[1] is used both for the area of study and for a family of sets with certain properties. From ch.1, recall that a family of open sets is induced by a metric on a nonempty set X. This family of open sets in a metric space, known as metric topology, satisfies a few simple properties. In defining a topology on an abstract set X, these properties are used as axioms. Thus a family τ of subsets of a set X is called a topology on X, provided

Properties of a topological space:

($\boldsymbol{\tau}$.1) X and $\emptyset$ are in τ,

($\boldsymbol{\tau}$.2) Unions of members of τ are in τ,

($\boldsymbol{\tau}$.3) Finite intersections of members of τ are in τ.

[1]The word *topology* is from Gk. *topos* (place) and *-logy* (speaking, discourse, treatise, theory, science, study), from Gk. *logia*, from root of *legein* (to speak), literally *science of place.*

The pair (X, τ) is called a **topological space**. Members of τ are called **open sets**.

There are alternate ways of beginning a study of topological spaces. In a manner similar to Theorem 1, Section 1.3, one begins with a set X together with a nearness relation between points and subsets of X satisfying the following conditions.

(**K**.1) No point in X is near the empty set,
(**K**.2) A point that is in a set $B \subset X$ is near B,
(**K**.3) A point is near the union of two sets, if and only if, it is near at least one of them.
(**K**.4) If a point x is near the set B and every point of B is near the set C, then x is near C.

Conditions (K.1)-(K.4) are usually written in an equivalent form using the Kuratowski closure operator. Beginning the study of topological spaces with the nearness form of (N.1)-(N.4) naturally leads to the study of L-proximity and Efremovič proximity in Ch. 3.

Let $\overline{E}$ denote the closure of set E. Then (K.1)-(K.4) are equivalent to

(**K'**.1) $\overline{\emptyset} = \emptyset$,
(**K'**.2) For each set E, $E \subset \overline{E}$,
(**K'**.3) For all sets E, F, $\overline{E \cup F} = \overline{E} \cup \overline{F}$,
(**K'**.4) For each set E, $\overline{\overline{E}} = \overline{E}$.

Recall from Sect. 1.3 that conditions (K.1)-(K.4) were arrived at via the metric on X. It is common practice in mathematics to use important results in analysis or metric spaces to define concepts in an abstract setting. Here, (K.1)-(K.4) satisfied by the family of open sets in a metric space are used to define a topology on a set X. In this generalisation, many, but not all of the previous results obtained in a metric space, remain valid. Thus, every metric space is a topological space with the metric topology.

2.2 Examples

We now give examples satisfying (K.1)-(K.4) with one exception in which (K.4) is not satisfied. Some of the examples will be useful later to illustrate concepts or as counterexamples. Students are urged to verify these and similar examples mentioned later. Such verifications are essential to an understanding of abstract concepts.

(**Ex** 2.2.1) Let X be a nonempty set and define $x \in X$ to be near a subset B of X, if and only if, $x \in B$ (this is the discrete case). Alternatively, the **discrete topology** on X contains every subset of X, including X and $\varnothing$.

(**Ex** 2.2.2) Let X be a nonempty set and define $x \in X$ to be near a subset $B \subset X$, if and only if, $B \neq \varnothing$ (indiscrete case). Alternatively, the **indiscrete topology** contains only X and $\varnothing$.

(**Ex** 2.2.3) Let X be an infinite set and define $x \in X$ to be near a subset $B \subset X$, if and only if, $x \in B$ or B is infinite. Alternatively, this topology consists of $X, \varnothing$ and complements of finite sets. Naturally, it is called a **co-finite topology**.

(**Ex** 2.2.4) Let $\mathbb{N}^* = \mathbb{N} \cup \{\infty\}$, where ∞ is an element not in the set of natural numbers $\mathbb{N}$. Define a *nearness* between points and subsets of $\mathbb{N}^*$ in the following way: a point $x \in \mathbb{N}^*$ is near a subset $B \subset \mathbb{N}$, if and only if, $x \in B$ or $x = \infty$ and $B \subset \mathbb{R}$ is infinite.

(**Ex** 2.2.5) Let $\mathbb{R}^* = \mathbb{R} \cup \{-\infty\} \cup \{\infty\}$, where $-\infty, \infty$ are elements not in the set of real numbers $\mathbb{R}$ and $-\infty < x < \infty$ for all $x \in \mathbb{R}$. Define a *nearness* between points and subsets of $\mathbb{R}^*$, provided $x \in B$ or
  (i) $x \in \mathbb{R}, B \subset \mathbb{R}$ and x is near B in the metric topology, or
  (ii) $x = -\infty$ and $B \subset \mathbb{R}$ is not bounded below, or
  (iii) $x = \infty$ and $B \subset \mathbb{R}$ is not bounded above.

(**Ex** 2.2.6) Let $\mathcal{F}$ denote the set of functions on $\mathbb{R}$ to $\mathbb{R}$ and define a function f to be near a set $E \subset \mathcal{F}$, if and only if, there is a sequence of functions $(f_n) \in E$ that converges pointwise to f, *i.e.*, for each $x \in \mathbb{R}$, the sequence $(f_n(x))$ converges to $f(x)$. Prove that conditions (K.1)-(K.3) are satisfied but (K.4) is not satisfied. **Hint**: Consider C, the set of continuous functions on $\mathbb{R}$ to $\mathbb{R}$ with pointwise convergence.

2.3 Closed and Open Sets

From Sect. 2.1, one may begin the study of topology on a set X starting either with a family of open sets satisfying $(\tau.1)$-$(\tau.3)$ or with a nearness relation between points and subsets of X satisfying conditions (K.1)-(K.4). Since both approaches are used in this book, a proof of their equivalence is given. For such a proof, it is easier to deal with closed sets instead of open sets. A *closed* set is one whose complement is open. In the space $\mathbb{R}$ of real numbers, an open interval $(a, b) = \{x : a < x < b\}$ is open and an interval $[a, b] = \{x : a \leq x \leq b\}$ is closed. In a metric space (X, d), an open ball $S(x, \varepsilon) = \{y \in X : d(x, y) < \varepsilon\}$ for $\varepsilon > 0$, is open. The half-open and half-closed interval $(a, b] = \{x : a < x \leq b\}$ is neither open nor closed. Hence, *closed* is not a negation of *open*. Naturally, by DeMorgan's laws, the family $\mathcal{F}$ of closed sets satisfy conditions (**F**.1)-(**F**.3), which are complementary to $(\tau.1)$-$(\tau.3)$, *i.e.*,

(**F**.1) X and $\varnothing$ are in $\mathcal{F}$,
(**F**.2) Intersection of members of any subfamily of $\mathcal{F}$ is in $\mathcal{F}$,
(**F**.3) Finite unions of members of $\mathcal{F}$ are in $\mathcal{F}$.

> The passage from (K.1)-(K.4) to (**F**.1)-(**F**.3) starts with
> *A set E is closed, if and only if, E contains all points of X that are near E.*

To prove the equivalence, it is necessary to derive (**F**.1)-(**F**.3) from (K.1)-(K.4). If E is any set and x is any point, then if x is near E, then x is near every closed set that contains E. Hence, x belongs to the intersection of all closed sets containing E. Then it is necessary to prove that a point x is in $\overline{E}$, if and only if, the point x is near E.

Proof. Begin with a set X with a nearness between points and sets satisfying (K.1)-(K.4). Let $\mathcal{F}$ be the family of all sets each of which contains points near it. We show that $\mathcal{F}$ satisfies (**F**.1)-(**F**.3).

(i) Obviously, (K.1)-(K.2) $\Rightarrow$ (**F**.1).

(ii) Now, go from the family $\mathcal{F}$ satisfying (**F**.1)-(**F**.3) back to the nearness between points. Notice that $x \in \overline{E}$, closure of a set E, if and only if, x belongs to the intersection G of all closed sets G containing E. Hence, x belongs to $\overline{E}$. Conversely, if $x \in \overline{E}$, then x is near $\overline{E}$ and, by (K.4), each point of $\overline{E}$ is near E. Hence, x is near E.

(iii) Suppose $\{F_n : 1 \leq n \leq m\} \subset \mathcal{F}$ and a point x is near the union $\bigcup\{F_n : 1 \leq n \leq m\}$. Then, by (K.3), x is near some F_n and so x belongs to F_n. Hence, x is in the union. □

2.4 Closure and Interior

Again, observe that the closure $\overline{E}$ of a set E is the set of all points that are near E. Alternatively, $\overline{E}$ (or cl$E$) is the intersection of all closed sets that contain E. There is an analogous operator called *interior* related to open sets. The **interior** of a set E (denoted by $\overset{\circ}{E}$ or int$E$) is the union of all open subsets of E. Alternatively, $\overset{\circ}{E}$ is the set of all points that are far from E^c (**complement** of E). Obviously, a set E is open, if and only if, E equals its interior $\overset{\circ}{E}$. It is easy to check that the relation between the interior and closure operators, where

$$\overset{\circ}{E} = \overline{E^c}^c, \text{ where}$$

$$\overline{E^c}^c = ((\overline{E^c}))^c, \textit{ i.e.}, \text{ complement of the closure of the complement of } E.$$

Dually,

$$\overline{E} = \left(\overset{\circ}{\overbrace{(E^c)}}\right)^c.$$

The interior operator satisfies conditions (**int**.1)-(**int**.4) that are the dual of (K'.1)-(K'.4), namely,

(**int**.1) $\overset{\circ}{X} = X$,

(**int**.2) For each E, $\overset{\circ}{E} \subset E$,

(**int**.3) $\overset{\circ}{\overbrace{(E \cap F)}} = \overset{\circ}{E} \cap \overset{\circ}{F}$,

(**int**.4) $\overset{\circ}{\overbrace{\left(\overset{\circ}{E}\right)}} = \overset{\circ}{E}$.

A **neighbourhood** N of a set E is a set whose interior contains E, *i.e.*, $E \subset \overset{\circ}{N}$. Neighbourhoods of a point satisfy some simple conditions from the above four satisfied by the interior operator. These conditions will serve as motivation for defining the concept of a filter, later. The family $\mathcal{N}_x$ (called the *neighbourhood filter* of the point x of a nonempty set X) satisfies the following conditions.

(**nbd**.1) $\mathcal{N}_x$ is not empty, since $x \in X$ and $X \in \mathcal{N}_x$,
(**nbd**.2) each member N of $\mathcal{N}_x$ is not empty,
(**nbd**.3) $\mathcal{N}_x$ contains finite intersections of its members,
(**nbd**.4) if $N \in \mathcal{N}_x$ and $N \subset M$ implies $M \in \mathcal{N}_x$.

Conversely, given a nonempty set X and, for each point x, a neighbourhood filter $\mathcal{N}_x$ satisfies conditions (**nbd**.1)-(**nbd**.4), it is possible to define a topology τ on X in the following way.

Neighbourhood Definition of a Topology

$G \in \tau$, if and only if, G is a neighbourhood of each of its points. Alternatively, a point x is far from F, if and only if, there is a neighbourhood N of x disjoint from F.

2.5 Connectedness

In Sect. 1.12, connectedness in metric spaces is introduced. The definitions and results in Sect. 1.12 also hold in abstract topological spaces.

2.6 Subspace

Suppose (X, τ) is a non-trivial topological space and Y is a nonempty subset of X. Then there is a topology on Y induced by τ. Suppose the topology τ is induced by a nearness relation satisfying (K.1)-(K.4). Then there is a naturally induced nearness relation on Y, namely, point $y \in Y$ is near a subset $E \subset Y$,if and only if, y is near E in X. In other words, the topology τ' on Y induced by τ is the family $\{G \cap Y : G \in \tau\}$. That is, a set is open in Y, if and only if, the set is the intersection of Y with an open set in X. Similarly, a closed set in Y is the intersection of Y with a closed set in X.

An important class of problems concerns the topological properties that are inherited by subspaces. For example, compactness is not inherited but, (*cf.* Sect. 1.9), a *closed* subspace of a compact space is compact.

2.7 Bases and Subbases

In analysis on the space of real numbers $\mathbb{R}$, we do not need all open sets but only the open intervals. In most definitions and proofs, one uses open intervals such as $(x - \varepsilon, x + \varepsilon)$ with $x \in \mathbb{R}$ and $\varepsilon > 0$. The reason for this is that any open set in $\mathbb{R}$ is a union of open intervals. In a metric space (X, d), the family of open balls

$$\{S_d(x, \varepsilon) : x \in X, \varepsilon > 0\}, \text{ where } S_d(x, \varepsilon) = \{y \in X : d(x, y) < \varepsilon\},$$

suffices in dealing with most concepts and proofs.

These observations lead to the concept of a *base* for a topology. A subfamily $\mathcal{B}$ of a topology τ on X is called a **base**, if and only if, each member of τ is a union of members of $\mathcal{B}$. In other words, for each member $U \in \tau$ and a point $x \in U$, there is a member B of the base $\mathcal{B}$ such that $x \in B \subset U$. It is possible for a topology to have several different bases.

For example, in the space $\mathbb{R}^2$ with the usual topology, three bases are frequently used, depending on the situation at hand. The bases are open balls that arise from three compatible metrics. Let $(x, y), (x', y')$ be two points in $\mathbb{R}^2$. Then the metrics are

(**metric**.1) $d_1((x, y), (x', y')) = \sqrt{(x - x')^2 + (y - y')^2}$ (Euclidean),
(**metric**.2) $d_2((x, y), (x', y')) = |x - x'| + |y - y'|$ (Taxicab),
(**metric**.3) $d_3((x, y), (x', y')) = max\{|x - x'|, |y - y'|\}$.

When is a subfamily $\mathcal{B}$ of a topology τ on X a base for τ? It is easy to show that $\mathcal{B}$ is a base, if and only if, for each pair of members $U, V \in \mathcal{B}$ and each point $x \in U \cap V$, there is a member $W \in \mathcal{B}$ such that $x \in W \subset U \cap V$.

In the space of real numbers $\mathbb{R}$, the family of open intervals $\{(a, b) : a, b \in \mathbb{R}\}$ is a base for the usual topology. Each open interval (a, b) is the intersection of two open rays, *i.e.*, $(a, b) = (-\infty, b) \cap (a, \infty)$. This observation leads to the concept of a *subbase* for a topology. A family $\mathcal{S} \subset \tau$ is called a **subbase** for the topology τ, if and only if, the family of finite intersections of the members of $\mathcal{S}$ is a base for τ.

Thus, the family of open rays in $\mathbb{R}$ is a subbase for the usual topology in $\mathbb{R}$. If $\mathcal{S}$ is any nonempty family of sets, then the family of unions of finite intersections of members of $\mathcal{S}$ is a topology on $X = \bigcup \mathcal{S}$ and it is the smallest topology on X that contains $\mathcal{S}$. This observation is useful in practice, since, in some problems, we need certain sets to be open and we can generate the smallest topology that contains these sets.

A **limit point** (also called an accumulation point or cluster point) p of a subset A in a topological space is one that is near $A - \{p\}$ or $p \in \mathrm{cl}(A - \{p\})$. This is equivalent to: *each neighbourhood p intersects $A - \{p\}$*. Naturally, each neighbourhood of a point x intersects A, if and only if, either x is a point in A or x is a limit point of A. A set is closed, if and only if, the set contains all of its limit points. The set of all limit points of a set A is called the **derived set** of A (denoted by A'). It is easy to see that the closure of A equals the union of A and A'.

A **boundary point** x of a set A is one that is near A as well as near its complement A^c. So x belongs to the intersection of $\mathrm{cl}\, A$ and $\mathrm{cl}\, A^c$, which is called the **boundary** (**frontier**) of A, denoted by A^b.

2.8 More Examples

This section introduces some interesting examples of topological spaces that will be useful as counterexamples. The usual method is to describe a base or subbase. Frequently, one starts with a standard topological space such as the real or complex numbers with the usual topologies and alters the neighbourhoods of some points.

(**Ex** 2.8.1) Consider the set of real numbers $\mathbb{R}$ with the usual topology generated by open intervals $\{(a,b) : a, b \in \mathbb{R}\}$. Let each real number x different from 0 have the usual neighbourhood filter $\mathcal{N}_x$. Let a typical neighbourhood of 0 be $(-\varepsilon, \varepsilon) - A$, where $\varepsilon > 0$ and A is a subset of $\{n^{-1} : n \in \mathbb{N}\}$. Note that the closure of each neighbourhood of 0 contains n^{-1} for sufficiently large n. Hence, $V = (-1, 1) - A$ cannot contain any closed neighbourhood of 0.

(**Ex** 2.8.2) Let X be any linearly ordered set with the linear order $<$. For each point a in X, set the left ray $L_a = \{x : x < a\}$ and the right ray $R_a = \{x : x > a\}$. The topology generated by the subbase of the left and right rays (namely, $\{L_a : a \in X\} \cup \{R_a : a \in X\}$) is called the **order topology** on X. Real numbers with the usual topology provide a familiar example.

Another important topology can be introduced on any linearly ordered set X with a base $\{[a,b) : a < b\}$ for open sets. This is called the **right half-open interval** topology. When $X = \mathbb{R}$, this topological space is known as the Sorgenfrey Line. It is first countable, separable but not second countable (see Sect. 2.9).

(**Ex** 2.8.3) Here is an example of an ordered topological space that is useful as a counterexample. Begin with the set $\mathbb{R} - \mathbb{N}$, the set of real numbers without the natural numbers and well-order it with $<$. Let $z = \{\mathbb{R}\}$ and extend the order $<$ to $\mathbb{R} \cup \{z\}$ as follows.

(i) $1 < 2 < 3$ ..., the usual order on the natural numbers $\mathbb{N}$,
(ii) $n < x$ for all natural numbers n and x in $\mathbb{R} - \mathbb{N}$,
(iii) $x < z$ for all x in $\mathbb{R}$.

Let Ω denote the least element of $\mathbb{R} \cup \{z\}$ with uncountably many predecessors. Notice that Ω exists because z has uncountably many predecessors and $<$ well-orders $\mathbb{R} \cup \{z\}$. Let M be the set of predecessors of Ω and $M^* = M \cup \{\Omega\}$. Let ω denote the first element whose predecessors are all natural numbers. Assign the order topology to M^*. Observe that if A is a countable subset of M, then $\sup A \in M$. This shows that although Ω is near M, there is no sequence in M that converges to Ω.

(**Ex** 2.8.4) Let X denote the set of all points in the upper half of the plane, including the x-axis. Let the neighbourhoods of points with positive y-coordinates be the same as those in the usual topology of the plane. For each point P on the x-axis, let the open sets be those sets that contain a closed disk touching the axis at P.

2.9 First Countable, Second Countable and Lindelöf

In Sect. 1.18, first-countable, second-countable and Lindelöf properties of metric spaces were introduced. In this section, these properties are revisited in the context of topological spaces. Let (X, τ) be a topological space. Recall that a base for the topology τ is a subfamily $\mathcal{B}$ of τ such that each open set is a union of members of $\mathcal{B}$. A **local base** $\mathcal{L}$ at $x \in X$ is a subfamily of $\mathcal{B}$ such that each neighbourhood of x contains a member of $\mathcal{L}$.

For example, in a metric space, the family of open balls $\left\{S(x, \frac{1}{n}) : n \in \mathbb{N}\right\}$ is a local base at each $x \in X$ (note that this family is *countable*). When a topological space has a countable neighbourhood base at each point, it is called **first countable** (denoted C_{I}). Thus, every metric space is first countable. An uncountable set such as the set of real numbers $\mathbb{R}$ with co-finite topology, is not first countable.

If a space has a countable base for the topology τ, then it is called **second countable** (denoted C_{II}). Obviously, a second countable space is first countable.

The space of real numbers $\mathbb{R}$ is second countable, since it has a countable base $\{(x, y) : x, y \text{ rational}\}$. A simple example of topological space that is first countable but not second countable is an uncountable set with the discrete 0-1 metric. The space of real numbers with half-open interval topology generated by $\{[x, y) : x < y\}$ is another example of a space that is not second countable.

An interesting property of the space of real numbers is that it has a countable dense subset, namely, the set of rational numbers. A topological space is **separable**, if and only if, it has a countable dense subset. A simple example of a topological space that is not separable is an uncountable set with discrete metric. On the other hand, every second countable metric space is separable. To see this, take a set that contains one point from each member of the countable base (the countable set so obtained is dense). A valuable property satisfied by a separable metric space is that it is second countable! But this is not true in general. The half-open interval space is separable but has no countable base.

There is another property reminiscent of the Heine-Borel definition of compactness, namely, every open cover of X has a finite subcover. Lindelöf introduced an analogous concept: every open cover of X has a countable subcover. This property is satisfied by the space of real (complex) numbers but is not satisfied by an uncountable, discrete metric space. In metric spaces, it has already been observed that the three properties (separable, second countable and Lindelöf) are equivalent but this is not true in general topological spaces. Every second countable space is Lindelöf. The half-open interval space is not second countable but is Lindelöf.

2.10 Application: Topology of Digital Images

This section introduces topological structures in digital images. This approach to the study of digital images naturally leads to a recognition of different types of sets embedded in images. For example, an easy outcome of this approach is the discovery of spatially near sets of picture elements. Sets of picture elements are spatially near, provided the sets have picture elements in common. Another outcome of this approach is the discovery of descriptively near sets of picture elements. Sets of picture elements are descriptively near, provided the sets contain picture elements that resemble each other (*e.g.*, picture elements that have the same colour). Nearness in this case is based on the perception of the closeness of picture element feature values.

2.10.1 *Topological Structures in Digital Images*

The notion of a topological structure (*cf.*, [32]) provides a setting for the analysis of digital images. Let X denote a set of points (picture elements) in a digital image. A **topological structure** on X is a structure given by a set of subsets τ of X, having the following properties.

(**O**.1) Every union of sets in τ is a set in τ.
(**O**.2) Every finite intersection of sets in τ is a set in τ.

A **digital image topological space** is a digital image endowed with a topological structure. When a topology has been defined on a digital image X, this set is *underlying* the topological space X. This view of digital images is inspired by a more general view of topological spaces (see, *i.e.*, [158; 149; 51; 32]). In general, a set X endowed with a relation (*e.g.*, equivalence, proximity, or tolerance relation) is a structured set. In each case, we obtain a collection of subsets in a structured set. In terms of a digital image, the term *structure* means some meaningful collection of the parts of a structured image. By *collection*, we mean grouping the pixels in an image based on some scheme, *i.e.*, not relocating pixels but viewing particular image pixels as belonging to a grouping based on the given location of the pixels or on the features of the pixels. Such groupings of pixels lead to the discovery of patterns in a digital image.

Čech View of Structure

For E. Čech [229], a set X endowed with a relation α is a structure represented by the pair (X, α). The Čech view of structure leads to a study of patterns in digital images, *e.g.*, pairs of structures that are near or far (remote) in some sense. Structures that are far from each other are called **remote structures**.

2.10.2 *Visual Sets and Metric Topology*

This section gives a brief introduction to visual sets and metric topology for digital images. A **visual set** is a set of picture elements in a digital image. Examples of visual sets are given here as well as in the sequel to this chapter.

Fig. 2.1 Remote sets $\boldsymbol{A}, \boldsymbol{B}$, Near sets $\boldsymbol{B}, \boldsymbol{C}$

Str.1 Spatially Remote Sets.
In Fig. 2.1, let X be endowed with a proximity relation δ. Spatially remote sets have no points in common. Sets A and B in Fig. 2.1 are examples of remote sets, *i.e.*, $A \underline{\delta} B$ such that $A \cap B = \emptyset$.

Str.2 Spatially Near Sets.
Spatially near sets have points in common. Sets B and C in Fig. 2.1 are examples of near sets, *i.e.*, $B \ \delta \ C$ such that $A \cap B \neq \emptyset$.

Str.3 Metric Topology on a Digital Image.
Let X denote the nonempty set of points in a digital image. Let $d : X \times X :\rightarrow \mathbb{R}$ be a metric. A metric topology of a digital image is determined by considering the open neighbourhood of each point $x \in X$ (also called an open ball), which is the set of all points that are sufficiently close to x. Let $\varepsilon > 0$. A point y is sufficiently close to a point x, if and only if, $d(x, y) < \varepsilon$. Then, an **open neighbourhood** $N_\varepsilon(x)$ of a point $x \in X$ (for simplicity, we also write N_x) defined by

$$N_x = \{y \in X : d(x, y) < \varepsilon\}.$$

N_x is called an **open ball** with centre x and radius ε such that $d(x, y) < \varepsilon$. The set A in Fig. 2.1 is an example of an open neighbourhood of the point x.

In the case of descriptive neighbourhoods of a point, there are occasions where one must consider distances between point descriptions that lie between zero and some ε. A descriptive neighbourhood of a point x is the set of all points that are descriptively close to the description of x. In Fig. 2.2, the set C is an example of a descriptive neighbourhood with a fixed radius, *i.e.*, all points with matching description (zero visual intensity) within a fixed distance of the centre of C.

Again, for example, let Φ denote a set of probe functions $\phi : X \to \mathbb{R}$ representing features such as greylevel intensity, colour, shape or texture of a point p (picture element) in a digital image X and let $\Phi(p)$ denote a feature vector containing numbers representing feature values extracted from p. Then let δ_Φ denote a descriptive proximity relation between singleton sets $\{p\}, \{p'\}$, where p, p' are points in a proximity space X. Then $\{p\}\ \delta_\Phi\ \{p'\}$, provided points p, p' have matching descriptions, *i.e.*, $d(\Phi(p), \Phi(p)) = 0$. The choice of $\varepsilon \in (0, \infty)$ is important, since we want to consider neighbourhoods containing spatially distinct points that have matching descriptions, where $d(\Phi(p), \Phi(p)) = 0$. Then, the **descriptive neighbourhood** $N_{\Phi(x)}$ of a point $x \in X$ with fixed radius ε is defined by

$$N_{\Phi(x)} = \{y \in X : d(\Phi(x), \Phi(y)) = 0 \text{ and } d(x, y) < \varepsilon\}.$$

The notion of **spatially** distinct points that have matching descriptions is explained in detail in the sequel.

Let τ denote the set of all open neighbourhoods of points in X. By convention, the empty set $\emptyset$ is declared open, so that open sets are closed under finite intersections and X is open because it is the union of all open neighbourhoods of points in X. If $N_x, N_{x'}$ are open neighbourhoods in τ, then $N_x \cap N_{x'} \in \tau$, since every point y in the intersection is the centre of a neighbourhood in τ. Similarly, $N_x \cup N_{x'} \in \tau$, since every point y in the union is the centre of a neighbourhood in τ. Hence, τ is a metric topology on the digital image X.

2.10.3 *Descriptively Remote Sets and Descriptively Near Sets*

Descriptively remote and descriptively near sets are briefly introduced in this section.

Fig. 2.2 Descriptively remote open sets $\boldsymbol{A}, \boldsymbol{C}$, Near open sets $\boldsymbol{A}, \boldsymbol{B}$

Str.4 Descriptively Remote Sets.
In Fig. 2.2, let X be endowed with a descriptive proximity relation δ_Φ. **Descriptively remote sets** have no points with matching descriptions. Choose Φ to be a set containing a probe function representing greylevel intensity. Let A and C in Fig. 2.2 represent visual open sets. Sets A and C are examples of descriptively remote sets, since all pixels in C have zero intensity (the pixels are black) and all pixels in A have greylevel intensity greater than zero, *i.e.*, $A \ \underline{\delta}_\Phi \ C$.

Str.5 Descriptively Near Sets.
Descriptively near sets have points with matching descriptions. Again, choose Φ to be a set containing a probe function representing greylevel intensity. Let A and B in Fig. 2.2 represent visual open sets. Sets A and B in Fig. 2.2 are examples of descriptively near sets, since A and B have some pixels with greylevel intensities that are equal, *i.e.*, $A \ \delta_\Phi \ B$.

Spatially near sets have members in common. This observation leads to the following result.

Theorem 2.1. *Spatially near sets are descriptively near sets.*

Proof. See Problem 2.11.18. □

2.11 Problems

(2.11.1) A **lattice** is a partially ordered set (poset) in which any two elements have a supremum (lub) and an infimum (glb). Show that the family of all topologies on a nonempty set X forms a lattice under the partial order $\subset$.

(2.11.2) Do the following:

(a) Show that the co-finite topology on an infinite set X is the smallest topology in which singleton sets are closed. Which topology is the largest?

(b) For each subset $E \subset X$, find the closure, interior, boundary and derived sets.

(c) Show that there is a family of subsets $\{A_\lambda\}$ of X such that $\overline{(\bigcup A_\lambda)}$ is not equal to $\bigcup \overline{(A_\lambda)}$.

(2.11.3) Construct an example of a topology (different from discrete and indiscrete topologies) in which a set is open, if and only if, the set is closed.

(2.11.4) Suppose X is uncountable and has a co-countable topology τ, *i.e.*, the open sets are $X, \varnothing$ and complements of countable sets. Show that this space is separable but not second countable.

(2.11.5) Give examples to show that first countable and separable are not comparable.

(2.11.6) Consider the half-open interval space or Sorgenfrey line X generated by $\mathcal{B} = \{[a, b] : a, b \in \mathbb{R}(\text{reals})\}$. Show that

(a) X is not connected.

(b) X is separable but it is not second countable. However, X is Lindelöf.

(c) Every subspace of X is separable and Lindelöf.

(2.11.7) In Example 2.8.3, show that if A is a countable subset of $\mathcal{M}$, then $\sup A \in \mathcal{M}$.

(2.11.8) Show that (a) separable and (b) Lindelöf are not hereditary but (c) first countable and (d) second countable are hereditary.

(2.11.9) Do the following:

(a) Show that the closure of every set A is the union of its interior $\overset{o}{A}$ and its boundary A^b.

(b) Prove that (i) $A^{bb} \subset A^b$, (ii) $(A \cap B)^b \subset A^b \cup B^b$, (iii) $(A \cup B)^b \subset A^b \cup B^b$.

(2.11.10) Show that the closure of every set A is the union of A and its derived set A'. Hence, A is closed, if and only if, $A' \subset A$.

(2.11.11) Show that $(A \cup B)' = A' \cup B'$ and $(A')' = A'$. It is shown in this chapter that a topology can be defined via an interior operator or a closure operator. Is it possible to arrive at a topology via properties of the 'derived' operator?

(2.11.12) If finite sets are closed in a topological space, show that the derived set of any set is closed.

(2.11.13) Find all topologies that can be defined on a set of three elements (the problem of finding the number of topologies on a set of n elements seems to be open).

(2.11.14) Suppose X is a topological space and $X = A \cup B$. Show that $X = \overline{A} \cup \overset{o}{B}$.

(2.11.15) Suppose G is an open set and D is dense in a topological space X. Prove that G is a subset of the closure of $(G \cap D)$.

(2.11.16) **Kuratowski problem**: For a subset E in a topological space X, $\overline{E}, E^c$ denote the closure and complement of E, respectively. Show that if these operators are used successively on any subset of X, then at most 14 sets can be constructed. Give an example of a subset of the real numbers that yields exactly 14 sets.

(2.11.17) In Fig. 2.2, give an example spatially near sets that also descriptively near sets.

(2.11.18) Prove Theorem 2.1.

(2.11.19) Let (X, δ_Φ) be a descriptive proximity space (the nonempty set X is endowed with the descriptive proximity relation δ_Φ) and let $N_{\Phi(x)}$ denote a descriptive neighbourhood of a point with a fixed radius $\varepsilon > 0$. Prove $y \in N_{\Phi(x)}$, if and only if, $y\ \delta_\Phi\ x$.

(2.11.20) Let $(X, \delta), (X, \delta_\Phi)$ be spatial and descriptive proximity spaces, respectively. Let Φ be a set of probe functions representing features of $x \in X$. For $x, y \in X$, assume that N_x, N_y are spatially near neighbourhoods of x and y with fixed radius $\varepsilon > 0$. Prove that N_x, N_y are descriptively near neighbourhoods relative to the descriptive proximity δ_Φ.

Chapter 3

Symmetric Proximity

3.1 Proximities

In Ch. 1, metric and fine proximities are central topics. The focus of Ch. 2 is on *nearness of a point to a set.* In this Chapter, there is a natural transition to the following topics.
(**topic**.1) Nearness of two sets called **proximity**,
(**topic**.2) nearness of finitely many sets called **contiguity**, and
(**topic**.3) nearness of families of sets with arbitrary cardinality called **nearness**.
Additionally in this chapter, proximities are considered in relation to topologies that are not necessarily related to a metric. Obviously, *proximity* gives rise to a topology whenever the first set is a point or singleton set. In that case, the proximity and topology are called **compatible**. Just as there are many topologies on a nonempty set, there are, in general, a number of proximities on a set as well as a number of proximities compatible with a given topological space.

If a proximity is compatible with a topology, then it may be symmetric or non-symmetric. If the proximity is symmetric, then the topology must satisfy the $(*)$ condition, *i.e.*,
$(*)$ x is near $\{y\} \Rightarrow y$ is near $\{x\}$.

The $(*)$ condition is satisfied by $\mathbb{R}$ and $\mathbb{C}$. However, the topological space $X = \{a, b, c\}$ with topology $\tau = \{\varnothing, X, \{a\}, \{a, b\}\}$ does not satisfy condition $(*)$.

Proof. Not $(*)$, since a is near $\{b\}$ but b is not near $\{a\}$. □

Topological spaces satisfying $(*)$ are called R_0 or **symmetric** or **weakly regular**. It is possible for spaces to have compatible, non-symmetric proximities. Only symmetric proximities are considered in this book. Motivation for the axioms of proximity stems from two topics in Chs. 1 and 2:

(**topic**.1) Kuratowski closure operator,
(**topic**.2) metric proximity.

From the analogy of (topic.1), we get the conditions for a proximity first studied by S. Leader for the non-symmetric case and by M.W. Lodato for the symmetric case. Suppose X is a nonempty set. We often write x for $\{x\}$. For each point $x \in X$ and subsets $B, C \subset X$, the Kuratowski closure operator satisfies (K.1)-(K.4), where $x\ \delta\ B \Leftrightarrow x$ is near B.

(**K**.1) $x\ \delta\ B \Rightarrow B \neq \varnothing$,
(**K**.2) $\{x\} \cap B \neq \varnothing \Rightarrow x\ \delta\ B$,
(**K**.3) $x\ \delta\ (B \cup C) \Leftrightarrow x\ \delta\ B$ or $x\ \delta\ C$,
(**K**.4) $x\ \delta\ B$ and $b\ \delta\ C$ for each $b\ \delta\ B \Rightarrow x\ \delta\ C$.

Thus axioms (P.1)-(P.4) for symmetric proximity (called **L-proximity** or **symmetric generalized proximity**) are obtained by replacing x or $\{x\}$ by A in (K.1)-(K.4) and adding the symmetry axiom (P.0).

(**P**.0) $A\ \delta\ B \Leftrightarrow B\ \delta\ A$ (symmetry),
(**P**.1) $A\ \delta\ B \Rightarrow A \neq \varnothing, B \neq \varnothing$,
(**P**.2) $A \cap B \neq \varnothing \Rightarrow A\ \delta\ B$,
(**P**.3) $A\ \delta\ (B \cup C) \Leftrightarrow A\ \delta\ B$ or $A\ \delta\ C$,
(**P**.4) $A\ \delta\ B$ and $b\ \delta\ C$ for each $b\ \delta\ B \Rightarrow A\ \delta\ C$ (L-axiom).

Recall from §1.7 that metric proximity satisfies a stronger condition than (P.4), namely,

(**P**.5) $A\ \underline{\delta}\ B$ implies there is an E such that $A\ \underline{\delta}\ E$ and $B\ \underline{\delta}\ E^c$ (EF-axiom).

The proof of Lemma 1.3(a) in §1.7, (P.5)⇒(P.4) holds, since it does not use the gap functional. A proximity that satisfies (P.0)-(P.5) is called **Efremovič** or **EF-proximity**. Observe that up to the 1970s, *proximity* meant EF-proximity, since this is the one that was studied intensively. But in view of later developments, there is a need to distinguish between various proximities. A **basic proximity** or **Čech-proximity** δ is one that satisfies (P.0)-(P.3).

A question arises, if an R_0 topological space has a compatible L-proximity. The answer is Yes.

Theorem 3.1. *Every R_0 topological space (X, τ) has a compatible proximity δ_0 defined by*

$$A \ \delta_0 \ B \Leftrightarrow \operatorname{cl} A \cap \operatorname{cl} B \neq \varnothing \ (\textbf{Fine L-proximity } \delta_0)$$

Proof. It is easily verified that δ_0 satisfies (P.0)-(P.3).
(P.4) Suppose $A \ \delta_0 \ B$ and $\{b\} \ \delta_0 \ C$ for each $b \in B$. Then $\operatorname{cl} A \cap \operatorname{cl} B \neq \varnothing$ and, for each $b \in B$, there is an $x \in X$ such $x \in \operatorname{cl}\{b\} \cap \operatorname{cl} C$. Since X is R_0, $b \in \operatorname{cl}\{x\} \subset \operatorname{cl} C$. So, $B \subset \operatorname{cl} C$ and $\operatorname{cl}B \subset \operatorname{cl} C$. This implies $\operatorname{cl} A \cap \operatorname{cl} C \neq \varnothing$ and $A \ \delta_0 \ C$.

Compatibility follows from (P.4), *i.e.*, $x\delta_0 B \Leftrightarrow \operatorname{cl}\{x\} \cap \operatorname{cl} B \neq \varnothing$. Hence, there is a $y \in \operatorname{cl}\{x\} \cap \operatorname{cl} B$. Since X is R_0, $x \in \operatorname{cl}\{y\} \subset \operatorname{cl} B$. Conversely,

$$x \in \operatorname{cl} B \Rightarrow \operatorname{cl}\{x\} \subset \operatorname{cl} B \Rightarrow \operatorname{cl}\{x\} \cap \operatorname{cl} B \neq \varnothing \Rightarrow x\delta_0 B.$$

□

Remark 3.1. B^c = complement of B, $\underline{\delta}$ = *far*, $\operatorname{int} B$ = interior of B.
The following results proved for L-proximity also hold for EF-proximity.
(a) If δ is an L-proximity, $A\delta B \Leftrightarrow \operatorname{cl} A \ \delta \ \operatorname{cl} B$.

Proof.
⇒ By the union axiom, $A\delta B, A \subset C, B \subset D \Rightarrow C\delta D$. Hence $A\delta B \Rightarrow \operatorname{cl} A \ \delta \ \operatorname{cl} B$.
⇐ $x \in \operatorname{cl} B \Rightarrow x\delta B$. So, by the L-axiom, $\operatorname{cl} A \ \delta \ \operatorname{cl} B$, and each element $x \in \operatorname{cl} B$ is near $B \Rightarrow \operatorname{cl} A \ \delta \ B$. By symmetry, $B \ \delta \ \operatorname{cl} A$. Repeating the argument, $B\delta A$. Finally, by symmetry, $A \ \delta \ B$. □

(b) $A \ \underline{\delta} \ B \Rightarrow$ (i) $\operatorname{cl} A \subset B^c$, (ii) $A \subset \operatorname{int}(B^c)$.

Proof.
(i) $A \ \underline{\delta} \ B \Rightarrow \operatorname{cl} A \ \underline{\delta} \ B \Rightarrow \operatorname{cl} A \subset B^c$.
(ii) $A \ \underline{\delta} \ B \Rightarrow A \ \underline{\delta} \ \operatorname{cl} B \Rightarrow A \subset (\operatorname{cl} B)^c = \operatorname{int}(B^c)$ □

(c) If (X, δ) is an L-proximity space and $Y \subset X$, one may define for subsets $A, B \subset Y$, $A\ \delta'\ B$ in Y, if and only if, $A\ \delta\ B$ in X. It is easy to check that δ' is an L-proximity on Y, which is called the **subspace proximity** induced by δ. If δ is L or EF, then respectively δ' is L or EF. It is also easy to check that if δ is a proximity compatible with the topology τ on X, then the subspace proximity δ' is compatible with the subspace topology induced by τ on Y.

(d) It will be shown later that the *fine L-proximity* δ_0 *is the most important proximity*. This is so because every proximity, abstractly defined on a set X, is the subspace proximity induced by the fine L-proximity δ_0 on a suitable compactification of X.

(e) **Partial order**: Proximities (L or EF) on a set X are partially ordered in the following way.

$$\begin{aligned}\delta > \delta' &\Leftrightarrow \text{for all subsets } A, B \subset X, A\ \delta\ B \Rightarrow A\ \delta'\ B \\ &\Leftrightarrow \text{for all subsets } C, D \subset X, C\ \underline{\delta}'\ D \Rightarrow C\ \underline{\delta}\ D.\end{aligned}$$

We also write $\delta' < \delta$ for $\delta > \delta'$. If $\delta > \delta'$, then δ is said to be **finer** than δ' or δ' is **coarser** than δ. In this partial order, δ_0 is the finest compatible L-proximity on an R_0 topological space X. This is so because if δ is any compatible proximity on X, then

$$A\ \delta_0\ B \Leftrightarrow (\mathrm{cl}\, A \cap \mathrm{cl}\, B \neq \varnothing \Rightarrow \mathrm{cl}\, A\ \delta\ \mathrm{cl}\, B) \Rightarrow A\ \delta\ B.$$

So $\delta_0 > \delta$.

(f) The **coarsest compatible L-proximity** δ_C on an R_0 topological space (X, τ) is given by

$$A\ \delta_C\ B \Leftrightarrow \mathrm{cl}\, A \cap \mathrm{cl}\, B \neq \varnothing, \text{ or both } A \text{ and } B \text{ are infinite.}$$

Proof. It is easy to see that δ_C satisfies (P.0)-(P.2).
(P.3) The union axiom is satisfied, since $\mathrm{cl}(B \cup C) = \mathrm{cl}\, B \cup \mathrm{cl}\, C$ and the union of two sets is infinite, if and only if, at least one of the sets is infinite.
(P.4) To see that the L-axiom is satisfied, suppose $A\ \delta_C\ B$ and $\{b\}\ \delta_C\ C$ for each $b \in B$. Then $A\ \delta_C\ B \Rightarrow \mathrm{cl}\, A \cap \mathrm{cl}\, B \neq \varnothing$ or A, B are both infinite. For each $b \in B$,

$$\{b\}\ \delta_C\ C \Rightarrow B \subset \mathrm{cl}\, C \Rightarrow \mathrm{cl}\, B \subset \mathrm{cl}\, C.$$
$$\mathrm{cl}\, A \cap \mathrm{cl}\, B \neq \varnothing \Rightarrow \mathrm{cl}\, A \cap \mathrm{cl}\, C \neq \varnothing \Rightarrow A\ \delta_C\ C.$$
$$A, B \text{ are infinite} \Rightarrow A, C \text{ are infinite} \Rightarrow A\ \delta_C\ C.$$

Compatibility: $\{x\}\ \delta_C\ E \Leftrightarrow x \in \mathrm{cl}\, E$, since $\{x\}$ is finite. If δ is any compatible L-proximity, then $A\ \underline{\delta}_C B \Rightarrow \mathrm{cl}\, A \cap \mathrm{cl}\, B = \varnothing$ and $\mathrm{cl}\, A$ or $\mathrm{cl}\, B$

is finite. A finite set is far from a disjoint closed set in any compatible proximity and so $A \underline{\delta} B$. Hence, $\delta > \delta_C$. So δ_C is the coarsest compatible L-proximity. □

This strange looking proximity δ_C (called the **coarse proximity**) is useful in the construction of counterexamples.

3.2 Proximal Neighbourhood

In an L-proximity space (X, δ), assume A, B are subsets of X and $B^c = X - B$. Then B is called a **proximal neighbourhood** of A, if and only if, $A \underline{\delta} B^c$ (denoted $A \ll_\delta B$ or, simply, $A \ll B$, if the proximity is obvious). It is possible to express the axioms for a basic proximity in terms of proximal neighbourhoods in the following way. For all subsets $A, B, C \subset X$,

(PN.0) $A \ll B \Leftrightarrow B^c \ll A^c$,
(PN.1) $X \ll X$,
(PN.2) $A \ll B \Rightarrow A \subset B$,
(PN.3) $A \subset B \ll C \subset D \Rightarrow A \ll D$,
(PN.4) $A \ll B_k, 1 \le k \le n \Rightarrow A \ll \bigcap_{k=1}^{n} \{B_k\}$.

Further, the L-axiom can be expressed in the following way.
(**PN**.5) $A \ll B \Rightarrow \forall C, A \ll C$ or there is an $x \in C^c$ with $x \ll B$ (L-axiom).
The EF-axiom can be expressed as follows.
(**PN**.6) $A \ll B \Rightarrow \exists C$ with $A \ll C \ll B$ (EF-axiom).
Since the proof of the equivalence of this PN formulation of $\ll$ with the axioms of proximity is a routine set-theoretic exercise, it is omitted.

Lemma 3.1. *Let* (X, δ) *be a proximity space. Then*
(*a*) $A \ll B \Rightarrow \mathrm{cl}\, A \ll B$,
(*b*) $A \ll B \Rightarrow A \ll \mathrm{int}\, B \Rightarrow \mathrm{cl}\, A \ll \mathrm{int}\, B$.

3.3 Application: EF-Proximity in Visual Merchandising

EF-proximity provides a key to the underlying structure in what are known as planograms. A **planogram** is a visual representation of a store's products or services. A typical merchandising planogram specifies the arrangement in the display of similar products on neighbouring shelves.

Fig. 3.1 EF-proximity display: $\mathbf{A}\ \underline{\delta}\ \mathbf{B} \implies \mathbf{A}\ \underline{\delta}\ \mathbf{C}$ and $\mathbf{B}\ \underline{\delta}\ \mathbf{C^c}$

Planograms are tools in what is known as visual merchandising. Organizations use planograms to plan the appearance of their merchandise. For example, supermarkets use text and box-based planograms to optimise shelf space, presentation, inventory turns and profit margins. It is possible to view product displays in the context of Efremovič proximity. The motivation for doing this is an interest in not only arranging similar products together vertically and horizontally on display shelves but also include in the arrangement a separation within the same class of products. This can be achieved with a realization of the EF-proximity relation relative to sets of products within the same class of products (*e.g.*, bottles with different shapes and colours).

Recall that a binary relation δ over the family of sets $\mathcal{P}(X)$ of a nonempty set X is an **Efremovič proximity** (also called an **EF-proximity**), if and only if, δ satisfies the EF-axioms (P.0)-(P.5) in Section 3.1.

A representation of a descriptive EF-proximity relation between sets A, B and C^c is shown in Fig. 3.2. For example, a decriptive EF proximity relation δ_Φ on the sets A, B, C^c is shown in the sketch of a product display

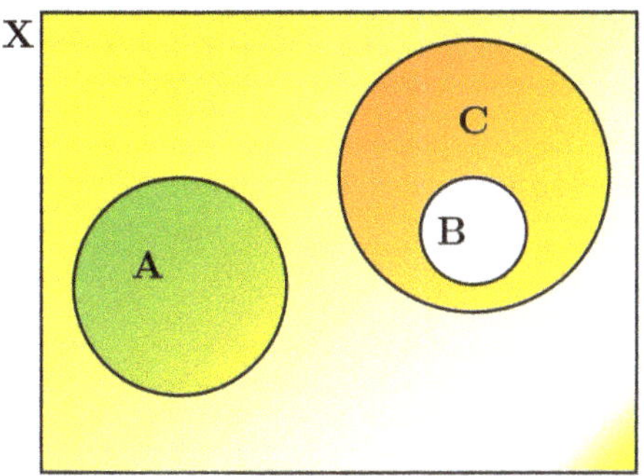

Fig. 3.2 $\mathbf{A}\ \underline{\delta}_{\Phi}\ \mathbf{B} \implies \mathbf{A}\ \underline{\delta}_{\Phi}\ \mathbf{C}$ and $\mathbf{B}\ \underline{\delta}_{\Phi}\ \mathbf{C^c}$

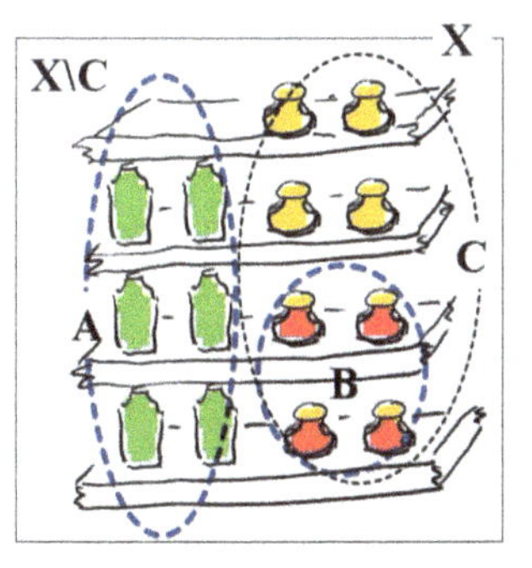

3.3.1: EF display

3.3.2: Thai display

Fig. 3.3 Sample EF-displays

in Fig. 3.3.1. The Thai display in Fig. 3.3.2 contains many instances of sets A, B, C^c such that $\mathbf{A}\ \underline{\delta}_{\Phi}\ \mathbf{B} \implies \mathbf{A}\ \underline{\delta}_{\Phi}\ \mathbf{C}$ and $\mathbf{B}\ \underline{\delta}_{\Phi}\ \mathbf{C^c}$ (see Problem 1.21.9). In Fig. 3.3.2, there are also many instances of sets A, B, C^c that exhibit spatial EF proximity such that $\mathbf{A}\ \underline{\delta}\ \mathbf{B} \implies \mathbf{A}\ \underline{\delta}\ \mathbf{C}$ and $\mathbf{B}\ \underline{\delta}\ \mathbf{C^c}$ (see Problem 1.21.10).

The EF-axiom (P.5) concerns remote sets. Sets $A, B \subset X$ are **remote** (denoted by $A\ \delta\ B$), if and only if, $A \cap B = \varnothing$ The import of an EF-proximity relation is extended rather handily to visual displays of products in a supermarket (see, *e.g.*, Fig. 3.1). The sets of bottles that have an underlying EF-proximity to each other is shown conceptually in the sets in Fig. 3.3. The basic idea with this application of topology is to extend the normal practice in the vertical and horizontal arrangements of similar products with a consideration of the topological structure that results when remote sets are also taken into account, representing the relations between these remote sets with an EF-proximity.

3.4 Problems

(**3.4**.1) Show that in any topological space

$$A \ \delta \ B \Leftrightarrow \text{cl}\, A \cap \text{cl}\, B \neq \varnothing$$

satisfies conditions (P.0)-(P.3) for basic or Čech proximity δ.

(**3.4**.2) Show that on $\mathbb{R}$

$$A \ \delta \ B \Leftrightarrow \text{cl}\, A \cap \text{cl}\, B \neq \varnothing,$$

or both A, B are uncountable, defines a compatible Čech proximity that is not an L-proximity.

(**3.4**.3) Suppose X is dense in a topological space αX. For subsets A, B of X, define

$$A \ \delta \ B \Leftrightarrow \text{cl}_{\alpha X} A \cap \text{cl}_{\alpha X} B \neq \varnothing$$

Verify that δ is a Čech proximity on X. Is it compatible?

(**3.4**.4) Construct a Čech proximity δ on $\mathbb{R}$ that is not an L-proximity, but satisfies the condition: $x \ \delta \ A, x \ \delta \ B \Rightarrow A \ \delta \ B$. Such a proximity is called a **Riesz proximity**.

(**3.4**.5) Let (x_n) be a sequence in a metric space (X, d) and put $P_n = \{x_m : m > n\}$. Define $A \ \delta \ B$, if and only if, there is a Cauchy sequence (x_n) in X such that $d(A, P_n) = 0 = d(B, P_n)$. Prove that δ is a compatible EF-proximity on X.

(**3.4**.6) **S-proximity**. On a nonempty set (in particular, on $\mathbb{R}$ or $\mathbb{C}$), define

$$A \ \delta \ B \Leftrightarrow (A \cap \text{cl}\, B) \cup (\text{cl}\, A \cap B) \neq \varnothing$$

Identify which axioms are satisfied by δ and show that δ is different from L-proximity [235].

(**3.4**.7) **Contiguity**. Using the analogy of L-proximity, give axioms for contiguity, *i.e.*, when finite subsets are near [99].

(**3.4**.8) **Nearness**. Using the analogy of L-proximity, give axioms for nearness, *i.e.*, when a family of subsets is near [88].

(**3.4**.9) **Descriptive EF Proximity**. Identify sets A, B, C^c in Fig 3.3.2 such that $\mathbf{A}\ \underline{\delta}_{\Phi}\ \mathbf{B} \implies \mathbf{A}\ \underline{\delta}_{\Phi}\ \mathbf{C}$ and $\mathbf{B}\ \underline{\delta}_{\Phi}\ \mathbf{C^c}$.

(**3.4**.10) **Spatial EF Proximity**. Identify sets A, B, C^c in Fig 3.3.2 such that $\mathbf{A}\ \underline{\delta}\ \mathbf{B} \implies \mathbf{A}\ \underline{\delta}\ \mathbf{C}$ and $\mathbf{B}\ \underline{\delta}\ \mathbf{C^c}$.

Chapter 4

Continuity and Proximal Continuity

4.1 Continuous Functions

In real and complex analysis, continuous functions play an important role. Recall from Ch. 1 that continuous functions on metric spaces which preserve nearness of points to sets as well as isometries that preserve distances between points, were considered. This chapter focuses on (1) continuous functions on topological spaces and (2) proximally continuous functions on proximity spaces, which preserve nearness of pairs of subsets.

Let X, Y be topological spaces and let $f : X \to Y$ denote a function. Then f is said to be **continuous at a point** $c \in X$, if and only if, for each subset $A \subset X$,

(**ContPt**.1) c is near A implies $f(c)$ is near $f(A)$, or
(**ContPt**.2) $c \in \text{cl}\, A$ implies $f(c) \in \text{cl}\, f(A)$, or
(**ContPt**.3) for each open set V containing $f(c)$, there is an open set U containing c, such that $f(U) \subset V$.

A function that is continuous at all points of X is said to be **continuous**. Clearly, this is equivalent to the assertion that a function is continuous, if and only if, for all subsets of $A \subset X, f(\text{cl}\, A) \subset \text{cl}\, f(A)$.

If f is a bijection and both f and f^{-1} are continuous, then f is called a **homeomorphism** or a **topological map**. Then the topological spaces X and Y are called homeomorphic spaces.

An important topic is that of continuous and homeomorphic invariants.

Let $(X,\delta),(Y,\delta')$ denote L-proximity spaces and let $f : X \to Y$ be a function. Then f is said to be **proximally continuous** or **p-continuous**, if and only if, for subsets $A, B \subset X$,

$$A\ \delta\ B \text{ implies } f(A)\ \delta'\ f(B).$$

If f is bijective and both f, f^{-1} are p-continuous, then f is called a **proximal homeomorphism** or **p-homeomorphism** or **p-isomorphism**. The following results from Lemma 4.1 follow easily[1].

Lemma 4.1. *Observe that*

(Lem.4.1.*1***)** *Constant functions are continuous (p-continuous),*

(Lem.4.1.*2***)** *Identity functions are continuous (p-continuous),*

(Lem.4.1.*3***)** *Compositions of continuous (p-continuous) functions are continuous (p-continuous),*

(Lem.4.1.*4***)** *Every p-continuous function is continuous with respect to compatible topologies. The converse does not hold, except in the case where the domain has the fine L-proximity.*

We now give some characterisations of continuous functions.

Theorem 4.1. *Let $(X,\tau),(Y,\tau')$ denote topological spaces and let $f : X \to Y$ denote a function. The following are equivalent.*

(*a*) *For each point $c \in X$ and each subset $A \subset X$, c is near A in X implies $f(c)$ is near $f(A)$ in Y,*

(*b*) *for all subsets $A \subset X$, $f(\mathrm{cl}\,A) \subset \mathrm{cl}\,f(A)$ in Y,*

(*c*) *for each open set $E \subset Y$, $f^{-1}(E)$ is open in X,*

(*d*) *for each closed set $F \subset Y, f^{-1}(F)$ is closed in X.*

Proof.
(c), (d) hold if the open (closed) sets are from a base or a subbase.
Obviously, (a) $\equiv$ (b) and (c) $\equiv$ (d).
(a)$\Rightarrow$ (c). Suppose E is an open subset of Y and $c \in f^{-1}(E)$. Then $f(c) \in E$ and, since E is open, $f(c)$ is far from E^c. By (a), c is far from $f^{-1}(E^c) = [f^{-1}(E)]^c$, showing thereby that $f^{-1}(E)$ is open in X.
(d)$\Rightarrow$ (a). Suppose $A \subset X$ and $c \in X$ is near A. We must show that $f(c)$ is near $f(A)$ in Y. If not, then $f(c) \notin \mathrm{cl}\,f(A)$, which is closed. By (d), $f^{-1}(\mathrm{cl}\,f(A))$ is closed in X and does not contain c. Since $A \subset f^{-1}(\mathrm{cl}\,f(A))$, it follows that c is not near A, a contradiction. □

[1] Leo Moser observed that a lemma creates a dilemma, if we cannot prove it.

4.2 Continuous Invariants

It was shown that continuous functions preserve compactness and connectedness (*cf.* Theorems 1.9 and 1.14). Also recall from §1.7 (Metric and Fine Proximities) that a space X is compact, if and only if, every open cover of X has a finite subcover. In the context of continuous functions, observe

Theorem 4.2. *Let X, Y denote topological spaces and assume $f : X \to Y$ is a continuous function. Then f preserves compactness and connectedness.*

Proof. Suppose E is compact in X and $\mathcal{G}$ is an open cover of $f(E)$. Since f is continuous, $f^{-1}(\mathcal{G}) = \left\{f^{-1}(G) : G \in \mathcal{G}\right\}$ is an open cover of E. Compactness of E implies that there is a finite subcover $\mathcal{H}$ of $f^{-1}(\mathcal{G})$. Finally, $\{f(H) : H \in \mathcal{H}\}$ is a finite subcover of $f(E)$. This shows that $f(E)$ is compact.

Next, suppose X is connected and $f(X) = Y$. If Y is not connected, then $Y = A \cup B$, where A, B are disjoint open sets in Y. Since f is continuous, $X = f^{-1}(A) \cup f^{-1}(B)$, a union of two disjoint open sets. This contradicts the connectedness of X. □

There is considerable literature on when the converse of the above results holds. In the above result, functions $f : X \to Y$ with the property that f^{-1} preserved open (closed) sets, are considered. Functions that take open sets into open sets as well as those whose graphs are closed are important in functional analysis. Some results are presented in the problems.

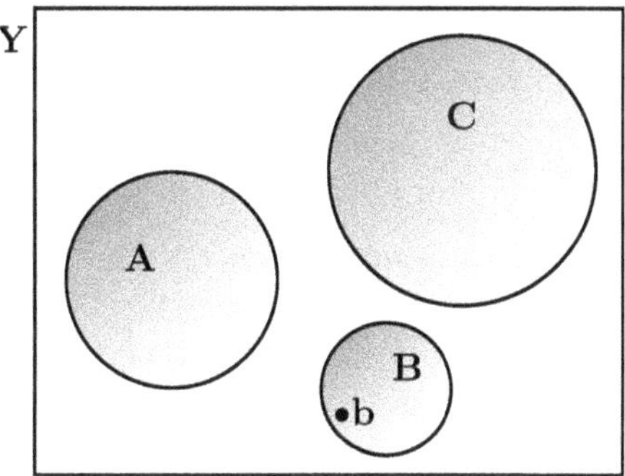

Fig. 4.1 $\mathbf{A}\ \delta_\Phi\ \mathbf{B}$ and $b\ \delta_\Phi\ \mathbf{C}$ for each $b \in B \implies \mathbf{A}\ \delta_\Phi\ \mathbf{C}$ (Descriptive L-space $\mathbf{Y}$)

4.3 Application: Descriptive EF-Proximity in NLO Microscopy

This section introduces an application of descriptive Efremovič proximity in non-linear optical (NLO) microscopy. In particular, descriptive EF-proximity provides a means of describing, analysing and classifying remote sets in NLO microscope images of the various stages in the development of plaque in the arterial tissue of the heart [121; 113]. This application begins with a brief overview of descriptive proximity (nearness) (see, *e.g.*, [183; 182; 186]).

In the study of microscope images, a descriptive form of EF-proximity is useful [183]. Let X be a nonempty set, x a member of X, $\Phi = \{\phi_1, \dots, \phi_n\}$ a set of probe functions that represent features of each x, where $\phi_i(x)$ equals a feature value of x. Let $\Phi(x)$ denote a feature vector for the object x, *i.e.*, a vector of feature values that describe x. A feature vector provides a description of an object. Let A, E be subsets of X. Let $\Phi(A), \Phi(E)$ denote sets of descriptions of members of A, E, respectively. The expression $A\ \delta_\Phi\ E$ means $\Phi(A)$ is near $\Phi(E)$, *i.e.*, the description of A contains points with descriptions that match the description of one or more points in E. By contrast, $A\ \underline{\delta}_\Phi\ E$ means $\Phi(A)$ is far from $\Phi(E)$, *i.e.*, the description of A does not contain points with descriptions that match the description of any point in E. The descriptive proximity of A and E is defined by

$$A\ \delta_\Phi\ E \Leftrightarrow \Phi(A) \cap \Phi(E) \neq \emptyset.$$

For example, let $\Phi(A), \Phi(C), \Phi(E)$ contain descriptions in terms of the colours of the members of A, E, respectively. Then $A\ \delta_\Phi\ E$ (A and E are descriptively near), but $A\ \underline{\delta}_\Phi\ C$ (A and E are descriptively remote, *i.e.*, A and C have no members with matching descriptions) in Fig. 4.2. The descriptive remoteness of A and C (denoted by $A\ \underline{\delta}_\Phi\ C$) is defined by

$$A\ \underline{\delta}_\Phi\ C \Leftrightarrow \Phi(A) \cap \Phi(C) = \emptyset.$$

Define the descriptive intersection $\underset{\Phi}{\cap}$ of A and E by

$$A \underset{\Phi}{\cap} E = \{x \in A \cup E : \Phi(x) \in \Phi(A) \text{ and } \Phi(x) \in \Phi(E)\}.$$

For $x \in A \cup E$, define the descriptive union of A and E by

$$A \underset{\Phi}{\cup} E = \{x \in A \cup E : \Phi(x) \in \Phi(A) \text{ or } \Phi(x) \in \Phi(E)\}.$$

Let $\mathcal{C}_{\Phi(x)}, \mathcal{C}'_{\Phi(y)}$ be covers of regions-of-interest in images X and X', respectively, for $\Phi(x) \in \Phi(X), \Phi(y) \in \Phi(X')$. Here is a nearness measure

$\mu(X, X')$ defined by

$$T = \sum_{\substack{A \in \mathcal{C}_{\Phi(x)}, \\ E \in \mathcal{C}'_{\Phi(y)}}} \left| A \underset{\Phi}{\cap} E \right|,$$

$$D = \sum_{\substack{A \in \mathcal{C}_{\Phi(x)}, \\ E \in \mathcal{C}'_{\Phi(y)}}} \left| A \underset{\Phi}{\cup} E \right|,$$

$$\mu(X, X') = \frac{T}{D}.$$

4.3.1 *Descriptive L-Proximity and EF-Proximity*

This section introduces a descriptive form of Lodato proximity space [133; 134; 135]. A binary relation δ_Φ is a *descriptive L-proximity*, provided the axioms (EF.1)-(EF.5) are satisfied for $A, B, C \in \mathcal{P}(X)$.

($\boldsymbol{EF}$.1) $A \ \delta_\Phi \ B$ implies $A \neq \emptyset, B \neq \emptyset$.
($\boldsymbol{EF}$.2) $A \underset{\Phi}{\cap} B \neq \emptyset$ implies $A \ \delta_\Phi \ B$.
($\boldsymbol{EF}$.3) $A \underset{\Phi}{\cap} B \neq \emptyset$ implies $B \underset{\Phi}{\cap} A$.
($\boldsymbol{EF}$.4) $A \ \delta_\Phi \ (B \cup C)$, if and only if, $A \ \delta_\Phi \ B$ or $A \ \delta_\Phi \ C$.
($\boldsymbol{EF}$.5) $A \underset{\Phi}{\cap} B$ and $b \underset{\Phi}{\cap} C$ for each $b \in B$ implies $A \underset{\Phi}{\cap} C$ (descriptive L-axiom).

The structure (X, δ_Φ) is a *descriptive L-proximity space* (or, briefly, *descriptive L space*), provided axioms (EF.1)-(EF.5) are satisfied. The relation δ_Φ is Efremovič (EF), provided δ_Φ satisfies axioms (EF.1)-(EF.5) as well as axiom (EF.6).

(EF.6) $A \ \underline{\delta}_\Phi \ B$ implies $A \ \underline{\delta}_\Phi \ C$ and $B \ \underline{\delta}_\Phi \ C^c$ for some $C \subset X$.

The structure (X, δ_Φ) is a *descriptive EF-proximity space* (or, briefly, *descriptive EF space*), provided axioms (EF.1)-(EF.6) are satisfied. Usually, a descriptive proximity space is Efremovič, provided axioms (EF.1)-(EF.4) and axiom (EF.6) are satisfied.

Example 4.1. Descriptive L space. Consider sets $A, B, C \in \mathcal{P}(Y)$ in Fig. 4.1 and Y is endowed with a descriptive proximity relation δ_Φ. Choose Φ to be a set containing a probe function that extracts the greyscale intensity of picture elements, *i.e.*, $\phi \in \Phi$ such that $\phi(x)$ equals the greyscale intensity of pixel $x \in Y$. Assume δ_Φ satisfies axioms (EF.1)-(EF.4). In this example, $A \ \delta_\Phi \ B$ and the description of each point in $b \in B$ is near the description of C, *i.e.*, $b\ \delta_{\Phi}\ C$ for each $b \in B$. Then, $A\ \delta_{\Phi}\ C$. Hence, δ_{Φ} is an L proximity relation, since it satisfies axiom (EF.5). ■

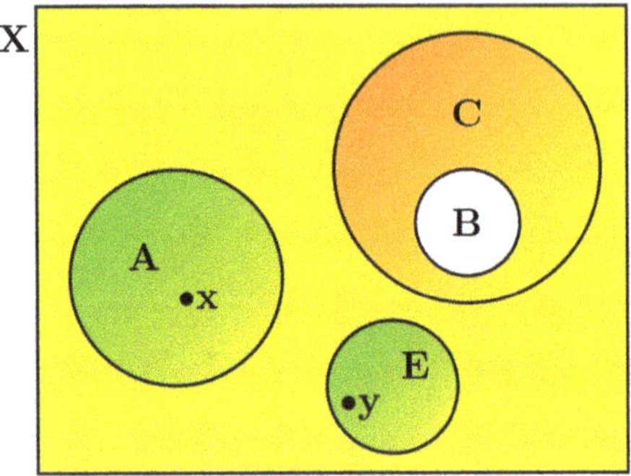

Fig. 4.2 $\mathbf{A}\ \underline{\delta}_{\Phi}\ \mathbf{B} \implies \mathbf{A}\ \underline{\delta}_{\Phi}\ \mathbf{C}$ and $\mathbf{B}\ \underline{\delta}_{\Phi}\ \mathbf{C}^c$ (Descriptive EF-space $\boldsymbol{X}$)

Example 4.2. Descriptive EF space. A representation of a set X endowed with a descriptive proximity relation δ_{Φ} is also shown in Fig. 4.2. Choose Φ to be a set of probe functions representing colour features of picture elements in X. Let $A, B, C \in \mathcal{P}(X)$ as shown in Fig. 4.2 and assume δ_{Φ} satisfies axioms (EF.1)-(EF.4). In this example, $B \subset C$ and the description of each point in B does not match the description of any point in C^c. There is no point in A that matches the description of any point in B. For this reason, $A\ \underline{\delta}_{\Phi}\ B$. Also notice that $A\ \underline{\delta}_{\Phi}\ C$ and that $B\ \underline{\delta}_{\Phi}\ C^c$. Hence, δ_{Φ} is EF, since it satisfies axiom (EF.6). ■

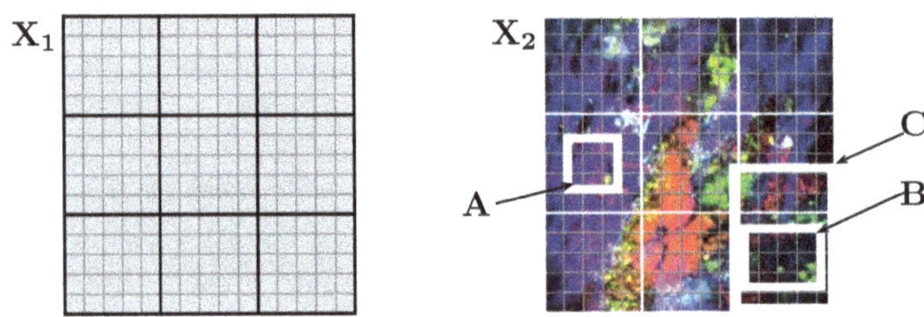

Fig. 4.3 Descriptive non-EF space $\boldsymbol{X_1}$ and descriptive EF-space $\boldsymbol{X_2}$

For distinct points $x, y \in X$, a descriptive EF-proximity δ_Φ is termed **descriptively separated**, if it satisfies

(EF.7) $\{x\}\ \delta_\Phi\ \{y\}$ implies $\Phi(x) = \Phi(y)$.

That is, $\{x\}$ is descriptively near $\{y\}$ implies the description of x matches the description of y, *i.e.*, $\Phi(x) = \Phi(y)$.

Theorem 4.3. *Let (X, δ_Φ) be a descriptive EF proximity space. The descriptive proximity δ_Φ is separated, if and only if, every point in X has the same description.*

Proof. See Problem 4.4.11. □

Example 4.3. Descriptively Near Separated Points.
Let (X, δ_Φ) be a descriptive EF space represented in Fig. 4.2. Let $x \in A, y \in E$ as shown in Fig. 4.2. Notice that $\{x\}\ \delta_\Phi\ \{y\}$. Hence, $\Phi(x) = \Phi(y)$. Points x and y are spatially separated. In addition, the descriptions of x and y are near, which implies x and y have matching descriptions. ■

Example 4.4. Descriptively Non-Near Separated Points.
Let $x \in A, c \in C$ as shown in Fig. 4.2. Notice that $\{x\}\ \underline{\delta}_\Phi\ \{c\}$, since $\Phi(x) \neq \Phi(c)$. That is, points x and c are spatially far and points x, y are descriptively not near. That is, $\Phi(x) \neq \Phi(c) \implies \{x\}\ \underline{\delta}_\Phi\ \{c\}$. ■

Example 4.5. Non-EF Space.
In Fig. 4.3, let the set X_1 be endowed with a descriptive L-proximity relation δ_Φ that satisfies axioms (EF.1)-(EF.5). All distinct points X_1 have matching descriptions. Let $x, y \in X_1$. Then δ_Φ is separated, since $\{x\}\ \delta_\Phi\ \{y\}$ implies $\Phi(x) = \Phi(y)$. Let $A \subset X_1$ and let $B \subset A^c$. The condition for axiom (EF.6) is not satisfied for any such A and B, since $A\ \delta_\Phi\ B$. Hence, δ_Φ is not EF. ■

Example 4.6. EF space.
In Fig. 4.3, let the set X_2 be endowed with a descriptive L-proximity relation δ_Φ that satisfies axioms (EF.1)-(EF.5). The set X_2 is represented by an NLO-microscope image with a grid superimposed on it. Choose a set of probe functions Φ representing features of points in X. Choose A, B, C in X such that $C \div A^c$ and $B \subset C$. Then, $A\ \underline{\delta}_\Phi\ C$ and $B\ \underline{\delta}_\Phi\ C^c$. Hence, axiom (EF.6) is satisfied and δ_Φ is EF. ■

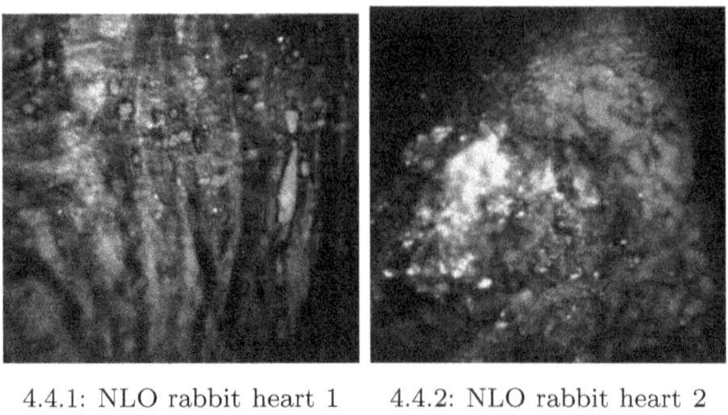

4.4.1: NLO rabbit heart 1 4.4.2: NLO rabbit heart 2

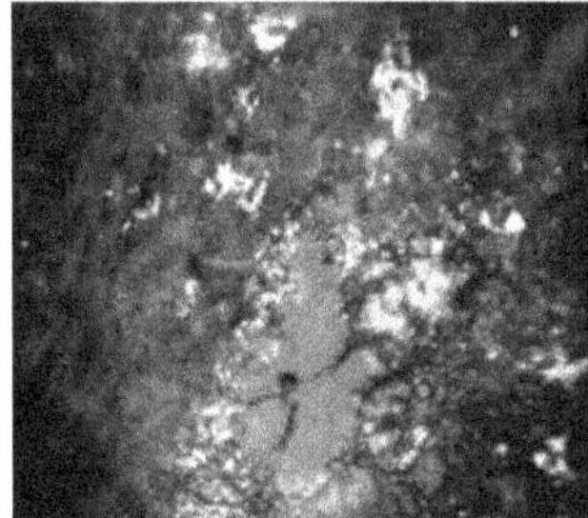

4.4.3: NLO rabbit heart 3

Fig. 4.4 Sample NLO microscope images of rabbit heart with plaque

4.3.2 *Descriptive EF Proximity in Microscope Images*

Next, consider the application of descriptive EF-proximity in NLO microscope images useful in the study of the early onset of heart disease. Atherosclerosis is the primary cause of heart disease, stroke and lower limb amputation worldwide. It is a progressive disease characterized by chronic inflammation of injured intima and is associated with fatty plaque deposits in the arteries [129; 113]. Recently, NLO microscopy has emerged as a powerful tool for tissue imaging. Several studies have demonstrated imaging of arterial tissue using NLO microscopy (see, *e.g.*, [121; 113; 190]). The images in Fig. 4.4 display different levels of plaque-laden rabbit heart tissue at the 50μm level. It is assumed that the points in each of these images are sets endowed with a descriptive proximity relation δ_Φ, satisfying axioms (EF.1)-(EF.4). In addition, depending on the choice of features represented by Φ, the points in the NLO images in Fig. 4.4 are descriptively distinct, *i.e.*, δ_Φ is separated. Hence, by Theorem 14.7, δ_Φ is EF.

Hence, the analysis of these NLO microscope images can be carried out in terms of a descriptive EF-proximity δ_Φ on remote sets of points in these images. In the NLO image in Fig. 4.5, let X denote the set of descriptively distinct points endowed with an EF-proximity relation δ_Φ. For example, it is clear that visual sets $A, B, C \subset X$ satisfy axiom (EF.5), *i.e.*

$$\mathbf{A} \ \underline{\delta}_\Phi \ \mathbf{B} \Rightarrow \mathbf{A} \ \underline{\delta}_\Phi \ \mathbf{C} \text{ and } \mathbf{B} \ \underline{\delta}_\Phi \ C^c.$$

At this point, it is then possible to analyse such an image in terms of descriptive EF-proximity for various combinations of remote sets and to quantify the degree of remoteness of such sets, gauging the severity of the plaque in different regions of this tissue sample.

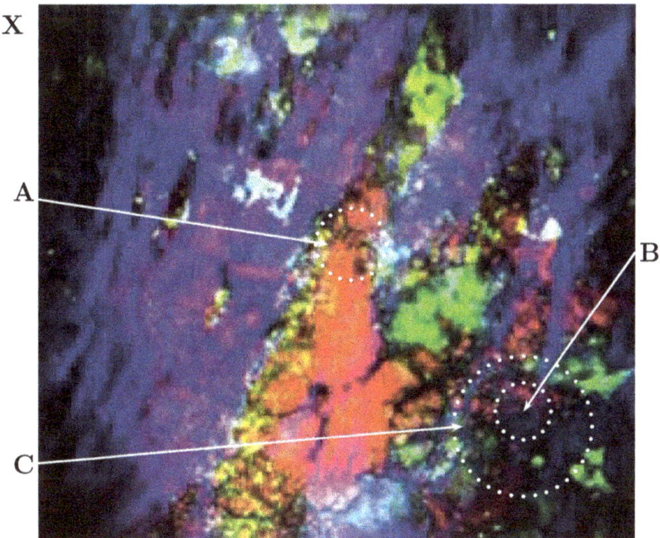

Fig. 4.5 Descriptive EF-proximity: $\mathbf{A} \ \underline{\delta}_\Phi \ \mathbf{B} \Rightarrow \mathbf{A} \ \underline{\delta}_\Phi \ \mathbf{C}$ and $\mathbf{B} \ \underline{\delta}_\Phi \ C^c$

4.4 Problems

(**4.4**.1) Show that $f : X \to Y$ is continuous, if and only if, for each member E of a base for open (closed) subsets of Y, $f^{-1}(E)$ is open (closed) in X. Show that the result from this Problem is true, if *subbase* is substituted for *base*. **Hint**: For the subbase problem, note that f^{-1} preserves unions and intersections.

(**4.4**.2) Show that $f : X \to Y$ is continuous, if and only if, for each subset $E \subset Y$, $\mathrm{cl}\, f^{-1}(E) \subset f^{-1}(\mathrm{cl}\, E)$.

(**4.4**.3) Show that $\mathbb{R}$ is homeomorphic to (0,1). Show that no two of the following are homeomorphic: (0,1), [0,1), and [0,1].

(**4.4**.4) A topological space is called T_2 (**Hausdorff space**), if and only if, it is a space in which distinct points belong to disjoint neighbourhoods. Show that
  (a) A closed subset of a compact Hausdorff space is compact.
  (b) Compact subsets of a Hausdorff space are closed.
  (c) A continuous function f from a compact space X to a Hausdorff space Y is closed, *i.e.*, if A is a closed subset of X, then $f(A)$ is closed in Y. Further, if f is a bijection, then f is a homeomorphism.

(**4.4**.5) Show that the set of all homeomorhphisms on a topological space onto itself form a group[2] under composition.

(**4.4**.6) Show that the set of all real-valued, continuous functions on a topological space X form a ring[3] under the operations
  (i) $(f + g)(x) = f(x) + g(x)$, and
  (ii) $(fg)(x) = f(x)g(x)$.

(**4.4**.7) Let X be a locally connected and first countable space. Let Y be a T_3 space[4] (if a point x does not belong to a closed set E, then x and E have disjoint neighbourhoods). Show that if a function f from X to Y preserves compact sets and connected sets, the f is continuous [111; 245].

(**4.4**.8) Suppose $f : \mathbb{R} \to \mathbb{R}$ takes connected sets into connected sets and, for each $y \in \mathbb{R}$, the fiber[5] $f^{-1}(y)$ is closed. Show that f is continuous [203].

(**4.4**.9) Suppose X is a topological space with a base $\mathcal{B}$ of connected open sets and Y is a T_3 space with a base $\mathcal{B}'$ of open sets. In addition, suppose $f : X \to Y$ is a function satisfying

---

[2]Recall that **group** is a nonempty set of elements G together with a binary operation (usually denoted $\cdot$) such that the operation is closed, associative, the set has an identify element and the operation is invertible (see, *e.g.*, [90]).

[3]There is a vast literature on rings of topological spaces in which there is an interesting interaction between algebra and topology (see, *e.g.*, [74]).

[4]Recall that a topological space X is $\boldsymbol{T_1}$, if and only if, for every pair of distinct points $x_1, x_2 \in X$, there exists an open set $E \subset X$ such $x_1 \in E$ and $x_2 \notin E$. Then a topological space X is called a $\boldsymbol{T_3}$ **space**, if and only if, X is a T_1 space and for every $x \in X$ and every closed set $E \subset X$ such that $x \notin E$, there exist open sets $U_1, U_2 \subset X$ such that $x \in U_1, E \subset U_2$ and $U_1 \cap U_2 \neq \varnothing$ (see, *e.g.*, [59].

[5]The **fiber** of a point $y \in Y$ for a function $f : X \to Y$ is $f^{-1}(\{y\}) = \{x \in X : f(x) = y\}$.

(a) For each $B \in \mathcal{B}, f(\mathrm{cl}\, B)$ is connected, and
(b) for each $B' \in \mathcal{B}', f^{-1}[\partial \mathcal{B}']$ is closed ($\partial \mathcal{B}'$ is the boundary of $\mathcal{B}'$).
Show that f is continuous [95].

(**4.4**.10) **Near and Quasi-Continuity** A function $f : X \to Y$ is *continuous*, if and only if, for each neighbourhood V of $f(x)$, there is a neighbourhood U of x such that $U \subset f^{-1}(V)$. By the changing the condition $U \subset f^{-1}(V)$ in different ways, we get weaker forms of continuity. Such weaker forms are valuable in functional analysis and in problems involving joint versus separate continuity.

(a) A function $f : X \to Y$ is **nearly continuous** at $x \in X$, if and only if, for each neighbourhood V of $f(x)$, there is a neighbourhood of U of x such that $U \subset \mathrm{cl}\, f^{-1}(V)$.
(b) A function $f : X \to Y$ is **quasi-continuous** at $x \in X$, if and only if, for each neighbourhood U of x and for each neighbourhood V of $f(x)$, there is a nonempty open set $W \subset U \cap f^{-1}(V)$.

A nearly continuous (respectively, quasi-continuous) function is one that is nearly continuous (respectively, quasi-continuous) at each $x \in X$. Prove that if Y is T_3, then a function $f : X \to Y$ that is nearly continuous and quasi-continuous is continuous [70].

(**4.4**.11) Prove Theorem 4.3.

Chapter 5

Separation Axioms

5.1 Discovery of the Separation Axioms

While working on a problem involving topological spaces, it is often necessary to add some extra condition called *separation axiom*. For example, in §3.1, the symmetric or R_0 axiom ((∗) a point a is near another point b implies b is near a) arose during a consideration of symmetric proximities compatible with a topology[1]. This axiom was discovered by A.S. Davis.

Earlier separation axioms had been discovered and were called *Trennungsaxiome* (German) by P. Alexandroff and H. Hopf. Hence, these axioms are known as $T_n, n = 0, 1, 2, 3, 4, 5$. Often these axioms have alternate names such as Hausdorff, normal, Tychonoff, and so on and there is no unanimity in the nomenclature. It is helpful to construct counterexamples to delineate various separation axioms.

> There are two types of separation, namely,
> (**Separation**.1) **Neighbourhood**: disjoint sets have neighbourhoods with distinct properties, and
> (**Separation**.2) **Functional**: disjoint sets can be separated by a continuous function.

In contrast to the symmetry axiom R_0, the anti-symmetric axiom T_0 (discovered by A. Kolmogorov) is defined as follows.

[1]See §1.20 for an example of an R_0 space, *i.e.*, the (∗) is satisfied due to the partition of an image into equivalence classes. Notice that the space formed by the partition with two or more classes in the example in §1.20 is not T_0, it is possible to have two distinct points a, b in separate classes such that $a \underline{\delta} b$ (a is far from b).

T_0: (a) For every pair of distinct points, at least one of them is far from the other, or
(b) For every pair of distinct points in a topological space X, there exists an open set containing one of the points but not the other point.

An indiscrete topological space is not T_0. A T_0 space that is not R_0 is $X = \{a, b\}$ with open sets $\varnothing, X, \{a\}$. An R_0 space that is not T_0 is $X = \{a, b, c\}$ with open sets $\varnothing, X, \{c\}, \{a, b\}$. Thus, R_0 and T_0 are independent[2]. The next separation axiom is $T_1 = T_0 + R_0$ (discovered by M. Fréchet and F. Riesz).

T_1: A topological space is T_1, if and only if, distinct points are not near.

In a T_1 space, singleton sets, and consequently, all finite sets are closed. If the topologies on a set are partially ordered with $\subset$, then the co-finite topology is the coarsest T_1 topology.

In analysis, it is shown that a sequence in the space of real or complex numbers has at most one limit. This result was extended to metric spaces in Ch. 1, starting in §1.4, Theorem 1.2 (see Problem 1.18.24). The proof depends on the fact that two distinct points have disjoint neighbourhoods. This fact is not necessarily true in topological spaces in general, a fact made plain by the space of an infinite set with co-finite topology.

While working in abstract spaces, it is necessary frequently to have this property of unique limits for nets and filters (see Ch. 6), which generalise sequences. Hausdorff observed this and used this axiom in his work. The corresponding space with *pairs of distinct points belong to disjoint neighbourhoods* is now named after him and is called the T_2 or Hausdorff space.

T_2: A topological space is T_2, if and only if, distinct points have disjoint neighbourhoods (distinct points live in disjoint *houses*[3]).

[2]Computable versions of the separation axioms T_0, T_1, and T_2 are considered in [243].
[3]In the application in §1.20, a partition is a T_2 space, if and only if, every class has no more than one point, *i.e.*, every class is single tenant 'house'.

In most problems in topology and functional analysis, this axiom is used. Obviously, T_2 implies T_1 and the reverse implication is not true. For example, an infinite set with co-finite topology is T_1 but not T_2.

It has also been observed that every T_1 space is a T_0 space. Putting these observations together, $T_2 \Rightarrow T_1 \Rightarrow T_0$.

The discovery of separation axioms continues. L. Vietoris defined a stronger axiom than T_2.

> T_3 (**regular**): A T_1 space is T_3 or **regular**, if and only if, a point a does not belong to a closed set B implies a and B have disjoint neighbourhoods. Since the complement of B is an open neighbourhood of a, an alternate description of T_3 can be made, *i.e.*, *each neighbourhood of a point contains a closed neighbourhood.*

Obviously, every T_3 space is Hausdorff or T_2. Example 2.8.1 in Ch. 2 shows that a Hausdorff space need not be regular. The next separation axiom (T_4 or **normal**) was discovered by H. Tietze in 1923.

> T_4 (**normal**): A T_1 topological space in which disjoint closed sets have disjoint neighbourhoods is called **normal**. Alternatively, if a closed set A is contained in an open set U, then there is an open set V such that
>
> $$A \subset V \subset \operatorname{cl} V \subset U.$$

Obviously, T_4 is stronger than all of the previous separation axioms and we have the following scheme exhibiting the relation between T_4 and the other axioms.

$$T_4 \Rightarrow T_3 \Rightarrow T_2 \Rightarrow T_1 \Rightarrow \begin{cases} R_0 \\ T_0 \end{cases}$$

An interesting result is that a T_1 topological space X is normal, if and only if, its compatible fine L-proximity δ_0 is Efremovič.

Proof. Recall that $A \ \delta_0 \ B$, if and only, $\operatorname{cl} A \cap \operatorname{cl} B \neq \emptyset$. Suppose A, B are disjoint closed sets of a normal space. Then they are far and so there is a subset $E \subset X$ such A is far from E and E^c is far from B. Then E^c and E are disjoint neighbourhoods of A, B, respectively. The converse follows easily. □

5.2 Functional Separation

Subsets A, B of a topological space X are **functionally separated** or **completely separated**, if and only if, there is a continuous function $f : X \to [0,1]$ such that $f(A) = 0$ and $f(B) = 1$. If A, B are thus separated, then $f^{-1}\left[0, \frac{1}{3}\right]$ and $f^{-1}\left[\frac{2}{3}, 1\right]$ are disjoint neighbourhoods of A and B. So functional separation implies neighbourhood separation.

A topological space X is termed **completely regular** or **Tychonoff**, if and only if, whenever a point a is not in a closed set B, a and B can be functionally separated. Occasionally, the symbol $T_{3.5}$ is used to distinguish a completely regular space from T_3. That the two spaces are not equivalent follows from the fact that a regular or T_3 space need not be completely regular. This was shown dramatically by the construction of a regular space in which every continuous function is constant [91].

One of the most beautiful, intricate and surprising results in topology was proved by P. Urysohn in 1925 by showing that, in a normal or T_4 space, neighbourhood separation of disjoint closed sets equals functional separation. This result (known as the Urysohn Lemma) is sometimes viewed as the first non-trivial fact of point set topology.

Lemma 5.1. (Urysohn Lemma) *If A, B are disjoint closed subsets of a normal space X, then there is a continuous function $f : X \to [0,1]$ with $f(A) = 0$ and $f(B) = 1$.*

Proof.
Clearly, $U_1 = X - B$ is an open set containing A. The trick is to construct, by induction, for each rational $r \in (0,1]$ an open set U_r such that (i) $A \subset U_r$ for each r and (ii) $r < s$ implies $\mathrm{cl}\, U_r \subset U_s$. This is done by repeated use of (T_4). Let the rationals in (0,1] be ordered $\{r_j : j \in \mathbb{N}\}$ with $r_1 = 1$. Suppose U_{rj} has been constructed for $j < n$. Let r_p, r_q be respectively the largest and the smallest elements of $\{r_j : j < n\}$ such that $r_p < r_n < r_q$. Since $\mathrm{cl}\, U_{rp} \subset U_{rq}$ by (T_4), there exists U_{rn} such that

$$\mathrm{cl}\, U_{rp} \subset U_{rn} \subset \mathrm{cl}\, U_{rn} \subset U_{rq}.$$

For each $x \in X$, define

$$f(x) = \begin{cases} inf\ \{r : x \in U_r\}, & \\ 1, & \text{otherwise.} \end{cases}$$

Clearly, f is a function on X to [0,1], $f(A) = 0$ and $f(B) = 1$. For each rational $t \in (0,1], f^{-1}([0,t)) = \bigcup\{U_r : r < t\}$ is open. Similarly, $f^{-1}((t,1])$ is open for each rational $t \in (0,1]$. Hence, f is continuous. □

It is obvious that a normal or T_4 space is Tychonoff and there are examples to show that the converse is not true. In an EF-proximity space, two far sets have closed neighbourhoods that are far from each other. This is akin to disjoint closed sets in a normal space.

So, in a T_1 space with compatible EF-proximity, two sets that are far are functionally separated. Hence, the topology is Tychonoff. Conversely, every Tychonoff space has a compatible EF-proximity δ_F given by $A \underline{\delta}_F B$, if and only if, A, B are functionally separated.

Proof. We need only prove the union axiom and the EF-axiom.

Suppose $A \underline{\delta}_F B$ and $A \underline{\delta}_F C$. Then there are continuous functions f, g on X to [0,1] such that $f(A) = 0, f(B) = 1, g(A) = 0$ and $g(C) = 1$. The function h defined by $h(x) = max\{f(x), g(x)\}$ is continuous and h separates A and $(B \cup C)$. So $A \underline{\delta}_F (B \cup C)$.

Suppose $A \underline{\delta}_F B$ and f separates A, B, Then $E = f^{-1}(\frac{1}{2}, 1]$ is far from A and E^c is far from B. It is clear that δ_F is the fine EF-proximity on a Tychonoff space, *i.e.*, if δ is any other compatible EF-proximity on X, then $A \, \delta_F \, B \Rightarrow A \, \delta \, B$. □

5.3 Observations about EF-Proximity

This section gives a brief summary of EF-proximity, the first proximity studied thoroughly. A topological space is **Tychonoff**, if and only if, it has a compatible EF-proximity. On a Tychonoff space, δ_F is the fine EF-proximity.

An EF-proximity space is akin to a normal or T_4 space in the sense that far (remote) sets correspond to sets with disjoint closures. So, what one can do in a normal space, can usually be done in a Tychonoff space by replacing sets with disjoint closures with sets that are far from each other. The fine L-proximity δ_0 in a T_1 space is EF, if and only if, the space is normal (Urysohn Lemma).

5.4 Application: Distinct Points in Hausdorff Raster Spaces

A **point** is an element of a space X [229], a subset of X having no proper subsets [122]. In the case of a digital image, picture elements (pixels) are considered points.

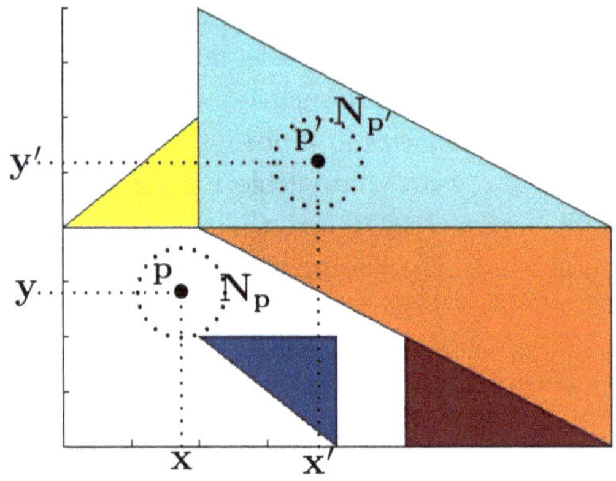

Fig. 5.1 Descriptively distinct points $\mathbf{p}, \mathbf{p}'$ in $\mathbf{N_p}, \mathbf{N_{p'}}$, resp.

Let X be a nonempty set, points $p, p' \in X$, $d : X \times X :\rightarrow \mathbb{R}$ a distance function that determines the distance between points in X, $\phi : X \rightarrow \mathbb{R}$ a real-valued probe function that represents a feature of $p \in X$, $\Phi(p) = [\phi_1(p), \ldots, \phi_n(p)]$ a feature vector that describes $p \in X$, and let $\Phi(p), \Phi(p')$ be descriptions of points p, p'. Then a **feature vector** is a vector of numbers, where each number is a feature value of a point extracted by a probe function.

5.4.1 *Descriptive Proximity*

Recall that **distinct points** belong to disjoint neighbourhoods in a Hausdorff T2 space. It is helpful to distinguish between spatial and non-spatial distinctness. Points p, p' are **spatially distinct**, provided x and y are in different locations (separated spatially). Points p and p' are considered **descriptively distinct**, provided p and p' have sufficiently different descriptions. Similarly, sets A and B in X are descriptively distinct, provided all points $a \in A, b \in B$ are descriptively distinct. For example, let X be a set of picture elements in Fig. 5.1 and let neighbourhoods $N_p, N_{p'}$ points p, p', respectively, be as shown in Fig. 5.1. Let Φ be a set of probe functions representing the colour and greyscale intensities of points in X. The white points in N_p are descriptively distinct from the turquoise points in $N_{p'}$. The notion of descriptive distinctness gives rise to a **descriptive proximity** relation denoted by δ_Φ.

Let $\mathfrak{I}, \mathfrak{I}'$ denote nonempty point sets, $p \in \mathfrak{I}, p' \in \mathfrak{I}'$. This gives rise to several different possible incarnations of descriptive proximities, namely,

(δ_Φ).1 **Descriptively near (remote) points.**

$$p\ \delta_\Phi\ p' \text{ **near** points or } p\ \underline{\delta}_\Phi\ p' \text{ **remote** points.}$$

(δ_Φ).2 **Descriptively near (remote) points and sets.**

$p\ \delta_\Phi\ \mathfrak{I}$ (p **near** $\mathfrak{I}$) or $p\ \underline{\delta}_\Phi\ \mathfrak{I}'$ (p **remote (far)** from $\mathfrak{I}'$).

(δ_Φ).3 **Descriptively near (remote) sets.**

$\mathfrak{I}\ \delta_\Phi\ \mathfrak{I}'$ (**near** sets) or $\mathfrak{I}\ \underline{\delta}_\Phi\ \mathfrak{I}'$ (**remote (far)** sets).

The descriptive proximity relation δ_Φ ($\underline{\delta}_\Phi$) is defined in a similar fashion in each of the three cases. Then

$$\mathfrak{I}\ \delta_\Phi\ \mathfrak{I}', \text{ if and only if, } D_\Phi(\mathfrak{I}, \mathfrak{I}') = 0 \text{ (near sets), where}$$
$$D_\Phi(\mathfrak{I}, \mathfrak{I}') = inf\{d(\Phi(p), \Phi(p')) : p \in \mathfrak{I}, p' \in \mathfrak{I}'\}, \text{ and}$$
$$\mathfrak{I}\ \underline{\delta}_\Phi\ \mathfrak{I}', \text{ if and only if, } D_\Phi(\mathfrak{I}, \mathfrak{I}') \neq 0 \text{ (remote sets).}$$

Recall that the pair (X, δ_Φ) is a **descriptive EF proximity space**, provided axioms (EF.1)-(EF.6) are satisfied. In terms of the nearness (remoteness) of points, consider two singleton sets $\mathfrak{I}, \mathfrak{I}'$, where $\mathfrak{I} = \{p\}$ and $\mathfrak{I}' = \{p'\}$. For simplicity, we also write p, p' instead of $\{p\}, \{p'\}$. Non-spatial distinctness of points is explained in terms of point descriptions. Distinct points have different descriptions. This is expressed by writing $p\ \underline{\delta}_\Phi\ p'$, *i.e.*, point p is **descriptively remote** (far from) point p'. By contrast, points p, p' are near, provided $p\ \delta_\Phi\ p'$, *i.e.*, point p is **descriptively near** point p', provided $d(\Phi(p), \Phi(p')) = 0$ (p and p' have matching descriptions).

Theorem 5.1. *In a descriptive EF proximity space, a nonempty set is descriptively near itself.*

Proof. Let (X, δ_Φ) be a descriptive EF proximity space, $A \in \mathcal{P}(X)$. $A\ \delta_\Phi\ A$, since $D(\Phi(A), \Phi(A)) = 0$. □

Theorem 5.2. *A point is descriptively near itself.*

Proof. Let $\mathfrak{I} = \{p\}$, a nonempty set containing a single point. The result follows from Theorem 5.1. □

Lemma 5.2. *In a descriptive EF proximity space, spatially distinct points can be descriptively near or remote.*

Proof. Let (X, δ_Φ) be a descriptive EF proximity space. Assume $p, p', p'' \in X$ are separated spatially and assume $d(\Phi(\{p\}), \Phi(\{p'\})) = 0$, *i.e.*, spatially distinct points p and p' have the same descriptions. Then $p\ \delta_\Phi\ p'$ (*i.e.*, p is descriptively near p'). Assume $d(\Phi(\{p\}), \Phi(\{p''\})) \neq 0$, *i.e.*, p and p'' have different descriptions. Then $p\ \underline{\delta}_\Phi\ p''$ (p is descriptively remote (far) from p''). □

Theorem 5.3. *Descriptively remote points are spatially remote points.*

Proof. Let (X, δ_Φ) be a descriptive EF proximity space. Assume $p, p' \in X$ are descriptively remote points but spatially not distinct. Hence, $p\ \underline{\delta}_\Phi\ p'$, *i.e.*, $d(\Phi(\{p\}), \Phi(\{p'\})) \neq 0$. Assuming p, p' are spatially not distinct, then $p = p'$, if and only if, $\Phi(p) = \Phi(p')$. From Theorem 5.2, $p\ \delta_\Phi\ p'$, a contradiction. □

Example 5.1. Spatially Distinct Points.
Points p and p' in the digital image in Fig. 5.1 are spatially distinct, since p and p' have different coordinates in the image. Hence,

$$p\ \underline{\delta}\ p', \text{ i.e., points } p, p' \text{ are remote.} \quad \blacksquare$$

Example 5.2. Descriptively Distinct Points.
Points p and p' in the digital image in Fig. 5.1 are descriptively distinct, since p (white pixel) and p' (turquoise pixel) have different descriptions in the image. That is, $p\ \underline{\delta}_\Phi\ p'$ (p is descriptively remote from p'). ■

Assume p, p' are spatially distinct points. Neighbourhoods $N_p, N_{p'}$ are **spatially distinct neighbourhoods**, provided each point $x \in N_p$ is spatially distinct from each point $y \in N_{p'}$ such that $p\ \underline{\delta}\ p'$. Hence, $N_p\ \underline{\delta}_\Phi\ N_{p'}$, since $D(\{x\}, \{y\}) \neq 0$ for each pair $x \in N_p, y \in N_{p'}$. From this and Lemma 5.2, we obtain the following result.

Theorem 5.4. *Spatially distinct neighbourhoods are remote.*

Theorem 5.5. *Spatially distinct neighbourhoods can be descriptively near or remote.*

Proof. Let $(X, \delta), (X, \delta_\Phi)$ be spatial EF and descriptive EF spaces, respectively. Assume $N_p, N_{p'}$ are spatially distinct neighbourhoods of distinct points p, p', respectively, *i.e.*, $\mathrm{cl}(N_p)\ \underline{\delta}\ \mathrm{cl}(N_{p'})$. Let $x \in N_p, y \in N_{p'}$. From Lemma 5.2, x and y can be descriptively near. Assume that there is a least one pair x, y such that $x\ \delta_\Phi\ y$. Hence, $N_p\ \delta_\Phi\ N_{p'}$. Again, from Lemma 5.2,

x and y can be descriptively remote. Assume every pair x, y is descriptively remote, *i.e.*, $x \ \underline{\delta}_{\Phi} \ y$ for a set of features Φ. Hence, $N_p \ \underline{\delta}_{\Phi} \ N_{p'}$. □

Let (X, δ_{Φ}) be a descriptive L space, $x \in X$. A **descriptive neighbourhood** of x (denoted by $N_{\Phi(x)}$) is defined by

$$N_{\Phi(x)} = \{y \in X : x \ \delta_{\Phi} \ y\}.$$

That is, if $y \in N_{\Phi(x)}$, then $d(\Phi(x), \Phi(y)) = 0$.

Theorem 5.6. *In a descriptive L-proximity space, descriptively distinct points belong to descriptively remote neighbourhoods.*

Proof. Let (X, δ_{Φ}) be a descriptive L proximity space. Assume $p, p' \in X$ are descriptively distinct (*i.e.*, $p \ \underline{\delta}_{\Phi} \ p'$) and consider neighbourhoods descriptive $N_{\Phi(p)}, N_{\Phi(p')}$. Assume $N_{\Phi(p)} \ \delta_{\Phi} \ N_{\Phi(p')}$. Then there is an $x \in N_p$ such that

$$p \ \delta_{\Phi} \ x \text{ and } x \ \delta_{\Phi} \ p' \Rightarrow p \ \delta_{\Phi} \ p' \text{ (Lodato axiom).}$$

A contradiction. Hence, $N_{\Phi(p)} \ \underline{\delta}_{\Phi} \ N_{\Phi(p')}$ (remote neighbourhoods). □

From the proof of Theorem 5.6, observe that descriptively remote neighbourhoods are disjoint neighbourhoods. That is, descriptively remote neighbourhoods have no points in common.

Corollary 5.1. *In a descriptive EF proximity space, descriptively distinct points belong to descriptively remote neighbourhoods.*

Proof. See Problem 5.5.9. □

Example 5.3. Descriptively Remote Neighbourhoods

In Fig. 5.1, $N_p, N_{p'}$ are not only spatially distinct neighbourhoods but also descriptively remote neighbourhoods, since these neighbourhoods have no points in common. In Fig. 5.1, $N_{\varepsilon,\Phi(p)}$ is the neighbourhood of p containing only white points, *i.e.*, all points with intensity equal to zero. When it is clear from the context what is meant, we write N_p instead of $N_{\Phi(p)}$. In Fig. 5.1, $N_p, N_{p'}$ are examples of bounded descriptive neighbourhoods. Notice that if we introduce $\varepsilon \in (0, \infty)$ in a metric space (X, d), then we obtain a **bounded descriptive neighbourhood** defined by

$$N_{\varepsilon,\Phi(p)} = \{p' \in X : p \ \delta_{\Phi} \ p' \text{ and } d(\Phi(p), \Phi(p')) < \varepsilon\}.$$

Notice, for instance, $N_{\varepsilon,\Phi(p)} \subset N_{\Phi(p)}$ and $N_{\varepsilon,\Phi(p')} \subset N_{\Phi(p')}$ in Fig. 5.1. ■

Lemma 5.3. *Descriptively distinct points in a descriptive EF proximity space belong to disjoint descriptive neighbourhoods.*

Proof. See Problem 5.5.11. □

Example 5.4. Spatially Distinct Neighbourhoods.
Neighbourhoods $N_p, N_{p'}$ in the digital image in Fig. 5.1 are spatially distinct, since every point $x \in N_p$ is spatially distinct from every $y \in N_{p'}$. Hence, $N_p \underline{\delta} N_{p'}$. ■

Example 5.5. Descriptively Distinct Neighbourhoods.
Neighbourhoods $N_p, N_{p'}$ in the digital image in Fig. 5.1 are descriptively distinct follows from Lemma 5.3, since it can be observed that points p, p' are descriptively distinct, *i.e.*, $N_p \underline{\delta}_\Phi N_{p'}$. ■

5.4.2 *Descriptive Hausdorff Space*

A **descriptive Hausdorff space** is defined in the context of a descriptive proximity space X, where each descriptive neighbourhood $N_{\Phi(p)}$ of a point p contains one or more distinct points that have matching descriptions. Given two such neighbourhoods $N_{\Phi(p)}, N_{\Phi(p')}$, assume that p and p' are descriptively remote, *i.e.*, $p \underline{\delta}_\Phi p'$. For each $x \in N_{\Phi(p)}, y \in N_{\Phi(p')}$,

$$p \ \delta_\Phi \ x \text{ and } p' \ \delta_\Phi \ y\,.$$

Since $p \underline{\delta}_\Phi p'$, neighbourhoods $N_{\Phi(p)}, N_{\Phi(p')}$ have no points in common. This leads to the following result.

Theorem 5.7. *A descriptive EF proximity space X is a descriptive Hausdorff (T_2) space.*

Example 5.6. Hausdorff Raster Space.
Recall that a **raster** is a rectangular pattern of scanning lines followed by an electron beam on a television screen or computer monitor. A **raster image** is a 2D array of numbers representing picture element (pixel) intensities. Each row of numbers in a raster image contains intensity values for the pixels in one of the lines in a raster. The points in a raster image are descriptively distinct by virtue of their individual intensities. From Corollary 5.3, descriptively distinct points in a raster image belong to disjoint descriptive neighbourhoods. Hence, from Theorem 5.7, every raster image is a descriptive Hausdorff space. This observation has important implications in the

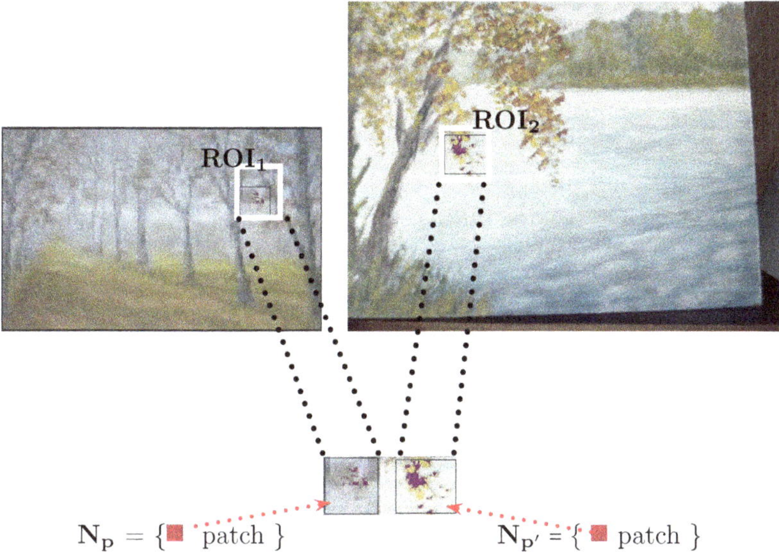

Fig. 5.2 Hausdorff raster spaces in Pawlak pictures

study of pictures (see, *e.g.*, [183; 51; 50; 156; 186; 178]). For recent application of the proposed approach in uncovering visual patterns in pictures, see the study of neighbourhoods in the paintings by Z. Pawlak [178], who introduced rough sets during the early 1980s [170; 171; 191; 174; 173; 172]. An example of a Hausdorff raster space is shown in Fig. 5.1. ■

An obvious application of descriptive Hausdorff spaces can be found in the study of various forms of pictures such as paintings. In fact, local descriptive Hausdorff spaces can be found in any digital image.

Example 5.7. Sample Hausdorff Spaces in a Pawlak Paintings Regions-of-interest (ROIs) in paintings of brightly-coloured autumn trees by Z. Pawlak are shown in Fig. 5.2. Each ROI is an example of a raster image. Hence, from Example 5.6, each ROI is a descriptive Hausdorff space. Assume that each ROI is selected manually. After identifying the ROIs, it is then possible to identify visual neighbourhoods of each point in the ROIs (see, *e.g.*, $N_p \in ROI_1, N_{p'} \in ROI_2$ in Fig. 5.2). To determine the nearness or apartness of the selected ROIs, it is necessary to identify the features of image pixels needed to extract the descriptions of the pixels in the form of feature vectors.

Detection of visual patterns (repetition of shapes or colors or textures like those found in Pawlak's paintings) in a visual neighbourhood then reduces to determining the perceptual distances between the neighbourhoods of ROI points. By perceptual distance, we mean determining the distance $D(\Phi(N_p), \Phi(N_{p'}))$, *i.e.*, the distance between the sets of descriptions $\Phi(N_p), \Phi(N_{p'})$ of neighbourhoods $N_p, N_{p'}$. $N_p \ \delta_\Phi \ N_{p'}$ ($\Phi(N_p)$ is near $\Phi(N_{p'})$), provided the description of N_p contains picture elements that match the description of one or more picture elements in $N_{p'}$. Visual patterns are discovered by searching for pairs of descriptive neighbourhoods that are near each other in descriptive Hausdorff spaces. ■

5.5 Problems

(**5.5**.1) Show that the following are equivalent:
  (a) X is R_0, *i.e.*, $a, b \in X, a$ is near b, implies b is near a.
  (b) $x, y \in X, \mathrm{cl}\{x\} = \mathrm{cl}\{y\}$ or $\mathrm{cl}\{x\} \cap \mathrm{cl}\{y\} = \varnothing$.

(**5.5**.2) Show that a topological space is T_0, if and only if, closures of distinct points are distinct.

(**5.5**.3) Show that $R_0, T_n, n = 0, 1, 2, 3$ are hereditary and topological invariants.

(**5.5**.4) Show that the following are equivalent for a topological space X to be T_1.
  (i) Singleton sets in X are closed.
  (ii) Each subset of X is the intersection of open sets containing it.

(**5.5**.5) Show that derived sets in a T_1 space are closed.

(**5.5**.6) Show that a space X is T_3, if and only if, each closed subset is the intersection of all of its closed neighbourhoods.

(**5.5**.7) Is it possible for two distinct metrics on a set to induce the same topology and yet induce different EF-proximities? If so, give an example.

(**5.5**.8) Show that the EF-axiom is equivalent to the assertion that *A pair of sets $A, B \subset X$ are near, if and only if, for every finite cover $\mathcal{G}$ of X, there is a set $G \in \mathcal{G}$ such that A is near G or B is near G.* Consider the case of countable covers.

(**5.5**.9) Give a proof of Corollary 5.1.

(**5.5**.10) Give a detailed proof of Corollary 5.3.

(**5.5**.11) Prove Lemma 5.3.

Chapter 6

Uniform Spaces, Filters and Nets

6.1 Uniformity via Pseudometrics

From Ch. 1, recall that a metric topology has, in general, at least two compatible EF-proximities, namely, (a) metric proximity and (b) fine L-proximity δ_0. This shows that proximity is a finer structure than topology. There is a structure finer than proximity that is called a **uniformity**. This was discovered by A. Weil. There are several ways to define a uniformity on a nonempty set X.

(**Uniformity**.1) family of pseudometrics, or
(**Uniformity**.2) family of covers, or
(**Uniformity**.3) set of entourages[1] that is a family of subsets of $X \times X$.

All of the approaches to defining a uniformity are in A. Weil's work with further developments by others (see Notes and Further Readings section for details). This book motivates (**Uniformity**.1) using pseudometrics. Recall that a pseudometric d on X satisfies all of the conditions for a metric with the possible exception of

$$d(x, y) = 0 \text{ implies } x = y.$$

For a pseudometric d, the distance between distinct points $x, y \in X$ could be zero.

Let X be a Tychonoff space with topology τ. For each disjoint pair $\{p, B\}$ (point p is disjoint from closed set B), there is a continuous function $f : X \to [0, 1]$ such that $f(p) = 0$ and $f(B) = 1$. This function induces the pseudometric d_f on X given by

$$d_f(x, y) = |f(x) - f(y)|, \text{ for } x, y \in X.$$

[1]French: neighbourhoods or *surroundings*.

Suppose $\mathcal{D}$ is a family of pseudometrics that induces a topology coarser than τ and contains at least one pseudometric for each disjoint pair $\{p, B\}$. Then $\mathcal{D}$ is called a **uniformity** on X.

Just as in the case of a metric, each pseudometric generates a topology on X. The *sup* of all topologies generated by members of $\mathcal{D}$ is called the uniform topology induced by $\mathcal{D}$. From the construction, it is apparent that this uniform topology is equal to the Tychonoff topology τ.

Using the family $\mathcal{D}$, one can also define a compatible EF-proximity δ on X. For each $d \in \mathcal{D}$, set the distance functional D_d by

$$D_d(A,B) = \begin{cases} inf\{d(a,b) : a \in A, b \in B\}, & \text{if } A, B \neq \emptyset, \\ \infty, & \text{otherwise.} \end{cases}$$

The EF-proximity $\delta = \delta_{\mathcal{D}}$ induced by $\mathcal{D}$ is defined by

$$A \ \delta \ B \text{ , if and only if, for each } d \in \mathcal{D}, D_d(A,B) = 0.$$

The proof that δ, thus defined, is a compatible EF-proximity is analogous to that of the metric proximity in §1.7. Hence,

$$\text{Uniformity } \mathcal{D} \Rightarrow \text{ EF-proximity } \delta_{\mathcal{D}} \Rightarrow \text{ Tychonoff topology } \tau.$$

This shows the importance of Tychonoff spaces among all topological spaces and also shows that important abstract concepts are related to the familiar real numbers. Tychonoff spaces occur naturally in topological groups, topological vector spaces, rings of continuous functions and so on.

In §1.6 (see Theorem 1.5), first a metric topology is defined and then an abstract topology on a set by isolating three important properties satisfied by the family of open sets. Similarly, using what has been learned so far about uniformity, define a uniformity on a nonempty set X by means of pseudometrics. Notice that the family of pseudometrics is not empty due to the existence of the identically zero pseudometric. A uniformity on a nonempty set X is a family of $\mathcal{D}$ of pseudometrics satisfying the following properties.

(**Uprop**.1) $d, d' \in \mathcal{D} \Rightarrow max\{d, d'\} \in \mathcal{D}$,

(**Uprop**.2) if a pseudometric e satisfies the condition

$$\text{for each } \varepsilon > 0, \exists d \in \mathcal{D}, \delta > 0 :$$
$$\forall x, y \in X, d(x,y) < \delta \Rightarrow e(x,y) < \varepsilon,$$

then $e \in \mathcal{D}$,

(**Uprop**.3) $x, y \in X$ and $x \neq y \Rightarrow \exists d \in \mathcal{D}:\ d(x,y) > 0$.
The pair $(X, \mathcal{D})$ is called a **uniform space**.

The following things can be observed about uniform spaces.

(**Obs**.1) Property (Uprop.2) says that if a pseudometric is 'uniformly continuous' relative to $\mathcal{D}$, then the pseudometric belongs to $\mathcal{D}$.

(**Obs**.2) Property (Uprop.3) implies the induced topology is Hausdorff.

(**Obs**.3) A subfamily $\mathcal{B}$ is a **base** for $\mathcal{D}$, if and only if, for each $e \in \mathcal{D}$ and $\varepsilon > 0$, $\exists d \in \mathcal{B}, \delta > 0 : \forall x, y \in X, d(x,y) < \delta$ implies $e(x,y) < \varepsilon$.

(**Obs**.4) Suppose $\mathcal{S}$ is any family of pseudometrics on a nonempty set X and suppose $\mathcal{S}$ has at least one member that distinguishes points of X as in Property (Uprop.3). Then the intersection $\mathcal{D}$ of all uniformities on X containing $\mathcal{S}$ is also a uniformity on X ($\mathcal{S}$ is called a **subbase** of $\mathcal{D}$. Moreover, the family of all suprema of finite subsets of $\mathcal{S}$ is a base for $\mathcal{D}$).

Not surprisingly from what has been observed so far, uniformity induces an EF-proximity. The proof of the result is analogous to that of Theorem 1.6 in §1.7.

Theorem 6.1. *Every uniform space $(X, \mathcal{D})$ has a compatible EF-proximity $\delta = \delta_{\mathcal{D}}$ defined by*

$$A\ \delta\ B \Leftrightarrow \textit{ for each } d \in \mathcal{D},\ D_d(A,B) = 0.$$

The following result is now obvious from the above motivation and results from Ch. 5.

Corollary 6.1. *A topological space (X, τ) is Tychonoff, if and only if, it has a compatible uniformity.*

If d is pseudometric on X, then $d' = min\{d, 1\}$ is a member of $\mathcal{D}$. Further, if $\mathcal{D}$ is a uniformity, $\mathcal{D}' = \{d' : d \in \mathcal{D}\}$ is a base for $\mathcal{D}$. If the uniformity $\mathcal{D}$ has a countable base $\{d_j : j \in \mathbb{N}\}$, then we can construct a metric e compatible with the topology τ that is given by

$$e(x,y) = \sum \{2^{-j} d'_j(x,y) : j \in \mathbb{N}\}.$$

Corollary 6.2. *A uniform space $(X, \mathcal{D})$ is metrisable, if and only if, uniformity $\mathcal{D}$ has a countable base.*

The definition of a uniformly continuous function is analogous to that in metric spaces (see §1.11). Let $(X, \mathcal{D})$ and $(Y, \mathcal{E})$ denote uniform spaces. A function $f : X \to Y$ is called **uniformly continuous**, if and only if, for each $e \in \mathcal{E}$ and $\varepsilon > 0$, there is a $d \in \mathcal{D}$ and $\delta > 0$ such that $d(x, y) < \delta$ implies $e(f(x), f(y)) < \varepsilon$. Every uniformly continuous function is continuous with respect to the induced topologies.

> Albert Einstein observed that *The significant problems we face cannot be solved by the same level of thinking that created them.* Similarly, there are many purely topological problems whose solution requires the use of proximity and/or uniformity, which are finer structures.

6.2 Filters and Ultrafilters

Recall that a metric topology can be studied with sequences (see, *e.g.*, starting in §1.6). For example, a point x is near a subset B in a metric space X, if and only if, there is a sequence of points in B that converges to x. This result is not always true in general topological spaces.

Recall from §2.8, Example 2.8.3 that although Ω is near M (set of predecessors of Ω)[2], there is no sequence in M that converges to Ω. So we need a generalisation of a sequence.

This was done in two ways: with nets by E.H. Moore and H.L. Smith and filters by H. Cartan. Both nets and filters are useful. Although they are logically equivalent, in specific problems and applications, one of them is easier to handle than the other. Both are used in this book.

By way of motivation for the study of filters, consider the following special cases.

(**filter**.1) **Neighbourhood filter**. Consider $\mathcal{N}(x)$, the family of all neighbourhoods of a point x in a topological space. Then
 (i) $\mathcal{N}(x)$ is a nonempty family of nonempty sets.
 (ii) $\mathcal{N}(x)$ is closed under finite intersections.

---

[2] Ω denotes the least element of $\mathbb{R} \cup \{z\}$ with uncountably many predecessors.

(iii) A subset of X that contains a member of $\mathcal{N}(x)$ is also in $\mathcal{N}(x)$.

(**filter**.2) **Principal filter**. Let x be a point in a set X. Then $\mathcal{V}(x)$ (the family of all subsets of X containing x) satisfies conditions (filter.1)(i)-(iii).

(**filter**.3) **Elementary filter**. Suppose (a_n) is a sequence in X. For each $n \in \mathbb{N}$, set $A_n = \{a_m : m \geq n\}$. Let

$$\mathcal{F} = \{E \subset X : E \supset A_n \text{ for some } n \in \mathbb{N}\}.$$

Then $\mathcal{F}$ satisfies conditions (filter.1)(i)-(iii).

A **filter** on a nonempty set X is
(i) a nonempty family of nonempty subsets of X,
(ii) closed under finite intersections,
(iii) closed under supersets.

Consider a family $\mathcal{S}$ of nonempty subsets of X in which finite intersections of its members are nonempty. Then $\mathcal{S}$ generates a filter $\mathcal{F}(\mathcal{S})$ that is made up of all subsets of X that contain finite intersections of members of $\mathcal{S}$. The family $\mathcal{S}$ is called a **filter subbase**. A **filter base** $\mathcal{B}$ is a filter subbase in which every finite intersection contains some member of $\mathcal{B}$.

For example, the family of all open sets containing a point x in a topological space X is a filter base for the neighbourhood filter $\mathcal{N}(x)$. The set of all filters on a set X can be partially ordered by set inclusion $\subset$. If $\mathcal{F} \subset \mathcal{G}$, then $\mathcal{F}$ is called a **coarser filter** and $\mathcal{G}$ is called a **finer filter**. Maximal filters are called **ultrafilters**. A standard Zorn's Lemma[3] argument shows that every filter is contained in an ultrafilter.

In analysis, two important concepts are related to a sequence:

---

[3]Recall that a nonempty set X is **totally ordered** under a binary relation $\leq$, if and only if, for all $x, y, z \in X$,

(**TO**.1) $x \leq y$ and $y \leq b \Rightarrow x = y$ (antisymmetry),
(**TO**.2) $x \leq y$ and $y \leq z \Rightarrow x \leq z$ (transitivity),
(**TO**.3) $x \leq y$ or $y \leq x$ (totality).

By contrast, a **partial order** requires antisymmetry, transitivity, and reflexivity $x \leq x$ (instead of totality). **Zorn's Lemma**: Assume a partially ordered set X has the property that every totally ordered subset (chain) has an upper bound in X. Then the set X contains at least one maximal element [248]. A version of Zorn's lemma was also proved by C. Kuratowski [117]. See P. Campbell's illuminating article [35] on the tangled origin of Zorn's Lemma.

(**seq**.1) Limit l. Each neighbourhood of a limit l of a sequence (a_n) contains some A_n. For neighbourhood filter $\mathcal{N}(l)$ and elementary filter $\mathcal{F}$, this is equivalent to

$$\mathcal{N}(l) \subset \mathcal{F} \text{ generated by } \{A_n\}.$$

(**seq**.2) Adherence point a of a sequence. An adherence point a of a sequence (a_n) belongs to the closure of each A_n.

These examples concerning sequences lead to analogous definitions for a filter $\mathcal{F}$ in a topological space X.

(a) Limit of $\mathcal{F} = \mathit{lim}\,\mathcal{F} = \{x \in X : \mathcal{N}(x) \subset \mathcal{F}\}$. In symbols, $\mathcal{F}$ converges to x or $\mathcal{F} \to x$. In the case of a filter base, $\mathcal{B} \to x$, if and only if, each member $G \in \mathcal{N}(x)$ contains a $B \in \mathcal{B}$.

The limit of a filter equals the limit of its base.

(b) Adherence of $\mathcal{F} = \mathit{adh}\,\mathcal{F} = \bigcap\{\mathrm{cl}\, F : F \in \mathcal{F}\}$.

Theorem 6.2. *In a topological space X, a point x is near a subset A, if and only if, there is a filter base $\mathcal{B}$ in A converging to x.*

Proof. Obviously, if $\mathcal{B} \to x$, then x is near A. Conversely, if x is near A, then $N \in \mathcal{N}(x)$ intersects A. Hence, $\mathcal{B} = \{N \cap A : N \in \mathcal{N}(x)\}$ is a filter base $\mathcal{B}$ in A converging to x. □

In a metric space, any sequence has at most one limit. But in an infinite set with cofinite topology, the filter of cofinite subsets converges to every point. To have unique limits, the Hausdorff axiom T_2 is needed.

Theorem 6.3. *A topological space X is Hausdorff, if and only if, every filter in the space has at most one limit.*

Proof. If $x, y \in X$ are distinct elements of a Hausdorff space X, then they have disjoint neighbourhoods. So the neighbourhood filters $\mathcal{N}(x), \mathcal{N}(y)$ cannot both be in a filter. If X is not Hausdorff, then there are distinct points x and y whose neighbourhoods intersect. So, $\mathcal{N}(x) \cup \mathcal{N}(y)$ generates a filter that converges to both x and y. □

Let X, Y be metric spaces. It was shown that a function $f : X \to Y$ is continuous, if and only if, for each sequence $(x_n) \to x$ in X, the image sequence $(f(x_n)) \to f(x)$ in Y. We now show that an analogous result holds in topological spaces for filters.

Theorem 6.4. *Let X, Y denote topological spaces. A function $f : X \to Y$ is continuous, if and only if, for each filter $\mathcal{F} \to x$ in X, the filter $\mathcal{G}$ generated by $f(\mathcal{F})$ converges to $f(x)$ in Y.*

Proof. First note that $f(\mathcal{F})$ is a filter base in Y and generates the filter

$$\mathcal{G} = \{E \subset Y : f^{-1}(E) \in \mathcal{F}\}.$$

Also note that f is continuous at x, if and only if, $f^{-1}(\mathcal{N}(f(x))) \subset \mathcal{N}(x)$. Also, $\mathcal{F} \to x$, if and only if, $\mathcal{N}(x) \subset \mathcal{F}$. The result follows. □

6.3 Ultrafilters

Now consider what are known as ultrafilters that have no analogues in sequences. An **ultrafilter** $\mathcal{L}$ is a maximal filter. That is, if $\mathcal{L}$ is contained in a filter $\mathcal{F}$, then $\mathcal{L}$ equals $\mathcal{F}$. A simple example of an ultrafilter is the principal filter $\mathcal{V}_x$, the family of all subsets of X containing x. Here are some useful characterisations of ultrafilters.

Theorem 6.5. *A filter $\mathcal{F}$ in a set X is an ultrafilter, if and only if, $(A \cup B) \in \mathcal{F}$ implies $A \in \mathcal{F}$ or $B \in \mathcal{F}$.*

Proof.
($\Rightarrow$) Suppose $(A \cup B) \in \mathcal{F}$ and $A \notin \mathcal{F}$. Let $\mathcal{H} = \{E \subset X : (A \cup E) \in \mathcal{F}\}$, which is not empty, since $B \in \mathcal{H}$. It is easy to verify that $\mathcal{H}$ is a filter finer than $\mathcal{F}$, *i.e.*, $\mathcal{F} \subset \mathcal{H}$. Since $\mathcal{F}$ is an ultrafilter, $\mathcal{F} = \mathcal{H}$, which contains B.
($\Leftarrow$) Suppose $(A \cup B) \in \mathcal{F}$ implies $A \in \mathcal{F}$ or $B \in \mathcal{F}$. Then, since $X \in \mathcal{F}$ for each subset E, $E \in \mathcal{F}$ or $E^c \in \mathcal{F}$. Suppose $\mathcal{F} \subset \mathcal{H}$. If $E \in \mathcal{H}$, then E^c cannot be in $\mathcal{F}$. So, $E \in \mathcal{F}$ and, hence, $\mathcal{F}$ is an ultrafilter. □

Corollary 6.3. *A filter $\mathcal{F}$ in X is an ultrafilter, if and only if, for each subset $E \subset X, E \in \mathcal{F}$ or $E^c \in \mathcal{F}$.*

Corollary 6.4. *A filter $\mathcal{F}$ in X is an ultrafilter, if and only if, E intersects every member of $\mathcal{F}$ implies $E \in \mathcal{F}$.*

The following result on ultrafilters is useful in the study of proximity spaces.

Theorem 6.6. *A family $\mathcal{F}$ of subsets of a set X is an ultrafilter, if and only if,*
(a) any two sets in $\mathcal{F}$ intersect,
(b) if a set E intersects every member of $\mathcal{F}$, then $E \in \mathcal{F}$, and

(c) $(A \cup B) \in \mathcal{F}$, *if and only if,* $A \in \mathcal{F}$ *or* $B \in \mathcal{F}$.

The following observation is useful in applications.

> If X contains at least two elements, there are at least two distinct ultrafilters on X. Hence, the ordered set of filters on X has no greatest member.

Theorem 6.7. *Every filter $\mathcal{F}$ on a set X is the intersection of the ultrafilters finer than $\mathcal{F}$.*

Proof. See Problem 6.7.4. □

6.4 Nets (Moore-Smith Convergence)

Looking at different limits in analysis, E. Moore and H. Smith generalised them and this generalisation fitted like a well-worn glove in topology. We use this approach in motivating nets. Observe

(**analysis**.1) The domain of a sequence is the set $\mathbb{N}$ of natural numbers, which is linearly ordered.

(**analysis**.2) In the limit of a function as $x \to a$, the domain of x is the family of neighbourhoods of a, which are partially ordered by $\subset$ and any two neighbourhoods of a contain their intersection.

(**analysis**.3) In Riemann integration on an interval I of $\mathbb{R}$, one works with partitions of I. Here, for every pair of partitions, there is a common refinement of both.

So, to include the above three analysis cases, we define a **directed set** $\mathcal{D}$ as a set with a partial order $\geq$ such that, for each pair of elements $d_1, d_2 \in \mathcal{D}$, there is an element $d \in \mathcal{D}$ satisfying $d \geq d_1, d \geq d_2$. Notice that an important example of a directed set is a filter with reverse set inclusion $\supset$ as the partial order.

A **net** in a set X is a function on a directed set $\mathcal{D}$ to X. As in the case of a sequence, a net is denoted simply by (x_d) or by $(x_d : d \in \mathcal{D})$. Analogous to the terminology in sequences, observe

(**netterms**.1) A net (x_d) is **eventually** in a subset $G \subset X$, if and only if, there is a $d_0 \in \mathcal{D}$ such that, for all $d \geq d_0$, $x_d \in G$.

(**netterms**.2) A net (x_d) is **frequently** in a subset $G \subset X$, if and only if, for every $d_0 \in \mathcal{D}$, there is a $d \geq d_0$ such that $x_d \in G$.

Next, consider nets in a topological space (X, τ). A net (x_d) is said to **converge to a limit** $x \in X$, if and only if, for every neighbourhood G of x, the net (x_d) is eventually in G. The notation for convergence of a net (x_d) to a limit x is $(x_d) \to x$.

Remark. Nets vs. Convergence in Topological Spaces:
This approach to nets and convergence in topological spaces resembles the notion of the limit of a sequence, but there are two major differences. First, sequences have a common domain, namely, the set of natural numbers but nets may have different directed sets. Second, a sequence in a metric space has at most one limit but a net in a topological space may converge to many points. A net cannot be eventually in two disjoint sets. So, to get unique limits, the topological space must be Hausdorff.

A point $x \in X$ is an **adherence point** of a net (x_d), if and only if, the net (x_d) is frequently in each neighbourhood of x. The notation $adh(x_d)$ denotes the set of all adherence points of (x_d). Obviously, as in the case of a sequence, a limit of a net is also an adherence point of the net but not conversely.

There are results about nets that are the counterparts of those about filters. We collect the results about nets here for reference and suggest that these results be treated as problems. Let X, Y denote topological spaces. Observe

(**net**.1) A point x is near a set A, if and only if, there is a net in A that converges to x.
(**net**.2) A point x is a limit point of a set A, if and only if, there is a net in $A - \{x\}$ that converges to x.
(**net**.3) A subset $E \subset X$ is closed, if and only if, no net in E converges to a point in E^c.
(**net**.4) X is Hausdorff, if and only if, each net in X has at most one limit.
(**net**.5) A function $f : X \to Y$ is continuous if, and only if, for each net $(x_d) \to x$ in X, the image net $(f(x_d)) \to f(x)$ in Y.

A concept analogous to an ultrafilter can be defined in any set X and is called a universal net. A **universal net** (x_d) in X is one which has the property: for every subset $A \subset X$, the net (x_d) is eventually in A or in A^c. In a universal net, $adh(x_d) = lim(x_d)$.

6.5 Equivalence of Nets and Filters

It is obvious that both filters and nets are adequate generalisations of sequences and fit like a glove in topological spaces. Their equivalence has been shown by R.G. Bartle. We just give the main points of this technical result and leave the details for readers as problems to explore.

Let $\mathcal{F}$ be a filter in a topological space (X, τ). We use $\mathcal{F}$ to get a directed set $(\mathcal{D}, \geq)$ as follows: $\mathcal{D} = \{(x, F) : x \in F \in \mathcal{F}\}$ and $(x, F) \geq (x, G) \Leftrightarrow G \supset F$. Then a net (x_d) (where $d = (x, F) \in \mathcal{D}$) in X can be defined on the directed set $\mathcal{D}$. It can be shown that

Proposition 6.1. *For a filter $\mathcal{F}$ in a topological space (X, τ) (a) $lim\mathcal{F} = lim(x_d)$, and (b) $adh\mathcal{F} = adh(x_d)$.*

Proof. Let (x_d) be a net on the directed set $(\mathcal{D}, \geq)$ in the topological space (X, τ). For each $d \in \mathcal{D}$, let $A_d = \{x_e : e \geq d\}$. Then the family $\{A_d : d \in \mathcal{D}\}$ has the finite intersection property and so is a base for filter $\mathcal{F}$. The results (a) and (b) follow. □

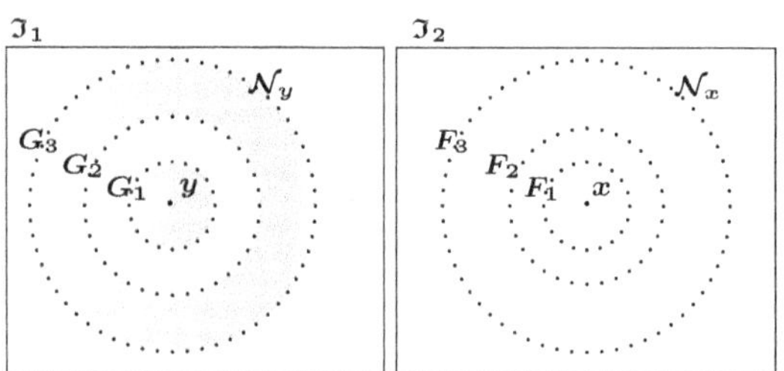

Fig. 6.1 Near neighbourhood filters $\mathcal{N}_y \ \delta_\Phi \ \mathcal{N}_x$

6.6 Application: Proximal Neighbourhoods in Camouflage Neighbourhood Filters

Recall that a filter on a set X is a nonempty collection $\mathcal{F}$ of $\mathcal{P}(X)$ has the following properties: $\varnothing \notin \mathcal{F}$; if $A \in \mathcal{F}$ and $A \subseteq B$, then $B \in \mathcal{F}$; and if $A \in \mathcal{F}$ and $B \in \mathcal{F}$, then $A \cap B \in \mathcal{F}$. The set of all neighbourhoods of a point $x \in X$ is called a **neighbourhood filter** of x (denoted by $\mathcal{N}(x)$). Two examples of spherical neighbourhood filters $\mathcal{N}_x, \mathcal{N}_y$ are shown in Fig. 6.1, *i.e.*,

$$\mathcal{N}_x = \{F_1, F_2, F_3 : F_1 \subset F_2 \subset F_3\},$$
$$\mathcal{N}_y = \{G_1, G_2, G_3 : G_1 \subset G_2 \subset G_3\}.$$

A maximal filter on X is called an *ultrafilter* on X. A subcollection $\mathcal{B}$ in a filter $\mathcal{F}$ is a **filter base**, provided every member of $\mathcal{F}$ contains some element of $\mathcal{B}$. For example, in Fig. 6.1, F_1 is a filter base for the filter $\mathcal{N}_x$ and G_1 is a filter base for $\mathcal{N}_y$.

Let δ_Φ denote a descriptive EF-proximity defined in terms of a set of features Φ. Neighbourhood filters $\mathcal{N}_x, \mathcal{N}_y$ are **near neighbourhood filters** (denoted $\mathcal{N}_x\ \delta_\Phi\ \mathcal{N}_y$), if and only if, $D(A, B) = 0$ for some $A \in \mathcal{N}_x, B \in \mathcal{N}_y$ (*cf.*, [177; 180; 183; 186]). For example, in Fig. 6.1,

$$\mathcal{N}_x\ \delta_\Phi\ \mathcal{N}_y, \text{ since, } D(G_2, F_2) = 0, \text{for } G_2 \in \mathcal{N}_y, F_2 \in \mathcal{N}_x.$$

That is, the description of the outer rim of G_2 matches the description of all of the points in F_2. Hence, $\mathcal{N}_x\ \delta_\Phi\ \mathcal{N}_y$.

A subset B of an EF-proximity space (X, δ) is a **proximal neighbourhood** of A (denoted $A \ll B$), if and only if, $A\ \underline{\delta}\ (X - B)$ [163], [50]. This means that A is *strongly contained* in B.

Theorem 6.8. Disjoint Proximal Neighbourhoods [50].
Any two remote sets have disjoint proximal neighbourhoods.

A subset B in an EF-proximity space is a **descriptive proximal neighbourhood** of a subset A (denoted $A \ll_\Phi B$), provided $A\ \underline{\delta}_\Phi\ (X - B)$, *i.e.*, A is descriptively remote (far) from the complement of B.

Theorem 6.9. Disjoint Descriptive Proximal Neighbourhoods.
Any two descriptively remote sets have disjoint descriptive proximal neighbourhoods.

Proof. Immediate from Theorem 6.8. □

Example 6.1. Let $X = \mathfrak{I}_1 \cup \mathfrak{I}_2$ in Fig. 6.1 and let δ, δ_Φ denote an EF-proximity relations on X. Choose Φ to be a set of probe functions that measure average greyscale intensity and various colours. Then, for example, $G_3\ \underline{\delta}_\Phi\ F_3$ in Fig. 6.1 by virtue of the difference between the descriptions of the points in G_1 and F_1. Further, G_3, F_3 are descriptively remote sets, $G_1 \ll_\Phi G_3$ and $F_1 \ll_\Phi F_3$. From Theorem 6.9, G_3, F_3 are disjoint descriptive proximal neighbourhoods. ■

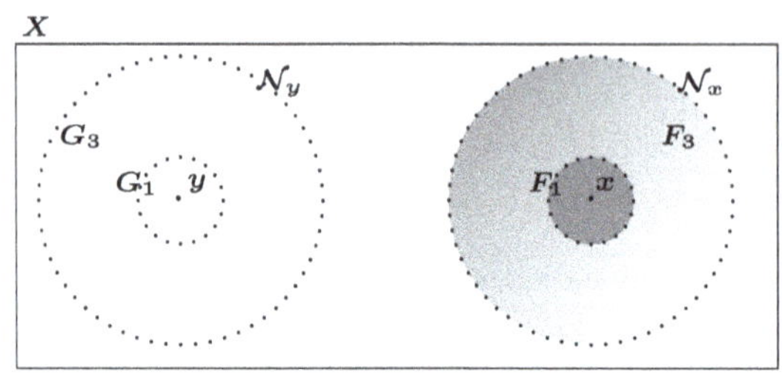

Fig. 6.2 Remote neighbourhood filters $\mathcal{N}_y \underline{\delta} \mathcal{N}_x$

Fig. 6.3 Natural camouflage example: Dragonfly

Example 6.2. Remote Neighbourhood filters.
In Fig. 6.2, $G_1 \in \mathcal{N}_y$ and $F_1 \in \mathcal{N}_x$ are disjoint sets. From Theorem 6.8, G_1 and F_1 have disjoint proximal neighbourhoods. In this example, $G_1 \ll G_3$ and $F_1 \ll F_3$. It can also be observed that G_3 and F_3 are descriptively remote sets in $\mathcal{N}_y, \mathcal{N}_x$, respectively. Then, from Theorem 6.9, G_3 and F_3 have disjoint descriptive proximal neighbourhoods. This can be seen from the fact that $G_3 \underset{\Phi}{\cap} F_3$ is empty. ∎

Descriptive proximal neighbourhoods play an important role in detecting discrepancies in comparing camouflage neighbourhood filters. We illustrate this in terms of a comparison between a camouflaged dragonfly centered in neighbourhood filter $\boldsymbol{\mathcal{N}_x}$ and a similar setting favoured by dragonflies centered in $\boldsymbol{\mathcal{N}_y}$ (see Fig. 6.4).

Fig. 6.4 Camouflage neighbourhood filters

The first step in the detection of camouflage in Fig. 6.3 is to consider a pair of disjoint neighbourhood filters $\mathcal{N}_x$, $\mathcal{N}_y$ superimposed on the zoomed versions of camouflaged object and non-camouflaged object (in this case, image of a dragonfly on a tree branch and image of a similar tree branch). This is done in Fig. 6.4. Next, let $X = \mathcal{N}_x \cup \mathcal{N}_y$ be an EF-proximity space and let δ_Φ be an EF-proximity on X.

Let Φ consist only of colour features (for simplicity, shape descriptors are omitted from Φ). Neighbourhoods $C_1 \in \mathcal{N}_x, G_1 \in \mathcal{N}$ are descriptive remote sets. From Theorem 6.9, we can expect to find a pair of descriptive proximal neighbourhoods. Observe that C_2 is a descriptive proximal neighbourhood of C_1, *i.e.*, $C_1 \ll_\Phi C_2$. In addition, G_2 is a descriptive proximal neighbourhood of G_1 ($G_1 \ll_\Phi G_2$). These proximal neighbourhoods are descriptively disjoint in the sense that the descriptions of points in one proximal neighbourhood are not the same as the descriptions of the points in the other proximal neighbourhood. From a camouflage point of view, this is an indication that a camouflaged object has been detected in the region occupied by the coarser subsets in the neighbourhood filters.

6.7 Problems

(**6.7**.1) Show that a uniform topology of a uniform space $(X, \mathcal{D})$ can be expressed in the following ways:

(a) $\operatorname{cl} A = \bigcap \{x \in X : D_d(x, A) = 0, d \in \mathcal{D}\}$.

(b) A base for the topology is $\{B_d(x, \varepsilon) : x \in X, \varepsilon > 0, d \in \mathcal{D}\}$, where $B_d(x, \varepsilon)$ is an open d-ball with centre x and radius ε.

(**6.7**.2) Let $A \subset X$ be a subset in a uniform space $(X, \mathcal{D})$. Show that the function $f(x) = d(x, A)$ is continuous for each $d \in \mathcal{D}$.

(**6.7**.3) Let A be a Tychonoff space, $\mathcal{C}(X)$ the family of real-valued continuous functions, $\mathcal{C}^*(X) \subset \mathcal{C}(X)$ the subfamily of bounded functions. Each member $f \in \mathcal{C}(X)$ induces a pseudometric d_f given by $d_f(x, y) = |f(x) - f(y)|$. Let $\mathcal{C}, \mathcal{C}^*$ denote the uniformities generated by pseudometrics induced by members of $\mathcal{C}(X), \mathcal{C}^*(X)$, respectively. Show that they are distinct. **Hint**: Use $X = \mathbb{R}$.

(**6.7**.4) Show that every filter in a set equals the intersection of all ultrafilters that contain it.

(**6.7**.5) Let X be an infinite set and let $\mathcal{F}$ denote the family of all subsets whose complements are finite. Show that $\mathcal{F}$ is a filter. Is it an ultrafilter?

(**6.7**.6) For any ultrafilter $\mathcal{L}$ in a topological space, show that $adh\mathcal{L} = lim\mathcal{L}$.

(**6.7**.7) Let $\mathcal{F}$ be a convergent filter in a Hausdorff space. Show that $adh\mathcal{F}$ contains only one point and $adh\mathcal{F} = lim\mathcal{F}$.

(**6.7**.8) Show that x is an adherence point of a filter $\mathcal{F}$, if and only if, there is a filter finer than $\mathcal{F}$ that converges to x.

(**6.7**.9) Let X, Y denote nonempty sets and let $\mathcal{F}$ be a filter in X. Show that $f(\mathcal{F}) = \{f(F) : F \in \mathcal{F}\}$ is a base for a filter $\mathcal{G}$ in Y and show that $\mathcal{G} = \{E \subset Y : f^{-1}(E) \in \mathcal{F}\}$. If $\mathcal{F}$ is an ultrafilter, then show that $\mathcal{G}$ is also an ultrafilter.

(**6.7**.10) Let $\mathcal{F}$ be a filter in X and $A \in \mathcal{F}$. Show that $\{F \in \mathcal{F} : F \subset A\}$ is a filter on A called **trace filter** $\mathcal{F}_A$. Further, show that if $\mathcal{F}$ is an ultrafilter, so is $\mathcal{F}_A$.

(**6.7**.11) Give a detailed proof of Theorem 6.9.

Chapter 7

Compactness and Higher Separation Axioms

> It has been observed by J.L. Kelley that the notion of a compact topological space is an abstraction of certain important properties of the set of real numbers. What is also known as the Heine-Borel-Legesgue theorem asserts that every open cover of a closed and bounded subset of the space of real numbers has a finite subcover. Like most good theorems, the conclusion of the Heine-Borel theorem has become a definition.

7.1 Compactness: Net and Filter Views

The importance of compactness is brought forward in §1.9. In fact, compactness is viewed as one of the most valuable concepts in analysis and topology.

Recall that in a metric space, the following are equivalent.

(**Equiv**.1) Every sequence in a metric space has a convergent subsequence (sequentially compact).

(**Equiv**.2) Every open cover of the space has a finite subcover (Heine-Borel Theorem).

(**Equiv**.3) Every infinite subset has a limit point in it (Bolzano-Weierstrass Theorem).

(**Equiv**.4) Every countable open cover of the space has a finite subcover (countably compact).

Notice that (Equiv.1)-(Equiv.4) are not necessarily equivalent in a topological space. A lot of experimentation was made and it was found that the most suitable definition in abstract topological spaces is (**Equiv**.2) (Heine-

Borel Theorem). In sum, a topological space is **compact**, if and only if, every open cover of the space has a finite subcover. Nomenclature for compactness differs among mathematicians, *e.g.*, Bourbaki added a Hausdorff axiom to the definition for a compact space and Soviet mathematicians called such spaces bicompact. In this chapter, we consider equivalent formulations of compactness in terms of filters and nets as well as its influence on other concepts. We also briefly explore generalisations of compactness.

Examples of compact and non-compact spaces are known from analysis. For example, closed and bounded subsets of Euclidean spaces are compact but the Euclidean spaces themselves are not compact. An infinite set with cofinite topology is compact but not Hausdorff.

With each family $\mathcal{O}$ of open sets of topological space X, there is an associated family $\mathcal{C}$ of closed subsets that are complements of the members of $\mathcal{O}$. The family $\mathcal{O}$ is a cover of X, if and only if, members of $\mathcal{C}$ have empty intersection. A family $\mathcal{F}$ is said to have the **finite intersection property**, if and only if, every finite subfamily of $\mathcal{F}$ has nonempty intersection.

By De Morgan's law, the definition

Every open cover of X has a finite subcover.

is equivalent to

Every family of closed sets in X with the finite intersection property has nonempty intersection.

Given a filter $\mathcal{F}$ in X, the closures of members of $\mathcal{F}$ have the finite intersection property. So compactness is equivalent to

Every filter has an adherent point.

and that is equivalent to

Every ultrafilter converges.

Naturally, there is an analogue in nets. This discussion is summarised as follows.

Theorem 7.1. *The following assertions are equivalent in a topological space X.*

(**compact**.*1*) *Every open cover of X has a finite subcover.*

(**compact**.*2*) *Every family of closed sets of X with the finite intersection property has a nonempty intersection.*
(**compact**.*3*) *Every filter in X has an adherent point.*
(**compact**.*4*) *Every ultrafilter in X converges.*
(**compact**.*5*) *Every net in X has an adherent point.*
(**compact**.*6*) *Every universal net in X converges.*

7.2 Compact Subsets

A subset E of a topological space X is **compact**, if and only if, it is compact in the subspace topology. It is easy to verify that this is equivalent to the assertion: *Every cover of E with open sets in X has a finite subcover.* Characterisation of compact subsets is similar to those in Theorem 7.1. Commonly used characterisations are given in Theorem 7.2.

Theorem 7.2. *Let E denote a subset of a topological space X. Then the following assertions are equivalent.*
(**subspace**.*1*) *E is compact.*
(**subspace**.*2*) *Every filter in E has an adherent point in E.*
(**subspace**.*3*) *Every ultrafilter in E converges in E.*
(**subspace**.*4*) *Every net in E has an adherent point in E.*
(**subspace**.*5*) *Every universal net in E converges in E.*

Compactness is inherited by closed subsets. Suppose E is a closed subset of a compact space X. Then every filter $\mathcal{F}$ in E is also a filter in X and, since X is compact, $\mathcal{F}$ has an adherent point p. Since E is closed, $p \in E$.

Theorem 7.3. *A closed subset of a compact space is compact.*

The result in Theorem 7.3 raises a question: *Is a compact subset of a topological space closed?* This need not be true. For example, a proper subset of an infinite indiscrete space is compact but not closed. However, the answer is *Yes* for Hausdorff spaces.

Theorem 7.4. *A compact subset E of a Hausdorff space X is closed. Hence, in a compact Hausdorff space, a set is closed, if and only if, it is compact.*

Proof. Suppose p is a closure point of E. Then there is an ultrafilter $\mathcal{L}$ in E that converges to p. E is compact implies $\mathcal{L}$ converges to a point $q \in E$.

Since X is Hausdorff, $\mathcal{L}$ has only one limit $p = q \in E$, showing thereby that E is closed. □

An alternative proof of Theorem 7.4 to show that a point $x \in (X - E)$ and E have disjoint neighbourhoods.

Proof. For each point $y \in E$, y and x have disjoint neighbourhoods U_y and V_x, respectively. Since E is compact, the open cover $\{U_y : y \in E\}$ has a finite subcover $\{U_j : 1 \leq j \leq n\}$, whose union is disjoint from the intersection of corresponding V_js, namely, $\bigcap\{V_j : 1 \leq j \leq n\}$. So, x is not a closure point of E. Thus, a compact subset behaves like a point. If there are disjoint closed subsets E, F of a compact Hausdorff space, then we may continue this argument to get disjoint neighbourhoods of E and F. □

7.3 Compactness of a Hausdorff Space

Thus, compactness of a Hausdorff space gives it normality, a higher separation axiom.

Theorem 7.5. *A compact Hausdorff space is normal.*

By Urysohn's Lemma 5.1 (see Section §5.2), a compact Hausdorff space is Tychonoff and has a compatible EF-proximity δ_0. It was shown that in an EF-proximity space, disjoint closed sets (one of which is compact) are far. So, in a compact Hausdorff space, any two disjoint closed sets are far. This shows that every compact Hausdorff space has a unique EF-proximity.

Theorem 7.6. *A compact Hausdorff space X has a unique compatible EF-proximity δ_0 (the fine proximity) given by*

$$A\ \delta_0\ B \Leftrightarrow \operatorname{cl} A \cap \operatorname{cl} B \neq \varnothing.$$

Hence, every continuous function on a compact Hausdorff space with proximity δ_0 to any L-proximity space Y is proximally continuous.

Notice that it is possible for a compact Hausdorff space to have two distinct compatible L-proxmities. For example, $[0,1]$ has, besides δ_0, the coarse L-proximity δ_C given by

$$A\ \delta_C\ B \Leftrightarrow \operatorname{cl} A \cap \operatorname{cl} B \neq \varnothing \text{ or } A \text{ and } B \text{ are infinite.}$$

In calculus, an important results is: *A real-valued continuous function on a compact subset of $\mathbb{R}$ is bounded and attains the bounds.* Theorem 7.7 is a generalisation.

Theorem 7.7. *Let $f : X \to Y$ denote a continuous function from a compact space X onto a topological space Y. Then Y is compact. Further, if Y is Hausdorff and f is injective*[1]*, then f is a homeomorhphism.*

Proof. If $\mathcal{O}$ is an open cover of Y, then $f^{-1}(\mathcal{O})$ is an open cover of X. Since X is compact, $f^{-1}(\mathcal{O})$ has a finite subcover $\mathcal{F}$. Then $\{E \in \mathcal{O} : f^{-1}(E) \in \mathcal{F}\}$ is a finite subcover of Y.

Suppose Y is Hausdorff and F is a closed subset of X. Then F is compact and, since f is continuous, $f(F)$ is closed. This shows that f is a homeomorphism, if f is injective. □

Corollary 7.1. *Suppose T is a compact Hausdorff topology on X, $T_1 \subset T \subset T_2$, T_1 is Hausdorff and T_2 is compact. Then $T_1 = T = T_2$. This shows that a compact Hausdorff topology is minimal Hausdorff and maximal compact.*

There is considerable literature on minimal Hausdorff spaces (see the Historical Notes Section).

7.4 Local Compactness

Compactness is an important concept but it is too strong. In real and complex analysis, we work with spaces that are not compact but are locally compact. In many situations, local compactness suffices. There are several forms of local compactness in the literature on topological spaces that are equivalent in the important case of Hausdorff spaces. Bourbaki defines local compactness only for Hausdorff spaces. We choose one form of local compactness and leave the others as problems.

> A topological space is **locally compact**, if and only if, each point of the space has a compact neighbourhood.

Every compact space is locally compact. Euclidean spaces are non-compact but locally compact. The space of rational numbers with the usual topology is *not* locally compact.

[1] Let f be a function whose domain is X. Then f is **injective**, if and only if, for all $x, y \in X$, $f(x) = f(y)$ implies $x = y$, *i.e.*, an injection preserves distinctness, which is another way of saying $x \neq y$ implies $f(x) \neq f(y)$. See, *e.g.*, [21].

Local compactness is not preserved by continuous functions. For example, the identity function on the rational numbers with discrete topology to the non-locally compact space of rational numbers with the usual topology is continuous. However, it is easy to show that the image of a locally compact space is locally compact under a function that is continuous and open, *i.e.*, a function that takes open sets into open sets.

In complex analysis, one-point-compactification of the space of complex numbers plays an important role. This was generalised by P.S. Alexandroff. Let (X,T) be a T_1-topological space and let ω be an element not in X, which is referred to as the *point at infinity*. Put $X^* = X \cup \{\omega\}$ with topology T^* in which (a) neighbourhoods of $x \in X$ are the same as in T and (b) open neighbourhoods of ω are complements of closed compact subsets of X in the topology T. Then X^* is compact and T_1.

Proof. If $\mathcal{O}$ is any open cover of X^*, the complement of a member of $\mathcal{O}$ containing ω is compact. Hence, $\mathcal{O}$ has a finite subcover, showing thereby that (X^*,T^*) is compact. The set $\{\omega\}$ (since it is the complement of the open set X) is closed and each $\{x\} \subset X$ is also closed. This shows that X^* is T_1. □

In the important case of a locally compact Hausdorff space X, each point $x \in X$ has a compact neighbourhood N_x that is closed. Hence, N_x and $X^* - N_x$ are disjoint neighbourhoods of x and ω. That is, (X^*,T^*) is Hausdorff in addition to being compact. So, (X^*,T^*) is normal and, hence, its subspace (X,T) is Tychonoff (completely regular).

Theorem 7.8. *Every locally compact Hausdorff space is Tychonoff, since it has Alexandroff one-point compactification that is Hausdorff.*

7.5 Generalisations of Compactness

The Heine-Borel theorem is now used as the definition of compactness in topology, functional analysis and elsewhere. Other formulations of compactness (*e.g.*, (a) sequential compactness and (b) Bolzano-Weierstrass property) that are equivalent to it in metric spaces have proved useful as counterexamples. There is a vast literature on this (see the Historical Notes section).

A generalisation *paracompactness* was discovered by J. Dieudonné that has proven to be important in topology and analysis. A family $\mathcal{P}$ of subsets of a topological space X is said to be **locally finite** (**locally discrete**), if and only if, each point of X has a neighbourhood that intersects at most *finitely* many members (one member) of $\mathcal{P}$.

A family of sets $\mathcal{Q}$ is called a **refinement** of the family $\mathcal{P}$, if and only if, each member of $\mathcal{Q}$ is contained in some member of $\mathcal{P}$.

> A topological space X is called **paracompact**, if and only if, each open cover of X has an open, locally finite refinement that is a cover of X. Usually, the Hausdorff axiom is added in defining a paracompact space.

Obviously, a finite cover is locally finite and so a compact space is paracompact. An important result was proved by A.H. Stone: *every metric space is paracompact.*

7.1.1: Ode by Keats 7.1.2: Forged copy

Fig. 7.1 Handwritten copies of line from J. Keats' *Ode on a Grecian Urn*

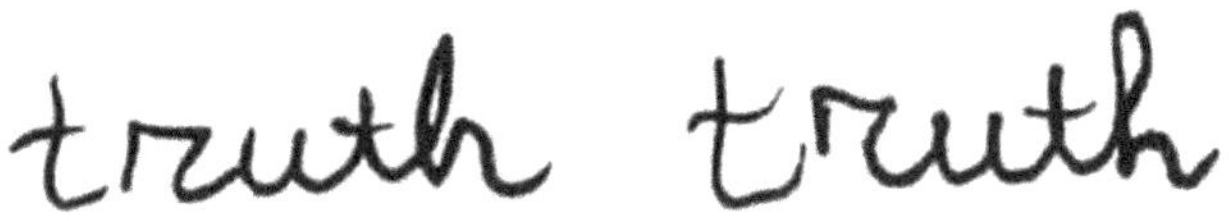

7.2.1: Word 7.2.2: Forged word

Fig. 7.2 Original and forged handwritten word

7.6 Application: Compact Spaces in Forgery Detection

The detection of forgery is based on the study of non-forged and possibly forged items (see, *e.g.*, [179; 185]). In this section, forgery detection is illustrated in terms of the EF-proximity of descriptive point clusters containing cursive characters in original handwriting and possibly forged handwriting samples (see, *e.g.*, Fig. 7.1 and handwritten versions of the word *truth* in Fig. 7.2).

In an EF-proximity space X, a **cluster** $\mathcal{C}$ is a collection of subsets of X satisfying the following conditions.

(a) If $A, B \in \mathcal{C}$, then A is near B.
(b) If A is close to every B in $\mathcal{C}$, then A belongs to $\mathcal{C}$.
(c) If $A \cup B$ belongs to $\mathcal{C}$, then either A or B belongs to $\mathcal{C}$.

Let X be a nonempty set. Then the collection of all points close to a fixed point x in X is a cluster. A **point cluster** $\mathcal{C}_x$ is the collection of all sets such that $x\ \delta\ A$ for each $A \in \mathcal{C}_x$. Let δ_Φ be the usual descriptive EF-proximity relation. The collection of all points descriptively close to the description of a fixed point x is a **descriptive point cluster** (denoted by $\mathcal{C}_{\Phi(x)}$) is the collection of all sets A such that $x\ \delta_\Phi\ A$ for each $A \in \mathcal{C}_x$ and $\mathcal{C}_{\Phi(x)}$ satisfies the following conditions.

(a_Φ) If $A, B \in \mathcal{C}_{\Phi(x)}$, then $A\ \delta_\Phi\ B$.
(b_Φ) If $A\ \delta_\Phi\ B$ for each B in $\mathcal{C}_{\Phi(x)}$, then A belongs to $\mathcal{C}_{\Phi(x)}$.
(c_Φ) If $A \underset{\Phi}{\cup} B$ belongs to $\mathcal{C}_{\Phi(x)}$, then either $\Phi(A)$ or $\Phi(B)$ belongs to $\mathcal{C}_{\Phi(x)}$.

This application takes advantage of the connection between compact EF proximity spaces and point clusters discovered by S. Leader [122] (see, also, [163; 50]), namely,

Theorem 7.9. Leader [122]. *An EF-proximity space X is compact, if and only if, every cluster from X is a point cluster.*

This leads to the following result for compact descriptive EF-proximiity spaces.

Theorem 7.10. *A descriptive EF-proximity space X is compact, if and only if, every cluster from X is a descriptive point cluster.*

Proof. See Problem 7.7.12. □

7.6.1 *Basic Approach in Detecting Forged Handwriting*

The basic approach in this application is to compare point clusters superimposed on the characters of handwriting specimens. In the proposed approach, points along the curvature of a handwritten character provide starting points for individual point clusters. Taking this approach a step further, we consider the set of points X along the curvature of handwritten characters to be a descriptive EF-proximity space, which is compact, since every cluster from X is a descriptive point cluster.

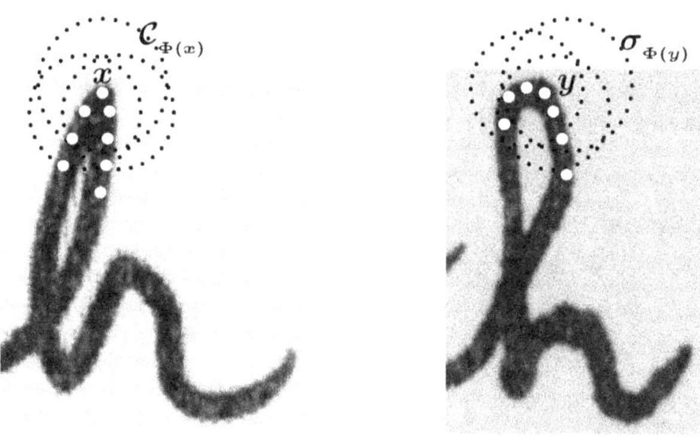

Fig. 7.3 Point clusters $\mathcal{C}_{\Phi(x)}$ and $\sigma_{\Phi(y)}$

Let X be a compact descriptive EF-proximity space and let $\mathfrak{I}, \mathfrak{I}'$ denote disjoint handwriting samples in X. In addition, let $\mathcal{C}_{\Phi(x)}, \sigma_{\Phi(y)}$ denote descriptive point clusters in $\mathfrak{I}, \mathfrak{I}'$, respectively (see Fig. 7.3 for examples). The handwriting samples in Fig. 7.3 are copies of the beginning of a line in J. Keats's *Ode on a Grecian Urn* [104]. We know that when the cursive characters in an original handwriting sample are compared with characters in a supposed duplicate, the duplicate is, in fact, a forgery, if point clusters containing the original and its copy are remote from each other. In other words, $\mathfrak{I}'$ is a forgery of the handwriting in $\mathfrak{I}$, provided a descriptive point cluster $\mathcal{C}_{\Phi(x)}$ in $\mathfrak{I}$ is remote (far from) the intersection of $\mathcal{C}_{\Phi(x)}$ and $\sigma_{\Phi(y)}$, *i.e.*,

$$\mathcal{C}_{\Phi(x)} \ \underline{\delta}_{\Phi} \left(\mathcal{C}_{\Phi(x)} \underset{\Phi}{\cap} \sigma_{\Phi(y)}\right).$$

That is, if the descriptive point clusters on points along the curvature of the original and another handwritten character contain corresponding points that do not have matching descriptions, then the points in the intersection $\mathcal{C}_{\Phi(x)} \underset{\Phi}{\cap} \sigma_{\Phi(y)}$ will not be a duplicate of the points in $\mathcal{C}_{\Phi(x)}$ and a forgery is detected. Using this idea, a forgery test reduces to checking if

$$\mathcal{C}_{\Phi(x)} = \left(\mathcal{C}_{\Phi(x)} \underset{\Phi}{\cap} \sigma_{\Phi(y)}\right).$$

The descriptive intersection $\mathcal{C}_{\Phi(x)} \underset{\Phi}{\cap} \sigma_{\Phi(y)}$ contains all points in $\mathcal{C}_{\Phi(x)}$ that match descriptions of points in $\sigma_{\Phi(y)}$. The descriptive intersection of point clusters is defined by

$$\mathcal{C}_{\Phi(x)} \underset{\Phi}{\cap} \sigma_{\Phi(y)} = \left\{A \underset{\Phi}{\cap} E : A \in \mathcal{C}_{\Phi(x)}, E \in \sigma_{\Phi(y)}\right\}.$$

7.6.2 *Roundness and Gradient Direction in Defining Descriptive Point Clusters*

To test whether one point cluster $\sigma_{\Phi(y)}$ contains a forgery of an original handwritten character in a point cluster in $\mathcal{C}_{\Phi(x)}$, it is necessary to define the set of features Φ in terms of one or more shape descriptors. A *shape descriptor* is a compact representation of a shape that is a closed subset in Euclidean space. Many shape descriptors are useful in the study handwriting, *e.g.*, barcode [44], Hu spatial moments [96], ridge [130], corner, junction, Zernike moment [247; 128; 169], roundness and gradient direction. In this introduction to the application of point clusters in discerning discrepancies in handwriting specimens, we consider only roundness and gradient direction shape descriptors. Let g denote a digital image, s a shape in g, a_s area of a shape s in g.

(**shape**.1) **Roundness**. Let S be a set of shapes (bordered 2D regions, a closed subset of Euclidean space) and let d_s (diameter) denote the length of a line between points on the perimeter of shape $s \in X$ in digital image g. A roundness function $rd : S \to \mathbb{R}$ [205] is defined by

$$rd(s) = \frac{4a_s}{\pi(max\{d_s\})^2}.$$

(**shape**.2) **Gradient Direction**. Let $\mathfrak{I}$ denote a digital image, (x, y) the coordinates of a pixel $g(x, y) \in \mathfrak{I}$ in the centre of a $n \times n$ neighbourhood N in $\mathfrak{I}$ (usually, n is an odd number), kernel K that is an $n \times n$ matrix containing weights. For example, using a Sobel kernel [210; 211] for the x and y directions, then the gradient direction ψ is the angle (in radians) from the x-axis to a point $g(x, y)$, defined by

$$\psi = tan^{-1}\left(\frac{\mathcal{D}_y}{\mathcal{D}_x}\right) = tan^{-1}\left(\frac{\frac{\partial g}{\partial y}}{\frac{\partial g}{\partial x}}\right).$$

Then, for example, obtain

$$N = \begin{bmatrix} z1 & z2 & z3 \\ z4 & z5 & z6 \\ z7 & z8 & z9 \end{bmatrix}, \quad K_x = \begin{bmatrix} -1 & -2 & -1 \\ 0 & 0 & 0 \\ 1 & 2 & 1 \end{bmatrix}, \quad K_y = \begin{bmatrix} -1 & 0 & 1 \\ -2 & 0 & 2 \\ -1 & 0 & 1 \end{bmatrix}.$$

$$\frac{\partial g}{\partial y} = (z3 + 2z6 + z9) - (z1 + 2z4 + z7),$$
$$\frac{\partial g}{\partial x} = (z7 + 2z8 + z9) - (z1 + 2z2 + z3).$$

Example 7.1. Forgery Test.
There are two handwritten h's in Fig. 7.3 from the word *truth* in Fig. 7.2. The enlargement of the individual handwritten letters is obtained using a stereo zoom microscope. It can be observed that $\sigma_{\Phi(y)}$ contains a forgery of the upper part of the h in the point cluster in $\mathcal{C}_{\Phi(x)}$ If we consider only roundness and gradient direction of the picture elements in the pair of descriptive point clusters $\mathcal{C}_{\Phi(x)}, \sigma_{\Phi(y)}$, the forgery test reveals that

$$\mathcal{C}_{\Phi(x)} \neq \left(\mathcal{C}_{\Phi(x)} \underset{\Phi}{\cap} \sigma_{\Phi(y)}\right).$$

The discrepancies between the top parts of the h's in the two point clusters is visually evident, if we consider the roundness shape descriptor (*e.g.*, the top of the h in $\mathcal{C}_{\Phi(x)}$ is less round than the top of the h in $\sigma_{\Phi(y)}$). Again, for example, the gradient (slope) directions of the pixels along the tops of the pair of h's are markedly different. Hence, the intersection of the two point clusters will not equal the set of descriptions in $\Phi(\mathcal{C}_{\Phi(x)})$. ■

7.7 Problems

(**7.7.1**) Show that the following are equivalent in any T_1 topological space X:
(a) Every sequence in X has an adherent point.
(b) Every infinite subset of X has a limit point.
(c) X is countably compact.

(**7.7.2**) Show that a sequentially compact space is countable compact. The converse is true in first countable spaces.

(**7.7.3**) Give an example of a non-Hausdorff space with a compact subset that is not closed.

(**7.7**.4) Show that if a filter in a compact space has only one adherent point, then the filter converges to the adherent point.

(**7.7**.5) Show that a finite union of compact sets is compact but the intersection of two compact sets need not be compact, unless the space is Hausdorff.

(**7.7**.6) Show that the closure of a compact set need not be compact, unless the space is regular.

(**7.7**.7) Show that a locally compact Hausdorff space has the family of closed compact neighbourhoods of each point as a base.

(**7.7**.8) Show that a dense locally compact subspace of a Hausdorff space is open.

(**7.7**.9) (Alexander) Suppose $\mathcal{S}$ is a subbase of open subsets of a space X. Show that if every open cover of X by members of $\mathcal{S}$ has a finite subcover, then X is compact. This is known as Alexander's Subbase Theorem [3] (see, also, [2])

(**7.7**.10) (Alexander) Prove that the following are equivalent for a space X:

(a) X is compact.

(b) There exists a subbase S for the closed subsets of X such that whenever $\mathcal{H}$ is a subfamily of S with the finite intersection property, then $\bigcap\{H : H \in \mathcal{H}\}$ is not empty.

This is another formulation of Alexander's subbase theorem [3], that can be used to prove Problem 7.7.11.

(**7.7**.11) Use the Alexander subbase theorem to prove that if $\{X_i : i \in \mathbb{N}\}$ is a family of compact spaces, then their product $X = \prod\{X_i : i \in \mathbb{N}\}$ is also compact. **Hint**: Let $\mathcal{S} = \{p_i^{-1}(A_i) : i \in \mathbb{N} \text{ and } A_i \text{ is closed in } X_i\}$, where p_i is the ith projective map. Start by proving that $\mathcal{S}$ is a subbase for the closed sets of X.

(**7.7**.12) Give a detailed proof of Theorem 7.10.

Chapter 8

Initial and Final Structures, Embedding

8.1 Initial Structures

Let X be a nonempty set, $\{(Y_\lambda, T_\lambda) : \lambda \in \Lambda\}$ a family of topological spaces, and $f_\lambda : X \to Y_\lambda$ functions. Then ask *Is there a topology on X that will make each function f_λ continuous*? Of course, there is a trivial solution, namely, the discrete topology on X.

A more interesting question is *Is there a coarsest topology on X such that each function f_λ is continuous*? Naturally, this topology $\mathcal{I}$ on X must contain the sets $f_\lambda^{-1}(T_\lambda)$ for $\lambda \in \Lambda$. So the answer to the second question is Yes and is the topology $\mathcal{I}$ (called the **initial topology** for the family $\{f_\lambda : \lambda \in \Lambda\}$) generated by the subbase $\bigcup \{f_\lambda^{-1}(T_\lambda) : \lambda \in \Lambda\}$. A useful result in this connection is

Theorem 8.1. *Let Z have the topology ζ and let $g : Z \to X$ be a function. Then g is continuous, if and only if, $f_\lambda g : Z \to Y_\lambda$ is continuous for each $\lambda \in \Lambda$.*

Proof. Clearly, if g is continuous, then, for each $\lambda \in \Lambda$, $f_\lambda g$ is continuous. Conversely, let $f_\lambda g$ be continuous for each $\lambda \in \Lambda$. Then, for each open set U in Y_λ, $f_\lambda^{-1}(U)$ is in the initial topology on X. Now $g^{-1}\left(f_\lambda^{-1}(U)\right) = (f_\lambda g)^{-1}(U)$ is open in Z. So g is continuous. □

Finite products of topological spaces are well known.

(product.1) The complex plane and Euclidean n-spaces are products of copies of the real line.

(product.2) In special relativity, 4D space is a product of three copies of real lines and $[0, \infty)$ for time.

(product.3) The product of a circle and $\mathbb{R}$ (interval) is an infinite (finite) cylinder.

The concept of an initial topology is useful in defining a Tychonoff topology on arbitrary products of topological spaces. Let $\{(X_\lambda, \tau_\lambda) : \lambda \in \Lambda\}$ be a family of topological spaces. Let $X = \prod\{X_\lambda : \lambda \in \Lambda\}$ be the product set, *i.e.*, X is the set of all functions $f : \Lambda \to \bigcup\{X_\lambda : \lambda \in \Lambda\}$ such that $f(\lambda) \in X_\lambda$.

One often uses x_λ for $f(\lambda)$. Let $p_\lambda : X \to X_\lambda$ be the projection mapping. The initial topology τ on X for the family $\{p_\lambda : \lambda \in \Lambda\}$ is called the (Tychonoff) **product topology**.

A typical open set in the product topology is a finite intersection of sets from the family

$$\left\{p_\lambda^{-1}(U_\lambda) : \lambda \in \Lambda, U_\lambda \text{ is an open set in } X_\lambda\right\},$$

i.e., where U_λ is an open set in X_λ and, for all except finitely many λs, $U_\lambda = X_\lambda$.

Another way of looking at the product topology is that it is the coarsest topology that makes all projections continuous. Moreover, in a product topology, each projection is an **open map**, *i.e.*, each projection p_λ takes an open set in X into an open subset of X_λ. This is obvious for basic open sets in X and this implies that it is true for all open sets of X.

A property P of topological spaces is called **productive**, if and only if, whenever each member of a family of topological spaces has P, then so does their product. For example, separation axioms R_0, T_0, T_1, T_2, T_3, completely regular are productive. But T_4 (normal) is not productive. Even a product of two normal spaces need not be normal. Compactness is productive (Tychonoff product theorem (see Theorem 8.2)), which is considered one of the most important results in general topology.

However, a projection need not preserve closed sets, even in finite products. For example, in $\mathbb{R}^2 = \mathbb{R} \times \mathbb{R}$, the hyperbola $\{xy = 1\}$ is a closed set in the plane but its projections are not closed maps. In case the coordinate spaces are all compact, then the projections are also closed maps. Let $X_\lambda, \lambda \in \Lambda$ be a family of topological spaces and let $X = \prod\{X_\lambda : \lambda \in \Lambda\}$, their product. Let $\mathcal{L}$ be an ultrafilter in X. Then, for each $\lambda \in \Lambda, \mathcal{L}_\lambda = p_\lambda(\mathcal{L})$ is an ultrafilter in X_λ. Since the projections are continuous, if $\mathcal{L}$ converges to $x = (x_\lambda) \in X$, then, for each $\lambda \in \Lambda$, $\mathcal{L}_\lambda$ converges to x_λ. Conversely, if, for each $\lambda \in \Lambda$, $\mathcal{L}_\lambda$ converges to x_λ, then every neighbourhood U_λ of x_λ in X_λ belongs to $\mathcal{L}_\lambda$. This implies that $p_\lambda^{-1}(U_\lambda) \in \mathcal{L}$. Since a typical neighbourhood of x is a finite intersection of the family $\{p_\lambda^{-1}(U_\lambda)\}$ and so it belongs to $\mathcal{L}$. Hence, $\mathcal{L}$ converges to x. This proves

Theorem 8.2. Tychonoff Product Theorem
A product of compact spaces is compact, if and only if, each coordinate space is compact.

J.L. Kelley observes that the Tychonoff product theorem on the product of compact spaces is the most useful theorem on compactness and is *probably the most important single theorem of general topology.* This theorem plays an important role in many results in topology as well as in applications, especially in functional analysis. A few examples that illustrate its utility are (1) Stone-Čech compactification, (2) Ascoli-Arzela theorem in compactness in function spaces (used in existence theorems in differential equations), (3) Gelfand representation theorem in Banach algebra, and (4) Alaoglu theorem on the weak compactness of the unit ball.

8.2 Embedding

It follows from Theorem 8.2 (Tychonoff Product Theorem) that a product of compact Hausdorff spaces is a compact Hausdorff space. We have seen that in a Tychonoff (completely regular) space X, a disjoint pair $\lambda = (x, \mathcal{C})$ of point x and closed set $\mathcal{C}$ can be separated by a continuous function on $X \to I = [0,1]$, a nice well-known space. If we take a sufficient number of copies of I, then it will be possible for us to embed X into a product of I, called a **cube**. We consider next a line of reasoning that leads to a general embedding theorem, namely, Theorem 8.3.

Suppose we wish to embed the space X into the product $Y = \prod\{Y_\lambda, \lambda \in \Lambda\}$. We begin with maps $f_\lambda : X \to Y_\lambda$. It is natural to consider the *evaluation* map

$$e : X \to Y \text{ given by } e(x) = (f_\lambda(x)).$$

We want $e : X \to e(X)$ to be a homeomorphism. For this to happen, obviously (a) the family $\{f_\lambda\}$ should be continuous and (b) points should be separated, *i.e.*, if $x, y \in X$, then there is a f_λ such that $f_\lambda(x) \neq f_\lambda(y)$, and (c) e should take open sets in X into open sets in $e(X)$. For this, it is sufficient that $\{f_\lambda\}$ should separate disjoint pairs $\{(x, \mathcal{C})\}$ of point x and closed set $\mathcal{C}$. It is easy to see that (a), (b) and (c) are also necessary. This gives

Theorem 8.3. Embedding Theorem
Let $X, \{Y_\lambda, \lambda \in \Lambda\}$ be topological spaces and let $f_\lambda : X \to Y_\lambda$ be a map for each $\lambda \in \Lambda$. Then X can be embedded in $Y = \prod\{Y_\lambda, \lambda \in \Lambda\}$, if and only if, $\{f_\lambda\}$ is a family of continuous functions that separate points as well as points and closed sets.

Corollary 8.1. *Every Tycnonoff space is homeomorhphic to a subspace of a cube. Every compact Hausdorff space is homeomorphic to a closed subspace of a cube.*

8.3 Final Structures

Next, consider what is known as a **final structure**, a concept that is the dual of initial structure. Suppose X is a set and $\{Y_\lambda, \lambda \in \Lambda\}$ is a family of topological spaces and, for each $\lambda \in \Lambda$, there is a function $f_\lambda : Y_\lambda \to X$. Then ask the question *Is there a topology on X that will make f_λ continuous*? A trivial answer is the indiscrete topology. A better question: *Is there a finest (largest) topology on X that will make each f_λ continuous*? The answer is Yes, the topology $\tau = \left\{U \subset X : f^{-1}(U) \text{ open in } Y_\lambda\right\}$ (verify that τ is indeed a topology). This topology τ is called the **final topology** for the family $\{f_\lambda\}$.

A special case of the final topology is the **quotient topology**. Suppose $\sim$ is an equivalence relation on a topological space Y and $X = Y/\sim$, the quotient set of equivalence classes $\{[y] : y \in Y\}$ (*i.e.*, $x \in Y$ belongs to $[y]$, provided $x \sim y$). Let f be the map on Y to X given by $f(y) = [y]$. The quotient topology on X is the final topology for f.

8.4 Application: Quotient Topology in Image Analysis

This section introduces an application of quotient topology in image analysis (IA). **Segmentation** is a generic process that subdivides a picture into its constituent regions. This process is often the first vital step that is taken before extracting features and attempting to classify a picture. The

8.1.1: Gleaner

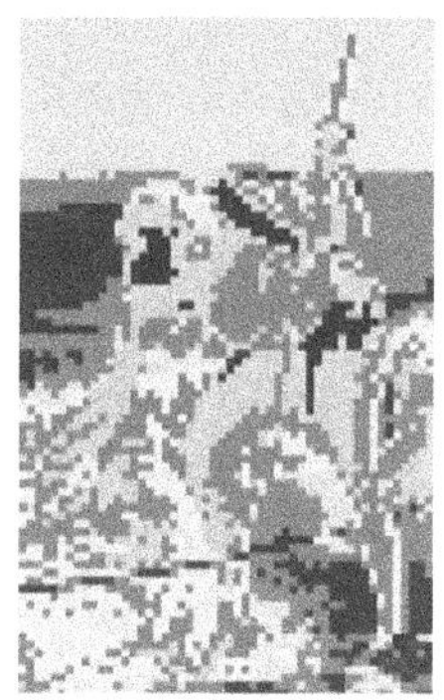

8.1.2: 10×10 partition

8.1.3: 10×10 class

Fig. 8.1 Sample image partition determined by relation $\sim_\Phi$

basic goal of picture segmentation is the partition of a picture into mutually exclusive regions so that one can assign meaningful labels [212].

A segmented digital image is a partition of an image into non-overlapping regions, where the objects in each region are similar in character. There are many segmentation methods (for a survey of segmentation methods, see, *e.g.* [76; 75], and for a detailed study of a topological view of segmentation properties, see, *e.g.*, [201]). This section introduces an approach to image segmentation considered in the context of metric spaces based on recent work on near sets in image analysis (see, *e.g.*, [187; 184; 81; 167]).

Let (X, η_Φ) denote a metric space, where X is a non-empty set of subimages of a digital image, Y is a set of non-empty set of n-dimensional feature vectors $\Phi(x)$ of reals that describe extracted subimages $x \in X$, and the set of probe functions $\Phi = \{\phi : \phi : X \to \mathbb{R}\}$. A *subimage* is an $n \times n$ array of image pixels (picture elements). Each probe $\phi \in \Phi$ represents a feature such as average greylevel, average local entropy, average gradient direction of the pixels in the subimage. A feature vector $\Phi(x)$ is defined by

$$\Phi(x) = (\phi_1(x), \phi_2(x), \ldots, \phi_i(x), \ldots, \phi_l(x)),$$

where $\phi_i \in \Phi$, l is the length of a feature vector. The function $f : X \to Y$ is defined by $f(x) = \Phi(x), x \in X$. The partition of X is determined by an equivalence relation denoted by $\sim_{\mathcal{B}}$ such that

$$\sim_\Phi = \{(x, y) \in X \times X : \ |\Phi(x) - \Phi(y)| = 0\}.$$

By way of illustration, let $\Phi(x) = (\phi_{grey}(x))$, where the feature vector has length $l = 1$, $\Phi = \{\phi_{grey}\}$, and ϕ_{grey} is defined by

$$\phi_{grey}(x) = \text{ average greylevel pixel intensity of subimage } x \in X.$$

Also assume that each subimage $x \in X$ is a 10×10 array of pixels. Consider, for example, the image in Fig. 8.1.1 showing gleaner (collector of plant samples). This gleaner is shown on the coast of Prince Edward Island. The relation $\sim_\Phi$ determines the partition shown in Fig. 8.1.2. Each segment in Fig. 8.1.2 consists of scattered 10×10 ■ subimages. A sample image segment (actually, an equivalence class) is shown in Fig. 8.1.3. This class consists of a scattered collection of 10×10 light grey boxes such as ■■■ that illustrate the scattered character of the subimages in the segment, *i.e.*, although the highlighted subimages belong to the same segment, many of the subimages are not adjacent to each other. This class can be found in the partition in Fig. 8.1.2 by observing the light grey patches along the extended arm and leg and the lower edge of hat of the gleaner. This example of a quotient topology also exhibits a form of nearness. To see this, let $x, y \in X, \Phi(x), \Phi(y) \in Y$ and define η_Φ using the taxicab metric, *i.e.*,

$$\eta_\Phi(\Phi(x), \Phi(y)) = \sum_{i=1}^{l} |\phi_i(x) - \phi_i(y)|.$$

Then, $f(x)$ is near $f(y)$, if and only if $D_{\eta_{\mathcal{B}}}(f(x), \{f(y)\}) = 0$. Further, given a segment (class) A in the partition of X, $f(x)$ is near A, if and only if, $D(f(x), A) = 0$. We know this to be true, since only subimages with matching feature vectors are part of an image segment in the partition determined by $\sim_\Phi$. The form of image segmentation presented in this section has been implemented in what is known as the Near system, fully described in [87; 86].

8.5 Problems

(**8.5**.1) Let $X_\lambda : \lambda \in \Lambda$ be a family of topological spaces and let $X = \prod\{X_\lambda : \lambda \in \Lambda\}$ have the product topology.

(a) If, for each $\lambda \in \Lambda$, $\mathcal{C}_\lambda$ is a closed set in X_λ, then show that $\mathcal{C} = \prod\{C_\lambda : \lambda \in \Lambda\}$ is closed in X.

(b) If, for each $\lambda \in \Lambda$, E_λ is a set in X_λ, then show that

$$\operatorname{cl} E = \operatorname{cl} \prod\{E_\lambda : \lambda \in \Lambda\} = \prod\{\operatorname{cl} E_\lambda : \lambda \in \Lambda\}.$$

Hence, show that E is closed, if and only if, each E_λ is closed

(c) Show that, if each X_λ satisfies the separation axiom $T_k, k = 1, 2, 3, 3\frac{1}{2}$, then so does X.

(d) Show that the Sorgenfrey line X is normal but $X \times X$ is not normal.

(**8.5**.2) Let $f : X \to Y$ be a function and let $G(f)$ be the graph $\{(x, y) : x \in X\} \subset X \times Y$. Show that f is continuous, if and only if, $G(f)$ is homeomorphic to X.

(**8.5**.3) Show that infinite products of discrete spaces are not discrete.

Note: The following problems require the use of one or more digital images[1].

(**8.5**.4) Let the space X contain the segments of a digital image. Then a subset $A \subset X$ contains subimages. Use Matlab® or Mathematica® to segment a digital image using the feature-based method described in Section 1.20. Give sample topologies τ on the set X contains subsets of segments in the segmentation.

(**8.5**.5) For a subset $A \subset X$ in a sample topology τ on X in Problem 4, replace the average greylevel of each subimage in A to a single colour such as □ and display all of the segments of A in the original image.

(**8.5**.6) Instead of the taxicab metric used for the metric $\eta_{\mathcal{B}}$ for image segmentation in Section 1.20, define $\eta_{\mathcal{B}}$ using a different metric such as

(**metric**.a) Euclidean metric $\rho_2 = \| \cdot \|_2$,

(**metric**.b) Discrete metric ρ_{01}, where

$$\rho_{01}(x, y) = \begin{cases} 1, & \text{if } x = y, \\ 0, & \text{otherwise.} \end{cases}$$

(**metric**.c) Ratio metric $\rho(x, y) = \frac{|x-y|}{1+|x-y|}$.

(**8.5**.7) Let (X, ρ) be a metric space with a set of subimages X and a metric ρ. Recall that a spherical neighbourhood $S_\rho(x_0, r) = \{y \in X : \rho(x, y) < r\}$ is the set of all points y in X with neighbourhood centre x_0 and radius $r \in \mathbb{R}$ such that $\rho(x, y) < r$. A spherical neighbourhood is also commonly called an open ball (see, *e.g.*, [59]). Then select r and let $B_r(x) = S_\rho(x, r)$ for every $x \in X$ to define a collection of subsets of X to generate a topology τ on X. Display

[1] For digital topology experiments, sample digital images are available at `http://www.vision.caltech.edu/html-files/archive.html`

one of the subsets $A \in X$ in a topology for a sample image by replacing each average greylevel with a dark grey box such as ▩ for each of subimages in A. Hint: for image segments $x \in X$, choose $r > 0$ for $r \in [0, 255]$, where 0 represents absence of light (black) and 255 represents maximum light intensity (white).

Chapter 9

Grills, Clusters, Bunches and Proximal Wallman Compactification

9.1 Grills, Clusters and Bunches

Grills, clusters and bunches that arise naturally in proximity spaces are covered in this chapter. Grills were introduced by G. Choquet in 1947, clusters by S. Leader in 1962 and bunches by Leader's student M.W. Lodato (see Historical Notes section for details).

A grill is a generalisation of an ultrafilter. Recall that an ultrafilter $\mathcal{L}$ on a nonempty set X is a family of nonempty subsets of X that are closed under finite intersections and satisfies the union property, namely,

Union property : $(A \cup B) \in \mathcal{L}$, if and only if, $A \in \mathcal{L}$ or $B \in \mathcal{L}$.

9.2 Grills

A **grill** $\mathcal{G}$ on a nonempty set X is defined as a family of nonempty subsets of X satisfying the union property. Consider proving the following statements.

(**grill**.1) Every ultafilter is a grill.

(**grill**.2) $\mathcal{G}$ is a grill, if and only if, $\mathcal{G}$ is a union of ultrafilters.

(**grill**.3) Let $\mathcal{F}$ be a filter on X and let $c(\mathcal{F})$ denote the family of all nonempty subsets of X that do not intersect some member of $\mathcal{F}$. Then $c(\mathcal{F})$ is a grill. This shows that filters and grills are duals of each other. Moreover, each ultrafilter is its own dual. That is, $c(c(\mathcal{F})) = \mathcal{F}$, $c(c(\mathcal{G})) = \mathcal{G}$, and $c(\mathcal{L}) = \mathcal{L}$ for an ultrafilter $\mathcal{L}$.

(**grill**.4) Let X be a dense subset of a topological space αX. For each point $p \in \alpha X$, let

$$\sigma(p) = \{E \subset X : p \in \operatorname{cl}_{\alpha X} E\}.$$

Then $\sigma(p)$ is a grill. Later, we will show that if αX is a T_1 space, then *subsets A, B of X are near, if and only if, their closures in αX intersect* gives a L-proximity on X. And, conversely, every L-proximity on X arises in the same manner from a T_1 compact space αX in which X is dense.

(**grill**.5) Every proximity space satisfies the union axiom. Hence, in an L-proximity space (X, δ), for each nonempty set $A \subset X$,

$$\delta(A) = \{E \subset X : E\delta A\} \text{ is a grill.}$$

(**grill**.6) In a topological space, the closure operator satisfies the union property. Hence, if $\mathcal{G}$ is a grill in an L-proximity space (X, δ), then

$$b(\mathcal{G}) = \{E \subset X : \text{ cl } E \in \mathcal{G}\} \text{ is a grill.}$$

If $\mathcal{G}, \mathcal{G}'$ are grills such that $\mathcal{G} \subset \mathcal{G}'$, then $b(\mathcal{G}) \subset b(\mathcal{G}')$.

9.3 Clans

Ultrafilters are useful in topology in studying convergence and concepts such as compactness. Important properties of an ultrafilter arise from (i) intersection of pairs and (ii) the union property.

> The relation *near* is a generalisation of intersection. Replacing intersection with near in a proximity space yields a useful generalisation. A **clan** in an L-proximity space (X, δ) is a grill in which any two members are near. We use the notation **δ-clan**, if it is necessary to specify proximity.

A standard Zorn's Lemma argument yields the following result.

Lemma 9.1. *In an L-proximity space X, every clan is contained in a maximal clan.*

In a nonempty set X, any two intersecting sets are contained in an ultrafilter. Its analogue holds in an L-proximity space (X, δ). Suppose $A\ \delta\ B$. Then $\delta(B) = \{E \subset X : E\ \delta\ B\}$ is a grill that contains A. Hence, there is an ultrafilter $\mathcal{L}_A$ such that $A \in \mathcal{L}_A \subset \delta(B)$.

$$B \in \delta(\mathcal{L}_A) = \{E \subset X : E\ \delta\ F \text{ for each } F \in \mathcal{L}_A\} \text{ is a grill.}$$

So there is an ultafilter $\mathcal{L}_B$ such that $B \in \mathcal{L}_B \subset \delta(\mathcal{L}_A)$. This shows that A, B are both in the clan $\mathcal{L}_A \cup \mathcal{L}_B$, which, in turn, is contained in a maximal clan. This yields

Theorem 9.1. *In an L-proximity space* (X, δ), $A\ \delta\ B$ *implies that there is a maximal* δ*-clan that contains both* A *and* B.

9.4 Bunches

In an L-proximity space, two sets are near, if and only if, their closures are near. Hence, it is conceivable that the concept of a clan which contains a set, if and only if, it contains its closure, is valuable. Such a clan is called a **bunch**. The following results are obvious.

Lemma 9.2. *A clan* $\mathcal{C}$ *is a bunch in an L-proximity space, if and only if,*

$$\mathcal{C} = b(\mathcal{C}) = \{E \subset X : \text{ cl } E \in \mathcal{C}\}.$$

Corollary 9.1. (*cf. Lemma 9.1*)
In an L-proximity space, every bunch is contained in a maximal bunch.

Corollary 9.2. *In an L-proximity space, every maximal clan is contained in a bunch.*

Theorem 9.2. (*cf. Theorem 9.1*)
In an L-proximity space (X, δ), $A\ \delta\ B$ *implies there is a maximal bunch* β *that contains both* A *and* B.

> It is now clear that in an L-proximity space (X, δ), every bunch is a union of ultrafilters.

If $\mathcal{L}$ is an ultrafilter in X, then $b(\mathcal{L})$ is a clan. Since $b(b(\mathcal{L})) = b(\mathcal{L})$, $b(\mathcal{L})$ is a bunch in (X, δ) and is contained in any bunch that contains $\mathcal{L}$. Thus, we have

Theorem 9.3. *In an L-proximity space* (X, δ), *for every ultrafilter* $\mathcal{L}$ *in* X, $b(\mathcal{L})$ *is a minimal bunch containing* $\mathcal{L}$.

9.5 Clusters

In an L-proximity space (X, δ), for each $x \in X$,

$$\sigma_x = \{E \subset X : x \ \delta \ E\} \text{ is a bunch with property } (^{**}).$$

$(^{**})$ $E \ \delta \ G$ for every $G \in \sigma_x$ implies $E \in \sigma_x$.

A bunch in an L-space (X, δ) is called a **cluster**, if and only if, it satisfies property $(^{**})$. Notice that $\sigma_x - (x)$ is a bunch that is not a cluster.

A family $\mathcal{L}$ of nonempty sets of a nonempty set X is an ultrafilter, if and only if, $\mathcal{L}$ satisfies the following three conditions.

(**ultrafilter**.1) $A, B \in \mathcal{L}$ implies $A \cap B \neq \varnothing$.
(**ultrafilter**.2) $(A \cup B) \in \mathcal{L}$, if and only if, $A \in \mathcal{L}$ or $B \in \mathcal{L}$.
(**ultrafilter**.3) $A \cap B \neq \varnothing$ for each $B \in \sigma$ implies $A \in \mathcal{L}$.

In the original definition of a cluster, S. Leader used the analogues of the ultrafilter properties (ultrafilter.1)-(ultrafilter.3), replacing intersection with the relation *near* to obtain

A family σ of subsets of an L-proximity space (X, δ) is a **cluster**, if and only if,
(**cluster**.1) $A, B \in \sigma$ implies $A \ \delta \ B$.
(**cluster**.2) $(A \cup B) \in \sigma$, if and only if, $A \in \sigma$ or $B \in \sigma$.
(**cluster**.3) $A \ \delta \ B$ for each $B \in \sigma$ implies $A \in \sigma$.

The equivalence of the two definitions of a cluster is left as an exercise. It will be seen in the sequel that bunches are useful in L-proximity spaces and clusters in EF-proximity spaces.

Remark 9.1. About Clusters, Bunches, Dense Subsets, & Grills
At this point, a number of useful observations can be made about clusters.

(**Obs**.1) In an L-space (X, δ), for each $x \in X$, σ_x is a cluster called **point cluster**. Point clusters are maximal bunches. They are useful in embedding an L-proximity space in the space of all maximal bunches, which is compact in a suitable topology.

(**Obs**.2) Let X be a dense subset of a T_1 topological space αX. Let X have proximity δ induced by the fine L-proximity δ_0 on αX. For each point $p \in \alpha X$, let

$$\sigma^p = \{E \subset X : p \in \mathrm{cl}_{\alpha X} E\}.$$

Then σ^p is a bunch in (X, δ).

(**Obs.**3) If $\mathcal{G}$ is a grill in an L-space (X, δ), we write

$$\delta(\mathcal{G}) = \{E \subset X : E\delta G \text{ for each } G \in \mathcal{G}\}.$$

If σ is a cluster, the $\delta(\sigma) = \sigma$ and so a cluster is both maximal and minimal. In particular, every cluster is a maximal bunch.

(**Obs.**4) In an L-space (X, δ), if $\{x\} \in \sigma$, a bunch, then $\sigma = \sigma_x$ is a point cluster.

We now show that ultrafilters and clusters are closely related in an Efremovič proximity space (X, δ). If $\mathcal{L}$ is an ultrafilter in X, then $\delta(\mathcal{L})$ is a grill and $\delta(\delta(\mathcal{L})) = \delta(\mathcal{L})$. We show that $\delta(\mathcal{L})$ is a cluster by showing that it is a clan, *i.e.*, any two members A, B of $\delta(\mathcal{L})$ are near. If not, then there is a $E \subset X$ such that $A \underline{\delta} E$ and $(X - E) \underline{\delta} B$. This shows that both E and $(X - E)$ are not in the ultrafilter $\mathcal{L}$, a contradiction.

Theorem 9.4.

(*a*) *Let (X, δ) be an EF-proximity space. If $\mathcal{L}$ is an ultrafilter in X, then $\delta(\mathcal{L})$ is a cluster. Conversely,*

(*b*) *if $\mathcal{L}$ is an ultrafilter in a cluster σ, then $\sigma = \delta(\mathcal{L})$.*

(*c*) *If $A \in \sigma$ is a cluster, then there is an ultrafilter $\mathcal{L}$ in X such that $A \in \mathcal{L} \subset \delta(\mathcal{L}) = \sigma$.*

Corollary 9.3. *If $A \ \delta \ B$ in an EF-proximity space (X, δ), then there is a cluster σ that contains both A and B.*

Corollary 9.4. *In an EF-proximity space (X, δ), σ is a cluster, if and only if, it is a maximal bunch. Hence, in an EF-proximity space, every bunch is contained in a unique cluster.*

Recall that a topological space is compact, if and only if, every ultrafilter in the space converges. This leads to an analogous result in EF-spaces. Let (X, δ) be an EF-proximity space. From Theorem 9.4(c), we get

Theorem 9.5. *An EF-proximity space (X, δ) is compact, if and only if, every cluster in the space is a point cluster.*

9.6 Proximal Wallman Compactification

We now pursue Wallman T_1 compactification of a T_1 space X. H. Wallman uses the family wX of *all* closed ultrafilters of X and assigns a topology on

wX (known as Wallman topology) that makes wX compact. X is embedded in wX via the map that takes a point $x \in X$ to the closed ultrafilter $\mathcal{L}_x$ containing $\{x\}$ (for details, see the Historical Notes section). Thus, Wallman gets just one T_1 compactification of any T_1 space.

To generalise Wallman compactification, we begin with an arbitrary compatible L-proximity δ on X and follow Wallman step-by-step replacing wX by X^*, the space of all maximal bunches in the L-proximity space (X, δ) and mapping x to the maximal bunch σ_x, the family of all subsets of X near x. The resulting **proximal Wallman compactification** of (X, δ) contains as special cases infinitely many T_1 compactifications as well as the Smirnov compatification, which includes all Hausdorff compactifications.

Let (X, δ) be a T_1 L-proximity space and let X^* denote the set of all maximal bunches in X. For each $E \subset X$,

$$E^* = \{\sigma \in X^* : E \in \sigma\}.$$

We use the family $\{E^* : E \subset X\}$ as a base for the closed subsets of the Wallman topology τ^* on X^*. Notice that, for subsets $E, F \subset X$, $(E \cup F)^* = E^* \cup F^*$. Since $E^* = (\text{cl } E)^*$, we may choose the family

$$\mathcal{B} = \{E^* : E \in \mathcal{C}, \text{ the family of closed subsets of } X\}$$

as the closed base for τ^*. Obviously, τ^* is T_1. Notice that by Theorem 9.2, $A \ \delta \ B$ implies that there is a maximal bunch containing A and B. Hence, $A^* \cap B^* \neq \emptyset$. Conversely, $A^* \cap B^* \neq \emptyset$ implies $A \ \delta \ B$. This proves

Proposition 9.1. *$A \ \delta \ B$ in X, if and only if, $A^* \cap B^* \neq \emptyset$ in X^*.*

Next, we show that (X^*, τ^*) is compact. Let $\mathcal{L}^*$ be a closed ultrafilter in X^* formed from members of $\mathcal{B}$ and $\alpha = \{E \subset X : E^* \in \mathcal{L}^*\}$. Then α is a bunch in (X, δ), which we verify as follows.

(a) $X \in \alpha$ implies α is not empty.

(b)

$$\begin{aligned} A, B \in \alpha &\Rightarrow A^*, B^* \in \mathcal{L}^* \\ &\Rightarrow A^* \cap B^* \neq \emptyset \\ &\Rightarrow A \ \delta \ B. \end{aligned}$$

(c)

$$\begin{aligned} A \cup B \in \alpha &\Leftrightarrow (A \cup B)^* \in \mathcal{L}^* \\ &\Leftrightarrow (A^* \cup B^*) \in \mathcal{L}^* \\ &\Leftrightarrow A^* \in \mathcal{L}^* \text{ or } B^* \in \mathcal{L}^* \\ &\Leftrightarrow A \in \alpha \text{ or } B \in \alpha. \end{aligned}$$

(d)

$$A^* = (\mathrm{cl}\, A)^* \text{ implies } A \in \alpha \Leftrightarrow \mathrm{cl}\, A \in \alpha\,.$$

The bunch α is contained in a maximal bunch σ and, obviously, $\{\sigma\} \in \mathcal{L}^*$ shows the compactness of τ^*.

Theorem 9.6. *Let (X, δ) be a T_1 L-proximity space and let X^* be the space of all maximal bunches with the topology τ^* generated by the closed base $\{\sigma \in X^* : E \in \sigma\}$ and fine L-proximity δ_0.*
(a) (X^*, τ^*) is a compact T_1 space.
(b) The map $\phi : X \to X^*$ given by $\phi(x) = \sigma_x$, the point cluster x is a proximal isomorphism on X to $\phi(X)$.
(c) $\phi(X)$ is dense in X^*.
(d) $A\ \delta\ B$, if and only if, $\mathrm{cl}\,\phi(A) \cap \mathrm{cl}\,\phi(B) \neq \varnothing$ in X^*.

Corollary 9.5. *Let (X, δ) be a T_1 EF-proximity space and let X^* be the space of all clusters with topology τ^* generated by the closed base $\{\sigma \in X^* : E \in \sigma\}$ and the fine L-proximity δ_0. Then δ_0 is EF and (X^*, τ^*) is a compact Hausdorff space called Smirnov compactification of (X, δ).*

9.7 Examples of Compactifications

This section briefly sets forth examples of compactifications.

(**Ex**.1) **Wallman compactification** The original Wallman compactification of a T_1 space X is obtained from Theorem 9.6 by choosing the fine proximity δ_0 on X and X^* to consist of all maximal bunches containing closed ultrafilters.

(**Ex**.2) **Wallman-Frink compactification** of a T_1 space. Frink generalised Wallman compactification by choosing a subfamily of all closed sets and this was put in proximal Wallman form. A family $\mathcal{R}$ of closed subsets of T_1 space X is called **separating**, if it satisfies

(a) $\mathcal{R}$ is a base for closed subsets of X.

(b) $\mathcal{R}$ is a ring that is closed under finite unions and finite intersections.

(c) If $x \notin A$ (a closed subset of X), then there are $P, Q \in \mathcal{R}$ such that $x \notin P, A \subset Q$, and $P \cap Q = \varnothing$. It is easy to show that δ defined by

(d) $A \underline{\delta} B$, if and only if, there are $P, Q \in \mathcal{R}$ such that $A \subset P, B \subset Q$, and $P \cap Q = \varnothing$ is a compatible L-proximity on X.

> The resulting proximal-Wallman compactification is T_1. Further, if X is Tychonoff, a separating family $\mathcal{R}$ is called a **normal base**. If A, B are disjoint members of $\mathbb{R}$, there are $P, Q \in \mathbb{R}$ with $A \subset X - P, B \subset X - Q$ and $P \cup Q = X$. In this case, δ defined by (d) is an EF-proximity on X and the resulting compactification is called **Wallman-Frink**. The question *Are all Hausdorff compactifications of the above type?* (Frink's problem) led to a considerable literature, finally ending in a counterexample.

(**Ex.**3) **Alexandroff one-point compactification** This is a generalisation of the construction of the Riemann sphere that compactifies the space of complex numbers. Let X be a non-compact T_1 space with L-proximity by

$$A \ \delta \ B \Leftrightarrow \text{cl}A,\ \text{cl}B \text{ intersect or both are non-compact.}$$

Then the space X^* of maximal bunches in (X, δ) with Wallman topology is a compact T_1 space. In this case, identifying X with $\phi(X), X^* - \phi(X)$ is just one point $\{\omega\}$. If X is compact, then ω is isolated.

> In analogy with the space of complex numbers, to get a Hausdorff one-point compactification of X, the space X must be Hausdorff and locally compact. Conversely, if the one-point compactification of X is Hausdorff, then X is locally compact.

(**Ex.**4) **Two-point compactification** of the space of real numbers $\mathbb{R}$. In measure theory on $\mathbb{R}$, it is customary to use the space $\mathbb{R}^* = \mathbb{R} \cup \{-\infty, \infty\}$, with order $-\infty < x < \infty$ for all $x \in \mathbb{R}$. We look at this space in two different ways.

(a) Define a topology T^* on $\mathbb{R}^*$ generated by the subbase

$$\{[-\infty, x) : x \in \mathbb{R}\} \cup \{(x, \infty] : x \in \mathbb{R}\}.$$

Then $\mathbb{R}^*$ is a compact Hausdorff space containing $\mathbb{R}$ as a dense subspace and compatible EF-proximity δ_0.

(b) Define EF-proximity δ on $\mathbb{R}$ by

> $A\ \delta\ B \Leftrightarrow \mathrm{cl}\,A$, $\mathrm{cl}\,B$ intersect or both A, B are unbounded above or unbounded below.

Then $(\mathbb{R}^*, \delta_0)$ is proximally isomorphic to the proximal Wallman compactification of (X, δ) in Corollary 9.5.

(**Ex**.5) **Stone-Čech compactification of a Tychonoff space**. Let X be a non-compact Tychonoff space with the fine EF-proximity $\underline{\delta}_F$, namely,

> $A\ \underline{\delta}_F\ B$, if and only if, there exists a continuous function $f : X \to [0,1] : f(A) = 0, f(B) = 1$.

Then, in Corollary 9.5, one gets a Hausdorff compactification βX called *Stone-Čech compactification* of X. It is considered the most important compactification and has many applications.

(**Ex**.6) **Freudenthal compactification of a Tychonoff rim-compact space**. Let X be a Tychonoff space with a base of neighbourhoods with compact boundaries. Let a proximity δ be defined by

> $A\ \underline{\delta}\ B$, if and only if, A, B are contained in disjoint closed subsets with compact boundaries.

Then δ is a compatible EF-proximity on X. The construction given in Corollary 9.5 gives a Hausdorff compactification ϕX called Freudenthal compactification.

Observe that

> Every abstract proximity is induced by the fine proximity on a compact space.

Remark 9.2. In Theorem 9.6, the abstractly defined T_1 L-proximity space (X, δ) is proximally embedded in X^*, the space of all maximal bunches with the Wallman topology T^*. Identifying X with its image $\phi(X)$, it follows that the proximity δ is the subspace proximity induced by the fine L-proximity δ_0 on the compact T_1 space X^* in which X is dense.

By Corollary 9.5, it follows that every EF-proximity δ is induced by the fine EF-proximity δ_0 on the compact Hausdorff space X^* of all clusters with Wallman topology.

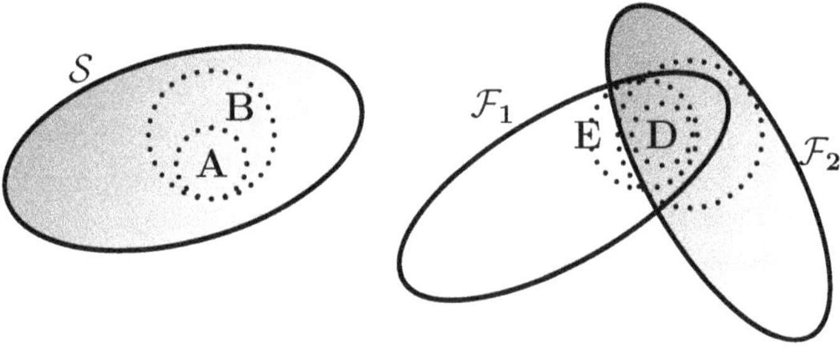

Fig. 9.1 Sample stack $\mathcal{S}$, grill $\mathcal{G} = \mathcal{F}_1 \cup \mathcal{F}_2$

9.8 Application: Grills in Pattern Recognition

This section uses results from [223; 162] in a study of grills and stacks in pattern recognition. Grill were introduced by G. Choquet in 1947 [42] and stacks by G. Grimeisen in 1960-1961 [77],[78], although the formulation by Grimeisen is slightly different from the usual one for stacks. Let X be a nonempty set and let $\mathcal{P}^2(X)$ be the set of all collections of subsets in X. Then a collection $S \in \mathcal{P}^2(X)$ is a **stack**, if and only if,

$$A \in \mathcal{S} \text{ and } B \supset A \text{ implies } B \in \mathcal{S}.$$

A **grill** $G \in \mathcal{P}^2(X)$ is a stack such that $A \cup B \in \mathcal{G}$, if and only if, $A \in \mathcal{G}$ or $B \in \mathcal{G}$. From an application point-of-view, an important result by Thron [223] is that *Every grill is the union of ultrafilters contained in it.*

Example 9.1. In Fig. 9.1, assume that filter $\mathcal{F}_1$ and $\mathcal{F}_2$ are ultrafilters, then the grill $\mathcal{G}$ in Fig. 9.1 is defined by

$$\mathcal{G} = \mathrm{cl}(\mathcal{F}_1) \cup \mathrm{cl}(\mathcal{F}_2).$$

The fact that grills can have many different arrangements of ultrafilters is advantageous, especially when one looks for patterns in, for example, fabric or layers of fossils in rocks. We show this next. ∎

In keeping with an interest in discovering similar and dissimilar patterns, we introduce a descriptive union operation. Let X be a nonempty Efremovič space with proximity relation δ_Φ defined on X, $A, E \subset X$, Φ a set features of members of x, $\Phi(x)$ a feature vector that describes x, $\Phi(A), \Phi(E)$ a set of descriptions of members of A, E, respectively. Let $\Phi(A), \Phi(E)$ denote the set of descriptions of members of A, E, respectively. For any two nonempty sets A, then for $x \in A$, descriptive union is defined by

$$A \underset{\Phi}{\cup} E = \{y \in A \cup E : \Phi(y) \in \Phi(A) \text{ or } \Phi(y) \in \Phi(E)\}.$$

A **descriptive filter** $\mathcal{F}_\Phi$ is a nonempty collection of subsets in $\mathcal{P}^2(\Phi(X))$ that satisfies the following conditions.

($\boldsymbol{\mathcal{F}_\Phi}$.1) $A, B \in \mathcal{F}_\Phi$ implies $A \underset{\Phi}{\cap} B \in \mathcal{F}_\Phi$.
($\boldsymbol{\mathcal{F}_\Phi}$.2) $\Phi(B) \supset \Phi(A) \in \mathcal{F}_\Phi$ implies $\Phi(B) \in \mathcal{F}_\Phi$.
($\boldsymbol{\mathcal{F}_\Phi}$.3) $\varnothing \notin \mathcal{F}_\Phi$.

A descriptive filter is called a **descriptive round filter**, if and only if, any set F that belongs to $\mathcal{F}_\Phi$ is δ_Φ (proximal) neighbourhood of some G belonging to $\mathcal{F}_\Phi$. A descriptive round filter that is not contained in another descriptive round filter is called an **descriptive end**.

The δ_Φ-**closure** of any subset of the EF-proximity space X is the set of all points x in X such that the singleton set $\{x\}$ is δ_Φ close to A, *i.e.*,

$$\mathrm{cl}_{\delta_\Phi}(A) = \{x \in X : \{x\}\ \delta_\Phi\ A\}.$$

We also write $\mathrm{cl}_\delta(A_\Phi)$ to denote the closure of a set of descriptions of points. A point $x \in X$ is descriptively close to a subset A in X, if and only if, x belongs to the descriptive topological closure of A. We write

$$\mathrm{cl}_\delta\left(\bigcap_{\Phi}\mathcal{F}_i\right) \text{ to denote } \mathrm{cl}_\delta\mathcal{F}_1 \underset{\Phi}{\cap} \dots \underset{\Phi}{\cap} \mathrm{cl}_\delta\mathcal{F}_i \dots \underset{\Phi}{\cap} \mathrm{cl}_\delta\mathcal{F}_n,$$

and

$$\bigcup_{\Phi}\mathrm{cl}_\delta\left(\mathcal{F}_i\right) \text{ to denote } \mathrm{cl}_\delta\mathcal{F}_1 \underset{\Phi}{\cup} \dots \underset{\Phi}{\cup} \mathrm{cl}_\delta\mathcal{F}_i \dots \underset{\Phi}{\cup} \mathrm{cl}_\delta\mathcal{F}_n.$$

A **descriptive grill** $\mathcal{G}_\Phi$ is a stack such that

$$\left(A \underset{\Phi}{\cup} E\right) \in \mathcal{G}_\Phi \Longleftrightarrow \Phi(A) \in \mathcal{G}_\Phi \text{ or } \Phi(E) \in \mathcal{G}_\Phi$$

Theorem 9.7. *Every descriptive grill $\mathcal{G}_\Phi$ is the union of all descriptive ultrafilters contained in $\mathcal{G}_\Phi$.*

Proof. Let $\mathcal{G}_\Phi$ be a descriptive grill and let $\mathcal{F}_1, \dots, \mathcal{F}_n$ be descriptive filters in $\mathcal{P}^2(\Phi(X))$. From the descriptive counterpart of Theorem 2.2 in [223], we obtain

$$\mathcal{G}_\Phi = \mathrm{cl}_\delta(\mathrm{cl}_\delta(\mathcal{G}_\Phi)) = \mathrm{cl}_\delta\left(\bigcap_{\Phi} \mathcal{F}_i\right) = \bigcup_{\Phi} \mathrm{cl}_\delta(\mathcal{F}_i) = \bigcup_{\Phi} \mathcal{F}_i.$$

□

A **pattern** is a repetition in the arrangement of the members of a nonempty set X. Each pattern is a partially ordered set[1]. In this work, a filter is a collection of ascending partially ordered sets. In other words, a filter is a composition of patterns and is likened to textile patterns where one textile pattern is a subpattern in a larger textile pattern or where one subpattern is descriptively similar to another subpattern that is part of a larger pattern.

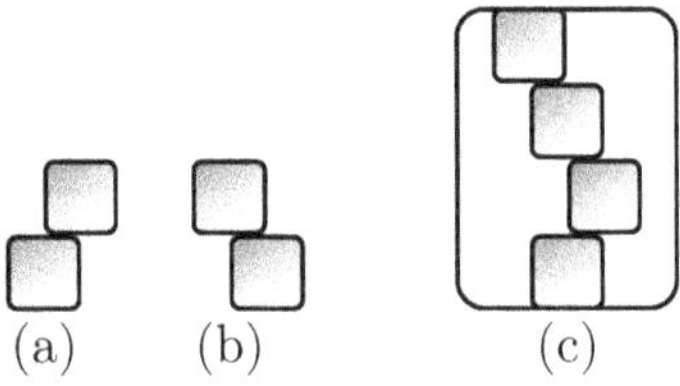

Fig. 9.2 Sample weave patterns

Example 9.2. Textile pattern Textile patterns are the focus of recent studies of weaving (see, *e.g.*, [219]). A fragment of a plain weave pattern and its reverse are shown in parts (a) and (b) in Fig. 9.2. These weave patterns are combined in a more complex pattern in part (c) in Fig. 9.2. A natural weave pattern can be detected in the arrangement of the surface

[1] For a, b, c in a nonempty set X, a partial ordering relation $\leq$ is reflexive ($a \leq a$), antisymmetric ($a \leq b$ and $b \leq a$ imply $a = b$), transitive ($a \leq b$ and $b \leq c$ imply $a \leq c$), and has the linearity property ($a \leq b$ or $b \leq a$).

fossil ammonoids on the surface of the rock shown in Fig. 9.3. In both cases, the weave patterns in Fig. 9.2 and the natural fossil pattern in Fig. 9.3 provide a basis for pattern-determined grills. ■

A **pattern-determined grill** is a grill such that the union of the filters of the grill contain subsets of a pattern. Let $\mathcal{G}_\Phi, \mathcal{G}'_\Phi$ denote pattern-determined descriptive grills, each containing a particular arrangement of the members of a set that belong to a pattern. Let $\mathcal{G}_\Phi$ contain ultrafilers $\mathcal{F}_i$ and $\mathcal{G}'_\Phi$ contain ultrafilers $\mathcal{F}'_i$. The patterns in the pair of grills are descriptively remote, if and only if,

$$(\bigcup \mathcal{G}_\Phi) \bigcap_\Phi (\bigcup \mathcal{G}'_\Phi) = \left(\bigcup_i \mathcal{F}_i\right) \bigcap_\Phi \left(\bigcup_j \mathcal{F}'_j\right) = \varnothing.$$

Otherwise, the patterns in the arrangement of the members in the grills are near. It is a straightforward task to quantify the degree of nearness of a pair of patterns (see, *e.g.*, [187; 86]).

The application of descriptive grills is illustrated in terms of the search for patterns in the arrangements of fossils on the surface of rocks found in the Rocky mountains (see Fig. 9.3).

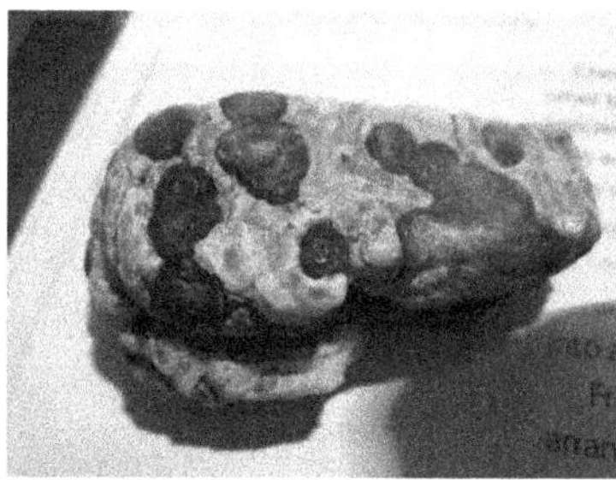

Fig. 9.3 Devonian era ammonoid fossil groupings

Example 9.3. Devonian Era Ammonoids Group Patterns. Ammonoids (also called ammonites) are derived the nautiloids, probably during the Devonian period, about 415 million years ago (MYA) [30]. Ammonoids were squidlike creatures that lived inside an external shell and are relatives of the modern squid and octopus. Ammonoids belong to the class of animals called cephalopods. They died out about 65 MYA. Ammonoid fossils are common in Cretaceous rocks found in the Alberta region of rocky mountains. The diversity of external shell form in ammonoids points to a wide range of adaptations in an aquatic environment.

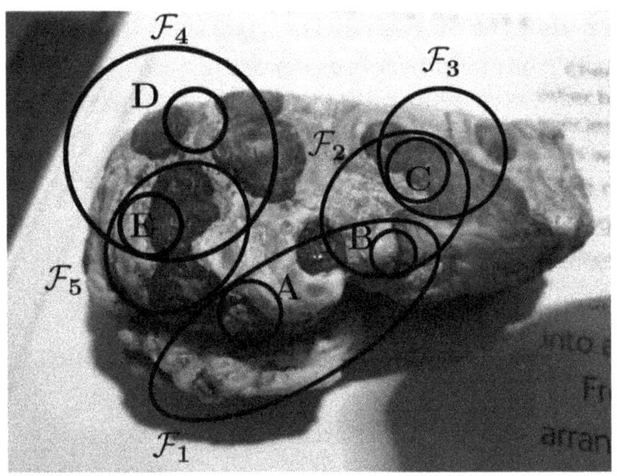

Fig. 9.4 Grills $\mathcal{G}_\Phi = \boldsymbol{F}_1 \cup \boldsymbol{F}_2 \cup \boldsymbol{F}_3$ and $\mathcal{G}'_\Phi = \boldsymbol{F}_4 \cup \boldsymbol{F}_5$

A number of groups of ammonoids are encrusted on the rock surface shown in Fig. 9.3. With an interest in the arrangements of groups of ammonoids on the same rock surface, two grills are superimposed on the rock surface in Fig. 9.4. In addition, to the roundness and gradient direction, the following texture features are useful in comparing the arrangement of picture elements in the study of fossil groups.

(**texture**.1) **Contrast**. Measure the intensity contrast $\mathfrak{C}$ between each pixel and its neighbour in an $n \times m$ image g, using

$$\mathfrak{C} = \sum_{i,j}^{m,n} |i - j| g(i,j).$$

(**texture**.2) **Correlation**. Measure the intensity correlation $\mathfrak{Co}$ between each pixel and its neighbour in an $n \times m$ image g, using

$$\mathfrak{Co} = \sum_{i,j}^{m,n} \frac{(i - \mu i)(j - \mu j)g(i,j)}{\sigma_i \sigma_j}.$$

(**texture**.3) **Local Entropy**. Expected value of a pixel intensity in a local image neighbourhood. Let X be a set of pixel intensities in an $n \times n$ neighbourhood of a pixel $p \in X$ in an image $\mathfrak{I}$, $Pr(x)$

the probability of $x \in X$. Then the local entropy $H(\{p\})$ is defined by

$$H(\{p\}) = \sum_{i=1}^{n} P(x_i) log_b \frac{1}{Pr(x_i)} = -\sum_{i=1}^{n} P(x_i) log_b Pr(x_i).$$

(**texture**.4) **Local range**. The local range $\mathfrak{r}$ of pixel p in an $n \times n$ neighbourhood X is defined by

$$\mathfrak{r} = max(X) - min(X).$$

(**texture**.5) **Edge frequency**. The ratio e_f of the number of edges n_e with the same gradient direction either in an entire image or in an image region relative to the total number of image edges n, where

$$e_f = \frac{n_e}{n}.$$

Then tailor the selection of features in the study of ammonoid groups and define the set of features Φ by

$$\Phi = [rd, \psi, \mathfrak{C}, \mathfrak{Co}, H, \mathfrak{r}, e_f],$$

where rd, ψ are shape roundness and pixel gradient direction, respectively. In Fig. 9.4, let $\mathcal{F}_1, \mathcal{F}_2, \mathcal{F}_3, \mathcal{F}_4, \mathcal{F}_5$ be ultrafilters. Then grills $\mathcal{G}_\Phi, \mathcal{G}'_\Phi$ are defined by

$$\mathcal{G}_\Phi = \mathcal{F}_1 \cup \mathcal{F}_2 \cup \mathcal{F}_3,$$
$$\mathcal{G}'_\Phi = \mathcal{F}_4 \cup \mathcal{F}_5.$$

It can be observed that there is considerable irregularity in the patterns formed by the arrangements of the ammonoids in Fig. 9.3. ■

Let $\mathcal{C}, \mathcal{D}$ be two nonempty collections of subsets in $\mathcal{P}^2(X)$. Recall that $\mathcal{C}$ is a refinement of $\mathcal{D}$ (denoted by $\mathcal{C} < \mathcal{D}$), if and only if, each set in $\mathcal{C}$ is contained in some set in $\mathcal{D}$.

Example 9.4. Let

$$\mathcal{C} = \{\text{patterns } (a), (b) \text{ in Fig. 9.2}\},$$
$$\mathcal{D} = \{\text{pattern } (c) \text{ in Fig. 9.2}\}.$$

Observe that $\mathcal{C}$ refines $\mathcal{D}$, since each set of points in $\mathcal{C}$ is contained in some set in $\mathcal{D}$, *i.e.*, $\mathcal{C} < \mathcal{D}$. ■

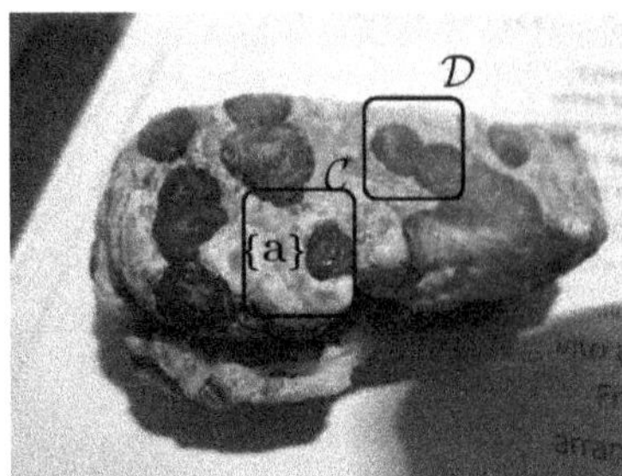

Fig. 9.5 Devonian era ammonoid fossil patterns

$\mathcal{C}$ is a **descriptive refinement** of $\mathcal{D}$ (denoted by $\mathcal{C} \underset{\Phi}{<} \mathcal{D}$), if and only if, the description of each set in $\mathcal{C}$ matches (effectively, is contained in) the description of some set in $\mathcal{D}$.

Example 9.5. The amonoid fossils on the surface of the rock in Fig. 9.5 can be viewed descriptively, where the shapes and colours of collections of single fossils that can be found repeated in the collections of fossil groupings. For simplicity, let $\mathcal{C}$ contain the single subset $A = \{a\}$, containing only the single fossil labelled a in Fig. 9.5. Let $\mathcal{D}$ consist of the fossils shown in Fig. 9.5. Assume that $\mathcal{C}, \mathcal{D}$ are disjoint. The shape and colour of the fossil in A is, for the most part, repeated in the fossils in $\mathcal{D}$. Hence, $\mathcal{C} \underset{\Phi}{<} \mathcal{D}$. ■

Both spatial as well as descriptive refinements are important in pattern recognition. Refinements can be found by comparing the descriptions of disjoint sets, since the members in disjoint sets can be descriptively identical.

Example 9.6. Let $\Phi = [\phi_g]$, where $\phi_g(x)$ returns the greylevel intensity of a picture element x in a picture X. In Fig. 9.1, $\Phi(A) = \Phi(B)$ for A, B in stack $\mathcal{S}$ and the description $\Phi(D)$ in filter $\mathcal{F}_2$ matches the descriptions of A and B in $\mathcal{S}$. Hence,

$$\mathcal{F}_2 \underset{\Phi}{<} \mathcal{S}, \text{ since } \Phi(D) = \Phi(A) = \Phi(B). \qquad ■$$

Let X be a nonempty EF-space X and let $\mathcal{A}, \mathcal{B} \in \mathcal{P}^2(X)$ denote collections of point patterns. Examples 9.5 and 9.6 illustrate two ways to define pattern nearness relations, namely,

Refinement-based pattern nearness relations:
(**refine**.1) $\mathcal{A}\ \delta_{<}\ \mathcal{B}$, if and only if, $\mathcal{A} < \mathcal{B}$ (pattern $\mathcal{A}$ refines pattern $\mathcal{B}$).
(**refine**.2) $\mathcal{A}\ \delta_{\underset{\Phi}{<}}\ \mathcal{B}$, if and only if, $\mathcal{A} \underset{\Phi}{<} \mathcal{B}$ ($\mathcal{A}$ descriptively refines $\mathcal{B}$).

A grill $\mathcal{G}$ is a **star refinement** of a grill $\mathcal{G}'$ (denoted by $\mathcal{G} *< \mathcal{G}'$), if and only if, for ultrafilter $F \in \mathcal{G}$ there is an ultrafilter $F' \in \mathcal{G}'$ such that $St(F, \mathcal{G}) \subset F'$, where

$$St(F, \mathcal{G}) = \bigcup \{E \in \mathcal{G} : E \cap F \neq \varnothing\}.$$

A pattern-determined grill $\mathcal{G}$ is remote from a second pattern-determined grill $\mathcal{G}'$, if and only if, $\mathcal{G} *\not< \mathcal{G}'$, *i.e.*, $St(F, \mathcal{G}) \not\subset F'$ for any $F' \in \mathcal{G}'$. On the other hand, the pattern represented by $\mathcal{G}$ resembles (is near) the pattern represented by $\mathcal{G}'$, if and only if, $St(F, \mathcal{G}) \subset F'$ for some $F' \in \mathcal{G}'$. The number of nonempty $St(F, \mathcal{G}) \subset F'$ for $F' \in \mathcal{G}'$ serves to quantify the extent of the resemblance between the patterns represented by $\mathcal{G}, \mathcal{G}'$.

Example 9.7. Since $\mathcal{G}, \mathcal{G}'$ are disjoint pattern determined grills in Fig. 9.4, then $\mathcal{G} *\not< \mathcal{G}'$. This is not the case, if we consider **descriptive star refinement** denoted by $\mathcal{G} *\underset{\Phi}{<} \mathcal{G}'$. That is, the two sets of ammonoids are descriptively near, even though the two sets spatially remote. In Fig. 9.4, it can be observed that the edge pattern in C in filter F_2 in grill $\mathcal{G}$ is descriptively near the pattern in E in filter F_2 in grill $\mathcal{G}'$, *i.e.*,

$$St(C, \mathcal{G}') \underset{\Phi}{\cap} E \neq \varnothing.$$

However, since C and E are spatially remote in Fig. 9.4, one can observe that

$$St(C, \mathcal{G}') \cap E = \varnothing.$$

In fact, there is no subpattern in $\mathcal{G}'$ that is near a subpattern in $\mathcal{G}$. ■

Let X be a nonempty EF-space and let $\mathcal{G}, \mathcal{G}'$ be pattern-determined grills in $\mathcal{P}^2(X)$. Example 9.7 suggests to two additional nearness relations based on star refinement, namely,

Star refinement-based pattern nearness relations:
(**star**.1) In an approach similar to defining an EF-proximity relation in [160], define

$$A \underset{*<}{\delta} B, \text{ provided } St(A, \mathcal{G}') \cap B \neq \varnothing, \text{ for some } A \in \mathcal{G}, B \in \mathcal{G}'.$$

(**star**.2) In terms of the descriptive nearness of pattern-determined filters in grills, define

$$A \underset{*<_\Phi}{\delta} B, \text{ provided } St(A,\mathcal{G}') \underset{\Phi}{\cap} B \neq \emptyset, \text{ for some } A \in \mathcal{G}, B \in \mathcal{G}'.$$

The above observations lead to the following result.

Theorem 9.8. *A filter subpattern A in one grill $\mathcal{G}$ is descriptively near a subpattern B in a filter in a grill $\mathcal{G}'$, if and only if, $St(A,\mathcal{G}')$ has a descriptive non-empty intersection with B in $\mathcal{G}'$.*

Proof. See Problem 9.9.7 □

With appropriate choices of features, it can be verified that this is the case in Example 9.7. Also observe that the definition of the EF-proximity relations (refine.2) and (refine.3) depend only on set operations. Hence, these star refinement-based proximity relations have utility across a wide spectrum of applications that are not dependent on a metric. A metric becomes useful, if we need to quantify the degree of nearness of patterns. This is essentially what is done in a number of recent studies (see, *e.g.*, [183; 186; 177]).

9.9 Problems

(**9.9**.1) Show that $\mathcal{G}$ is a grill, if and only if, it is union of ultrafilters.
(**9.9**.2) Show that $\mathcal{F}$ is a filter, if and only if, $\mathcal{G} = c(\mathcal{F})$ is a grill.
(**9.9**.3) Show that $c(c(\mathcal{F})) = \mathcal{F}$ and $c(c(\mathcal{G})) = \mathcal{G}$.
(**9.9**.4) Show that a grill $\mathcal{G}$ is an ultrafilter, if and only if, $c(\mathcal{G}) = \mathcal{G}$.
(**9.9**.5) Let X be a dense subset of a topological space αX. For each point $p \in \alpha X$, show that

$$\sigma(p) = \{E \subset X : p \in \text{cl}_{\alpha X} E\}$$

is a grill.
(**9.9**.6) Let X be a dense subset of a T_1 space αX. Show that

> Subsets $A, B \subset X$ are near, if and only if, the closures of A, B in αX intersect.

gives an L-proximity on X.
(**9.9**.7) Prove Theorem 9.8.

Chapter 10

Extensions of Continuous Functions: Taimanov Theorem

An important problem in topology is to find necessary or sufficient conditions for the existence of a continuous extension of a continuous function from dense subspaces. While most known results are special, A.D. Taimanov proved a beautiful *general* topological result. In this chapter, it is shown that the proximal avatar of Taimanov's result generalises all special cases.

10.1 Proximal Continuity

Let X, Y be dense in T_1 topological spaces $\alpha X, \lambda Y$, respectively. Let $f : X \to Y$ be a continuous function. An important problem concerns the necessary or sufficient conditions[1] for the existence of $\overline{f} : \alpha X \to \lambda Y$, a continuous extension of f. That is, for each $x \in X$, $\overline{f}(x) = f(x)$. Let δ_1, δ_2 be L-proximities on X, Y, respectively induced by fine L-proximities on αX and λY. That is,

$$A \ \delta_1 \ B, \text{ if and only if, their closures in } \alpha X \text{ intersect,}$$
$$C \ \delta_2 \ D, \text{ if and only if, their closures in } \lambda Y \text{ intersect.}$$

Suppose $\overline{f}$ is continuous and $A \ \delta_0 \ B$ in αX. Then $\mathrm{cl}\, A, \mathrm{cl}\, B$ intersect in αX. This implies $\overline{f}(\mathrm{cl}\, A), \overline{f}(\mathrm{cl}\, B)$ intersect in λY. Since $\overline{f}$ is continuous for each subset $E \subset \alpha X$, then $\overline{f}(\mathrm{cl}\, E) \subset \mathrm{cl}(\overline{f}(E))$. Hence, $\mathrm{cl}(\overline{f}(A), \mathrm{cl}(\overline{f}(B)$ intersect and so $\overline{f}(A), \overline{f}(B)$ are near in λY. So, a necessary condition

[1] Here, the expression 'necessary *or* sufficient condition' means a necessary condition is true, or sufficient condition is true, or both conditions are true. In general, *or* is known as an **inclusive disjunction** in mathematical logic.

for the continuity of $\overline{f}$ is that it should be proximally continuous and this implies the proximal continuity of $f : (X, \delta_1) \to (Y, \delta_2)$.

10.2 Generalised Taimanov Theorem

The first step toward a generalization of the Taimanov theorem is to note that, for each $p \in \alpha X$,

$$\sigma^p = \{E \subset X : p \in \mathrm{cl}_{\alpha X} E\} \text{ is a bunch in } (X, \delta_1).$$

Notice that if $p \in X$, then $\sigma^p = \sigma_p$ (point cluster). Thus, there is a map $\psi : \alpha X \to \Sigma_X$ (the space of all bunches in (X, δ_1)). To show that ψ is continuous, observe that a typical member of the base for closed sets in the Wallman topology on Σ_X is

$$E^* = \{\sigma^p : p \in \mathrm{cl}_{\alpha X} E\},$$

where $E \subset X$. Since $E^* = (\mathrm{cl}\, E)^*$ and $p \in \mathrm{cl}_{\alpha X} E$, it follows that $\psi^{-1}(E^*) = \mathrm{cl}\, E$, which is closed in αX.

Lemma 10.1. *Let X be dense in a T_1 space αX and let δ_1 be an L-proximity on X, defined by $A\ \delta_1\ B$ in X, if and only if, their closures in αX intersect. Let Σ_X denote the space of all bunches in (X, δ_1) with the Wallman topology. Then the map $\psi : \alpha X \to \Sigma_X$ given by $\psi(p) = \sigma^p$, is continuous.*

Step two in generalising Taimanov's theorem. Given a proximally continuous function $f : (X, \delta_1) \to (Y, \delta_2)$, construct a continuous function $f_\Sigma : \Sigma_X \to \Sigma_Y$, where the spaces of bunches Σ_X, Σ_Y are assigned the Wallman topology. For $\sigma \in \Sigma_X$, define

$$f_\Sigma(\sigma) = \left\{E \subset Y : f^{-1}(\mathrm{cl}\ E) \in \sigma\right\}.$$

We verify that $f_\Sigma(\sigma) \in \Sigma_X$.

Proof.

(i) Obviously, $A \in f_\Sigma(\sigma)$, if and only if, $\mathrm{cl}\, A \in f_\Sigma(\sigma)$.

(ii) Observe

$$
\begin{aligned}
(A \cup B) \in f_\Sigma(\sigma) &\Leftrightarrow f^{-1}(\mathrm{cl}(A \cup B) \in \sigma \\
&\Leftrightarrow f^{-1}(\mathrm{cl}(A) \cup \mathrm{cl}(B)) \in \sigma \\
&\Leftrightarrow f^{-1}(\mathrm{cl}(A)) \cup f^{-1}(\mathrm{cl}(B)) \in \sigma \\
&\Leftrightarrow f^{-1}(\mathrm{cl}(A)) \in \sigma \text{ or } f^{-1}(\mathrm{cl}(B)) \in \sigma \\
&\Leftrightarrow A \in f_\Sigma(\sigma) \text{ or } B \in f_\Sigma(\sigma).
\end{aligned}
$$

(iii) $A, B \in f_\Sigma(\sigma) \Rightarrow f^{-1}(\mathrm{cl}(A)) \in \sigma$ and $f^{-1}(\mathrm{cl}(B)) \in \sigma \Rightarrow f^{-1}(\mathrm{cl}(A))\ \delta_1\ f^{-1}(\mathrm{cl}(B))$. Since f is proximally continuous,

$$\mathrm{cl}(A)\ \delta_2\ \mathrm{cl}(B) \Rightarrow A\ \delta_2\ B.$$

f_Σ is continuous, since, for $E \subset Y$, $f^{-1}(E^*) = (f^{-1}(E))^*$. □

This leads to

Lemma 10.2. *Let X, Y be dense in T_1 spaces $\alpha X, \lambda Y$, respectively, and let X, Y be assigned subspace L-proximities δ_1, δ_2 induced by the fine proximity δ_0 on αX and λY. Let $f : X \to Y$ be a proximally continuous function. Also, let Σ_X, Σ_Y be spaces of all bunches in $(X, \delta_1), (Y, \delta_2)$, respectively, with the Wallman topology. Then the map $f_\Sigma : \Sigma_X \to \Sigma_Y$ given by*

$$f_\Sigma(\sigma) = \{E \subset Y : f^{-1}(\mathrm{cl}\, E) \in \sigma\}$$

is continuous (see arrow diagram in Fig. 10.1).

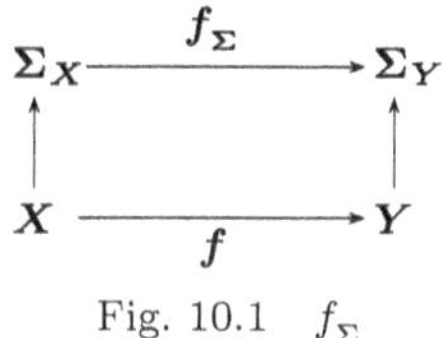

Fig. 10.1 f_Σ

Identifying $x \in X, y \in Y$ with σ_x, σ_y, the point clusters f_Σ can be considered an extension of f.

The third and final step in generalising Taimanov's theorem is to go from Σ_Y to λY. For this purpose, we take λY to be normal T_4, which makes the fine L-proximity δ_0 EF. In this case, every maximal bunch is a cluster. If $\beta \in \Sigma_Y$, then it is contained in a cluster σ. If $\mathcal{L}$ is an ultrafilter in β, then $\delta_2(\mathcal{L})$ is a cluster in (Y, δ_2). Since

$$\mathcal{L} \subset \beta \subset \delta_2(\mathcal{L}) = \sigma,$$

every bunch is contained in a unique cluster.

Thus, there is a map $\theta : \Sigma_Y \to Y^*$, the space of all clusters in (Y, δ_2) with Wallman topology. The map θ is continuous, since, for each member $E^*(E \subset Y)$ of the closed base in Y^*,

$$\theta^{-1}(E^*) = \{\beta \in \Sigma_Y : E \in \beta\}, \text{a member of the closed base for } \Sigma_Y.$$

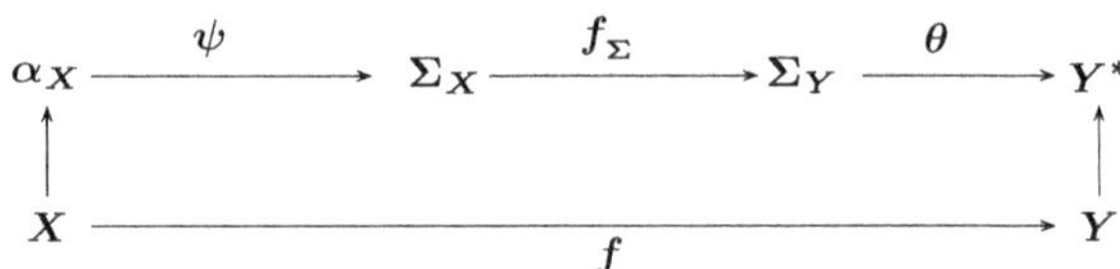

Fig. 10.2 $\alpha X \to \Sigma_X \to \Sigma_Y \to Y^*$

Lemma 10.3. *Let Y be dense in a normal space λY and let δ_2 be the EF-proximity on X induced by the fine proximity δ_0 on λY. Let Y^* be the space of all clusters in (Y, δ_s). Then the map $\theta : \Sigma_Y \to Y^*$ that takes each bunch into a cluster is continuous. Further, if λY is compact Hausdorff, then λY is homeomorphic to Y^*, the Smirnov compactification of Y. Hence, identifying λY with Y^*, θ may be considered a map from Σ_Y to λY.*

Combining Lemmas 10.1, 10.2 and 10.3, we obtain the diagram in Fig. 10.2, followed by the result in Theorem 10.1.

Theorem 10.1. Generalised Taimanov Theorem
Let X be dense in a T_1 space αX and let δ_1 be the L-proximity on X defined by $A\ \delta_1\ B$ in X, if and only if, their closures in αX intersect. Let (Y, δ_2) be an EF-proximity space and let Y^* be its Smirnov compactification. Then a function $f : X \to Y$ has a continuous extension $\overline{f} : \alpha X \to Y^*$, if and only if, f is proximally continuous.

Notice that $\overline{f}$ in Theorem 10.1 is a composition of three continuous functions, namely, ψ, f_Σ, θ. The original result by Taimanov is given in Theorem 10.2.

Theorem 10.2. Taimanov Theorem
Let X be dense in a T_1 space αX and let Y be compact Hausdorff. Then $f : X \to Y$ has a continuous extension $\overline{f} : \alpha X \to Y$, if and only if, for every

pair of disjoint closed sets $A, B \subset Y$, closures of $f^{-1}(A)$ and $f^{-1}(B)$ in αX are disjoint.

Corollary 10.1. *Let $(X, \delta_1), (Y, \delta_2)$ be Efremovič proximity spaces and let X^*, Y^* be their Hausdorff compactifications. Let X, Y be considered dense subspaces of their compactifications by identifying spaces with their point clusters. Then the following are equivalent.*
(a) $f : (X, \delta_1) \to (Y, \delta_2)$ is proximally continuous.
(b) f has a continuous extension to $\overline{f} : X^* \to Y^*$.

10.3 Comparison of Compactifications

It is obvious from Corollary 10.1 that if X is a Tychonoff space, then there is an ordered bijection from the set of compatible Efremovič proximities on X to its Hausdorff compactifications, *i.e.*, a finer proximity gives a larger compactification. The functionally separating proximity δ_F is given by $A \underline{\delta}_F B$, if and only if, A and B can be separated by a continuous function

$$f : X \to [0,1] \text{ with } f(A) = 0 \text{ and } f(B) = 1.$$

Compactification corresponding to δ_F is called Stone-Čech and is the largest Hausdorff compactification of X. In general, there is no smallest Hausdorff compactification. It exists and is the Alexandroff one-point compactification, if and only if, X is locally compact.

10.4 Application: Topological Psychology

Topological psychology was introduced by K. Lewin in 1936 [127]. Lewin was an experimental psychologist and a pioneer in social psychology. Lewin's studies initially focused on medicine and philosophy, then biology, and finally psychology. Early on, Lewis foresaw that psychology could beneficially utilize topology and the dynamics of social interaction viewed in the context of topological structures as a means of explaining and elucidating human behavior [73]. This section considers the topological implications of mathematical psychology in terms of the following topics.

(ψ.1) Generation of space by social interaction [168].
(ψ.2) Blind watchmaker selects best adapted generations of genes [48; 168].

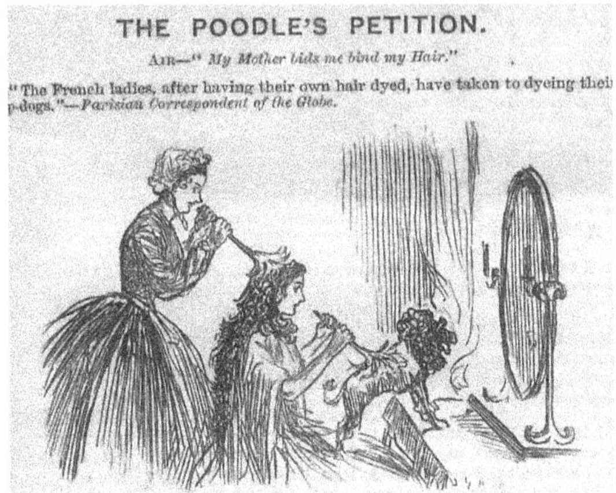

Fig. 10.3 Social behaviour, *Punch*, 1869

(ψ.3) Behaviour as a change in the psychological environment [127].
(ψ.4) EF-proximity space for social interaction (*cf.* [241; 196]).

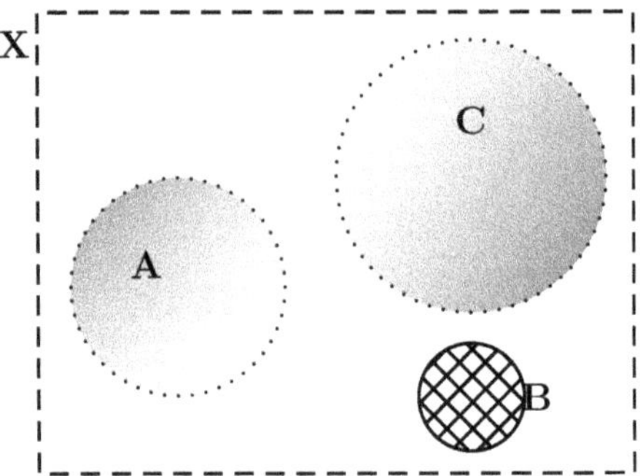

Fig. 10.4 Spatially remote but descriptively near sets in X

Example 10.1. (ψ.1): Generation of Space.
A quaint example of the **generation of space** by social interaction can be observed in the image[2] in Fig. 10.3, where a mother is braiding her daughter's hair while the daughter, in turn, is braiding her poodle's hair.

[2] A public domain image from *Punch, or the London Charivari*, vol. 57, July 1869.

The interaction in the Punch image is reminiscent of the symbiotic architecture of bacterial genomes adding layers upon layers, forming interacting neighbourhoods [168].

The *habitat evolutif* architectural designs of the late 1960s reflect this idea, where the movements and perceptions of dwellers acted as modalities of extensions, resulting in growing networked arrangements of physical relations [7]. An architecture evolves through aggregation, interconnection and juxtaposition of elements that fit together to form a habitable whole. This leads to a networked environment.

The mathematical model of a networked environment can be found in EF-proximity space, where spatially disjoint sets are brought together by their appearance in collections of descriptively near and descriptively remote sets. In fact, it is appearance such as the display of colour or rhythmic sounds by animals that are designed to attract or repel others. In Fig. 10.4, open sets A and B are spatially remote but descriptive near. By contrast, the closed set B in Fig. 10.4 is both spatially remote as well as descriptively remote from sets A and B, *i.e.*,

(1) **Spatially remote & descriptively near**: $A \underline{\delta} C$, A and δ_{Φ} C,
(2) **Spatially remote**: $B \underline{\delta} A$, and $B \underline{\delta} C$,
(3) **Spatially remote & descriptively remote**: $B \underline{\delta}_{\Phi} A$, and $B \underline{\delta}_{\Phi} C$.

> **Natural selection** is defined to be a continuous scrutinizing, throughout the world, every variation [in behaviour, in structure], even the slightest; rejecting that which is bad, preserving and adding up all that is good; silently and insensibly working, whenever and wherever opportunity offers, at the improvement of each organic begin in relation to its organic and inorganic conditions of life [46]. In social animals, [natural selection] will adapt the structure of each individual for the benefit of the community [46].

The configuration of sets in the environment represented by the open set X in Fig. 10.4 provides a succinct way to represent complex social interactions. The EF-space in Fig. 10.4 also provides a model for communities of individuals described by N. Tinbergen [225]. A detailed analysis of the learning that arises from social interaction is given in [181]. ■

Example 10.2. (ψ.2) Blind Watchmaker.
The **blind watchmaker** analogy is introduced by R. Dawkins [48] to describe natural selection, a blind, unconscious, automatic process discovered by Darwin.

For example, in the case of the social behaviour of animals represented by A, C in Fig. 10.4, scrutinizing each other, the EF-proximity relation between A and C will adaptively change in appearance. ∎

Example 10.3. (ψ.3) Topological Psychology.
Lewin defined **behaviour** as change in a psychological environment and introduced what he called field theory. Let B, S denote a set mental events and set of situations, respectively, where as a function $f : S \to B$ such that

$$f(s) = b, \text{ for } s \in S, b \in B.$$

In keeping with Darwin's natural selection as a form of scrutinizing by organisms in their environment, a set of situations S is decomposed into a set of persons P and a set of environments E. From this decomposition, social behaviour is then represented by a function $g : P \times E \to B$ defined by

$$g(p, e) = b, \text{ for } p \in P, e \in E, b \in B.$$

The function g provides a useful model of the social interaction between persons and their environment. ∎

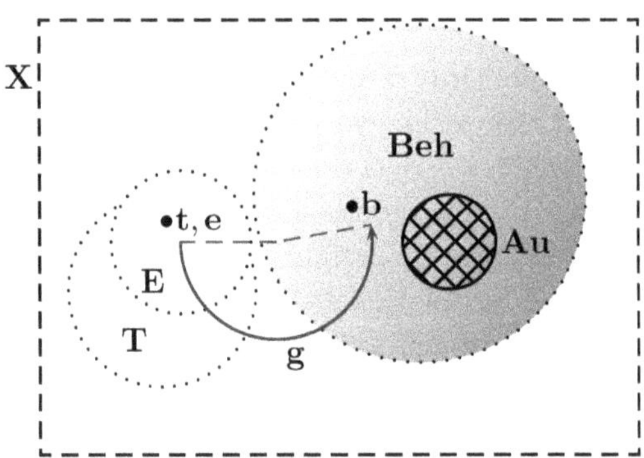

Fig. 10.5 Teacher-environment-student-audience EF-space

Example 10.4. (ψ.4) Social Interaction.
As a result of social interaction, the appearance of organisms in the environment tend to change either quickly or slowly and either dramatically

or almost imperceptibly. A recent example of social interaction can be observed in the continuous interaction between musician and pupil with Ravi Shankar teaching his daughter Anoushka to play the sitar[3]. Here Ravi Shankar mimes sitar sounds while his daughter listens or the daughter plays the sitar in short bursts, mimicking the sounds she has heard, while her father listens. The social interaction between teacher and student in an immediate, local environment (stage, instruments, sound equipment, lighting) in the presence of an audience can be viewed in the context of an EF-proximity space. In Fig. 10.5, let T, E, Beh, Au denote a set of teachers, set of surroundings and conditions that constitute the *local* environment for the teacher, set of student behaviours resulting from social interaction, and set of listeners that make up an audience, respectively. The assumption made here is that the behaviour of the audience (Au) is influenced in a vicarious way by the behaviour of the students. For simplicity, the relationship between the local environment for student behaviours is not shown in Fig. 10.5. This is an EF-proximity space, where the EF-proximity property is satisfied spatially as well as descriptively. That is, let

$$\begin{aligned}
T &= \{\text{Teachers}\} \\
E &= \{\text{Surroundings and conditions for teacher}\}, \\
Beh &= \{\text{student behaviours}\}, \\
Au &= \{\text{listeners in audience}\}, \\
X &= T \cup E \cup Beh \cup Au, \\
\Phi &= \{\text{probe functions representing features of } x \in X\}, \\
(X, \delta) &= \text{Spatial EF space}, \\
(X, \delta) &= \text{Descriptive EF space}.
\end{aligned}$$

Then, in terms of social interaction, observe that

$$\begin{aligned}
T \ \underline{\delta} \ Au \Rightarrow T \ \underline{\delta} \ Beh \text{ and } Au \ \underline{\delta} \ (Beh)^c \text{ for } T \subset X, \\
T \ \underline{\delta}_{\Phi} \ Au \Rightarrow T \ \underline{\delta}_{\Phi} \ Beh \text{ and } Au \ \underline{\delta}_{\Phi} \ (Beh)^c \text{ for } T \subset X.
\end{aligned}$$

Within the EF-proximity space described in Fig. 10.5, the fact that the behaviour of a student $b \in Beh$ depends on a teacher $t \in T$ and local environment $e \in E$ is represented by the mapping g from (t,e) to b. Social interaction leads to what has been described as cooperative coevolution [241; 196]. ■

[3] `http://www.youtube.com/watch?v=igDsu5QWhpo&feature=related`

10.5 Problems

Let X be dense in a T_1 space αX and let δ be the L-proximity on X defined by

$$A\ \delta\ B \Leftrightarrow \mathrm{cl}\,A, \mathrm{cl}\,B \text{ in } \alpha X \text{ intersect.}$$

Let Σ_X be the space of all bunches in (X, δ).

(**10.5**.1) Show that, for each $x \in \alpha X$, $\sigma^x = \{E \subset X : x \in \mathrm{cl}_{\alpha X} E\}$ is a bunch in (X, δ) and, for $x \in X$, $\sigma^x = \sigma_x$ (the point cluster).

(**10.5**.2) If αX is compact Hausdorff, show that σ^x is a cluster in (X, δ) and, hence, αX is homeomorphic to X^* (the space of clusters). **Note**: This shows the importance of a compact Hausdorff space, *i.e.*, such a space can be reconstructed from each of its dense subspaces.

(**10.5**.3) For $\sigma \in \Sigma_X$ and $\mathcal{E} \subset \Sigma_X$, define $\sigma \in \mathrm{cl}\,E$, if and only if, whenever $A \subset X$ belongs to each member of $\mathcal{E}$ (A absorbs $\mathcal{E}$), then $A \in \sigma$. Show that cl is a Kuratowski closure operator on Σ_X that generates the Wallman topology.

(**10.5**.4) Prove directly from Corollary 10.1 that Alexandroff one-point compactification of a locally compact Hausdorff space X is the smallest Hausdorff compactification.

(**10.5**.5) Suppose A, B are two closed subsets of a Tychonoff space X that are far in every compatible EF-proximity on X. Show that at least one of A, B is compact.

(**10.5**.6) If a Tychonoff space X is not locally compact, the Alexandroff proximity δ_A defined by

$$E\ \delta_A\ F \Leftrightarrow \mathrm{cl}\,E, \mathrm{cl}\,F \text{are disjoint, and both are non-compact}$$

is not EF. This shows that the Alexandroff one-point compactification space is not Hausdorff but is the infimum of all Hausdorff compactifications.

(**10.5**.7) Let $(X, \delta), (X, \delta_\Phi)$ be spatial and descriptive EF spaces, respectively. Give an example such that

(a) $(X, \delta), (X, \delta_\Phi)$ represent a networked environment containing spatial near as well as spatially remote sets and descriptively near as well as descriptively remote sets,

(b) $(X, \delta), (X, \delta_\Phi)$ represent natural selection,

(c) $(X, \delta), (X, \delta_\Phi)$ represent psychological environment,

(d) $(X, \delta), (X, \delta_\Phi)$ represent social interaction.

Chapter 11

Metrisation

11.1 Structures Induced by a Metric

Earlier, we generalized metric spaces to uniform spaces, proximity spaces and topological spaces. Three interesting problems arise, known as *metrisation problems*, *i.e.*, find necessary and sufficient conditions for

(**structure**.1) uniformity induced by a metric,
(**structure**.2) proximity induced by a metric,
(**structure**.3) topology induced by a metric.

These problems are studied in this chapter. We begin with a brief study of semi-metric spaces that form a bridge between a topological space and a metric space.

A **semi-metric** d on a T_1 space X is a function on $X \times X$ to $[0, \infty)$ such that d satisfies
(**semimetric**.1) $d(x, y) = d(y, x)$,
(**semimetric**.2) $d(x, y) = 0 \Leftrightarrow x = y$,
(**semimetric**.3) $x \in \text{cl}\, A \subset X \Leftrightarrow inf\{d(y, x) : y \in A\} = 0$.

First, notice that the range of semi-metric d need not be $[0, \infty)$. For instance, a set such as $\left\{\frac{1}{n} : n \in \mathbb{N}\right\} \cup \{0\}$ suffices, which shows that *countability* plays an important role in semi-metric spaces. In the absence of the triangle inequality in a semi-metric space, the 'open' ball

$$B(x, \varepsilon) = \{y \in X : d(x, y) < \varepsilon \text{ for } \varepsilon > 0\}$$

need not be open as in the case of a metric, but the family $\{B(x, \frac{1}{2^n}) : n \in \mathbb{N}\}$ is a countable base at $x \in X$. Let

$$\mathcal{U}_n = \bigcup\left\{B(x, \frac{1}{n}) : x \in X\right\}$$

be a cover of X. Then $\{\mathcal{U}_{n^+} : n \in \mathbb{N}\}$ is a countable family of covers of X and let

$$St(x,\mathcal{U}_n) = \bigcup\{U \in \mathcal{U}_n : x \in U\}.$$

Then $\{St(x,\mathcal{U}_n) : n \in \mathbb{N}\}$ is a neighbourhood base at $x \in X$.

Conversely, suppose we are given a countable family $\{\mathcal{U}_n : n \in \mathbb{N}\}$ of covers of X and $\{St(x,\mathcal{U}_n) : n \in \mathbb{N}\}$ is a neighbourhood base at $x \in X$. We may suppose here and in the sequel that $\mathcal{U}_{n+1}$ is a refinement of $\mathcal{U}_n$. If not, we can replace $\{\mathcal{U}_n\}$ by an equivalent family $\{\mathcal{V}_{n+1}\}$ defined by

$$\mathcal{V}_{n+1} = \{E \cap F : E \in \mathcal{U}_{n+1}, F \in \mathcal{U}_n\}.$$

Semi-metric: Now we can construct a semi-metric d defined by

$$(***) \qquad d(x,y) = \begin{cases} inf\{\frac{1}{2^n}\}, & \text{if } y \in St(x,\mathcal{U}_n), \\ 1, & \text{otherwise.} \end{cases}$$

Theorem 11.1. *A T_1 space X is semi-metrisable, if and only if, it has a countable family of covers $\{\mathcal{U}_n : n \in \mathbb{N}\}$ with $\mathcal{U}_{n+1}$ a refinement of $\mathcal{U}_n$ and $\{St(x,\mathcal{U}_n) : n \in \mathbb{N}\}$ is a neighbourhood base at each $x \in X$.*

An important special case occurs when the family $\{\mathcal{U}_n : n \in \mathbb{N}\}$ in Theorem 11.1 consists of open covers of X. In that case, the space X is called **developable**. It has been shown that a T_1 space X is developable, if and only if, the semi-metric **(***)** is upper semi-continuous.

11.2 Uniform Metrisation

We now consider the metrisation of a uniform space that has been considered in §6.1. Recall that uniformity on a nonempty set X is a family $\mathcal{D}$ of pseudometrics satisfying the following properties.

(**uniformity**.1) $d, d' \in \mathcal{D} \Rightarrow max\{d, d'\} \in \mathcal{D}$.

(**uniformity**.2) If a pseudometric e satisfies the condition

$$\forall \varepsilon > 0, \exists d \in \mathcal{D}, \delta > 0 \text{ such that } \forall x, y \in X, \quad d(x,y) < \delta \Rightarrow e(x,y) < \varepsilon,$$

then $e \in \mathcal{D}$.

(**uniformity**.3) $x, y \in X, x \neq y \Rightarrow \exists d \in \mathcal{D} : d(x,y) > 0$.

The pair $(X, \mathcal{D})$ is called a **uniform space**. A subfamily $\mathcal{B}$ of a uniformity $\mathcal{D}$ is called a **base** for $\mathcal{D}$, if and only if, for each $e \in \mathcal{D}$ and for each $\varepsilon > 0$, there is a $d \in \mathcal{B}$ and $\delta > 0$ such that for all $x, y \in X$, $d(x,y) < \delta$ implies $e(x,y) < \varepsilon$.

Notice that if d is a pseudometric on X, then it is uniformly equivalent to pseudometric e, bounded by 1 as defined by

$$e(x,y) = min\{d(x,y), 1\}.$$

Suppose uniformity $\mathcal{D}$ has a countable base $\mathcal{B} = \{d_n : n \in \mathbb{N}\}$ of pseudometrics bounded by 1. Using the well-known Weierstrass M-test, it follows that

$$d(x,y) = \Sigma\left\{\frac{1}{2^n} d_n(x,y) : n \in \mathbb{N}\right\}$$

is a pseudometric on X. It is easily verified that d is a actually a compatible metric on X. Thus, we obtain the **Alexandroff-Urysohn uniform metrisation** theorem.

Theorem 11.2. *Uniformity on X is metrisable, if and only if, it has a countable base.*

11.3 Proximal Metrisation

Let (X, δ) be an EF-proximity space and $(a_n), (b_n)$ be sequences in X. Then we define

$$\boxed{(a_n) \approx (b_n) \Leftrightarrow \text{for each infinite subset } M \subset \mathbb{N}, \{a_n : n \in M\}\ \delta\ \{b_n : n \in M\}.}$$

It is easily verified that $\approx$ is an equivalence relation. A cover $\mathcal{U}$ of X is called **admissible**, if and only if, $(a_n) \approx (b_n)$ implies $b_n \in St(a_n, \mathcal{U})$ for some $n \in \mathbb{N}$. If $\mathcal{F}$ is a family of all admissible covers, then it follows that $A\ \delta\ B$ in X, if and only if, for each $\mathcal{U} \in \mathcal{F}$, $B \cap St(A, \mathcal{U}) \neq \varnothing$. It follows that $\{St(x, \mathcal{U}_n : n \in \mathbb{N}\}$ is a neighbourhood base at $x \in X$ and the semi-metric d as defined by (***) is compatible with the topology of X. In fact,

$$A\ \delta\ B \text{ in } X \Leftrightarrow\ D(A,B) = inf\{d(a,b) : a \in A, b \in B\} = 0.$$

As can be guessed from Section §11.2 on uniform metrisation, the family $\mathcal{F}$ of all admissible covers must have a countable base $\{\mathcal{V}_n \in \mathcal{F} : n \in \mathbb{N}\}$. To show that d is a metric, we need only prove that d satisfies the triangle inequality. If d were a metric, then we would have the triangle inequality equivalent, for each $x \in X$,

$$B(B(x, \frac{1}{2^{n-1}}), \frac{1}{2^{n-1}}) \subset B(x, \frac{1}{2^n}).$$

If this is not true, then there exists $p \in \mathbb{N}$ and sequences $(a_n), (b_n), (c_n)$ for $n \in \mathbb{N}$ such that, for all $n > p$,

$$c_n \in B(b_n, \frac{1}{2^p}), b_n \in B(a_n, \frac{1}{2^p}), \text{ but } c_n \notin B(a_n, \frac{1}{2^n}).$$

This means

$$(a_n) \approx (b_n), (b_n) \approx (c_n) \text{ but } (a_n) \not\approx (c_n).$$

This contradicts the fact that $\approx$ is an equivalence relation. This leads to the **Efremovič-Švarc proximal metrisation** result in Theorem 11.3.

Theorem 11.3. *An EF-proximity space (X, δ) is metrisable, if and only if,*

(*i*) *$A\ \delta\ B$ in $X \Leftrightarrow$ there are sequences $(a_n) \in A, (b_n) \in B : (a_n) \approx (b_n)$.*
(*ii*) *The family of admissible covers has a countable base.*

So, in studying topological metrisation of a T_1 space of X, we need to consider which conditions are necessary and sufficient for a T_1 semi-metrisable space to be metrisable.

11.4 Topological Metrisation

Theorem 11.1 gives a characterization of a semi-metrisable topological space. So, in studying topological metrisation of a T_1 semi-metrisable space X, we need to consider which conditions are necessary and sufficient for a T_1 semi-metrisable space to be metrisable. To identify these likely conditions, we look at a metric space (X, d), a point $x \in X$, and sequences $(x_n), (y_n), (z_n)$. The following results stem from the triangle inequality.

(i) $d(x_n, y_n) \to 0, d(y_n, z_n) \to 0 \Rightarrow d(x_n, z_n) \to 0$.
(ii) $d(x_n, x) \to 0, d(x_n, y_n) \to 0 \Rightarrow d(y_n, x) \to 0$.
(iii) δ_d is an EF-proximity, where $A\ \delta_d\ B \Leftrightarrow D(A, B) = 0$.

(iv) If A is compact and disjoint from a closed set B, then $D(A,B) \neq 0$, *i.e.*, A, B are far in the metric proximity.

We will show that if (X,d) is a T_1 semi-metric space and d satisfies any one of conditions (i)-(iv), then X is metrisable.

First, consider a T_1 semi-metric space (X,d) that satisfies (i). The proof that X is metrisable requires verification of the triangle inequality and is similar to the proof of Theorem 11.2. If this is not true, then there exist sequences $(x_n), (y_n), (z_n)$ such that, for all $n > p$,

$$x_n \in B(y_n, \frac{1}{2^p}), y_n \in B(z_n, \frac{1}{2^p}), \text{ but } z_n \notin B(x_n, \frac{1}{2^n}).$$

This contradicts (i) and proves the **Chittenden metrisation** result in Theorem 11.4.

Theorem 11.4. *A T_1 space X is metrisable, if and only if, X has a compatible semi-metric d such that d satisfies*

$$d(x_n, y_n) \to 0, d(y_n, z_n) \to 0 \Rightarrow d(x_n, z_n) \to 0.$$

Similarly, we have the **Niemytzki metrisation** result in Corollary 11.1.

Corollary 11.1. *A T_1 space X is metrisable, if and only if, it has a compatible semi-metric d that satisfies*

$$d(x_n, x) \to 0, d(x_n, y_n) \to 0 \Rightarrow d(y_n, x) \to 0.$$

Next, suppose that a T_1 space X has a compatible semi-metric d with δ_d that is an EF-proximity. If d is not a metric, then, considering property (ii), assume there exists a point $x \in X$ and sequences $(x_n), (y_n)$ such that

$$d(x_n, x) \to 0, d(x_n, y_n) \to 0, \text{ but } d(y_n, x) \to 0 \text{ is not true.}$$

Without loss of generality, assume $x \notin \mathrm{cl}\, B$, where $B = \{y_n : n \in \mathbb{N}\}$. Since δ_d is EF, there is an $E \subset X$ such that $x \notin E$ and $(X - E)\ \underline{\delta}_d\ B$. This is a contradiction, since $d(x_n, x) \to 0$ implies (x_n) is eventually in $(X - E)$. Thus follows the **Arkhangle'skiĭ metrisation** result in Theorem 11.5.

Theorem 11.5. *A T_1 space X is metrisable, if and only if, X has a compatible semi-metric d such that δ_d is an EF-proximity, where $A\ \delta_d\ B \Leftrightarrow D(A,B) = 0$, for A, B in X.*

The following results are proved analogously.

Theorem 11.6. *A T_1 space X is metrisable, if and only if, X has a compatible semi-metric d such that disjoint closed sets, one of which is compact, are far in the proximity δ_d.*

Theorem 11.7. A.H. Stone Metrisation.
A T_1 space X is metrisable, if and only if, it has a countable family of open covers $\{\mathcal{U}_n : n \in \mathbb{N}\}$ such that for each $x \in X$ and neighbourhood U of x, there a neighbourhood V of x and an $n \in \mathbb{N}$ such that $St(V, \mathcal{U}_n) \subset U$.

11.1.1: Brasilian amethyst

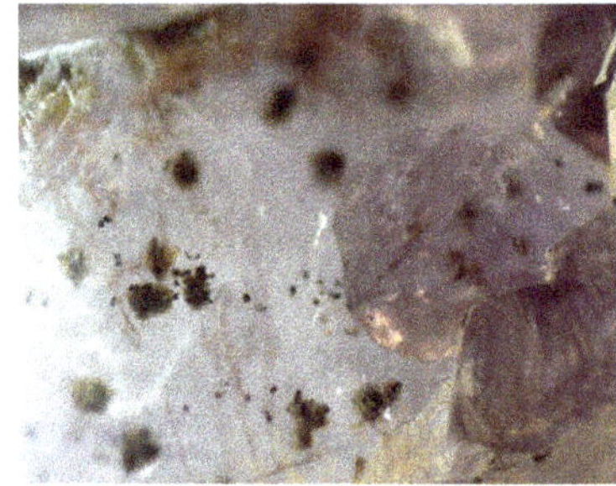

11.1.2: Surface microfossils

Fig. 11.1 Sample surface amethyst microfossils

11.5 Application: Admissible Covers in Micropalaeontology

Micropalaeontology is a branch of science concerned with fossil animals and plants at the microscopic level [13]. During the past three decades, significant advances have been made in the understanding of microscopic life and their fossil counterparts. The current completeness of the microfossil record means that the Phanerozoic era (0.01 MYA to 540 MYA) and parts of the Preterozoic era (540 MYA to 2500 MYA) can be dated using microfossils. In addition, microfossils are indispensable in the study of sedimentary basins, providing a biostratigraphical and palaeoecological framework as well as a measure of the maturity of hydrocarbon-prone rocks. A good overview of microfossils as environment indicators in marine shales is given by S.P. Ellison [58] and as indicators of glacial drift [102]. Sample microfossils near the surface of a Brasilian amethyst are shown in Fig. 11.1. A small group of surface ostracod microfossils [13], looking like bits of pepper and estimated to be from the Devonian era (350-415 MYA), are located in the north central part of the amethyst in Fig. 11.1.1. There are about 33,000 living and fossil ostracod species, originally marine and probably benthic (fauna and flora found on the bottom of a sea or lake). Some adapted to a semi-terrestrial life, living in damp soil and leaf litter. A small group of surface of the same microfossils can be seen more clearly on a 50μm scale in Fig. 11.1.2, obtained with an AxioCam Erc5s camera on a Zeiss Discovery V8 stereo zoom microscope using a Zeiss CL 6000 LED light.

This section is based on a recent study of the application of admissible covers in Micropalaeontology [188], which introduces two types metrisable EF-proximity spaces in the study of sequences of points in microfossil images.

(**cover**.1) **Spatial**. Metrisable EF-proximity space (X, δ), where

$$A\ \delta\ B \text{ in } X \Leftrightarrow (a_n) \approx (b_n) \text{ for } (a_n) \in A, (b_n) \in B.$$

(**cover**.2) **Descriptive**. An admissible cover of a descriptive EF-proximity space (X, δ_Φ), where

$$A\ \delta_\Phi\ B \text{ in } X \Leftrightarrow (a_n) \approx (b_n) \text{ for } (a_n) \in A, (b_n) \in B.$$

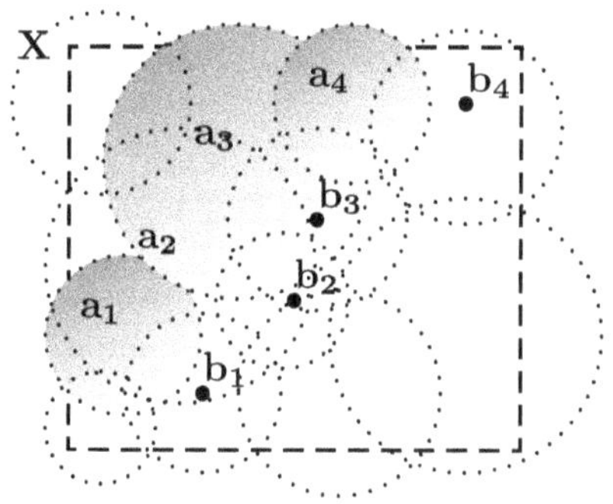

Fig. 11.2 $(a_n) \in A$ and $(b_n) \in B$, and admissible cover **X**

Example 11.1. Sample Admissible Cover.
Let (X, δ) be an EF-proximity space. Let A be a set of neighbourhoods of points N_a for $a \in (a_n)$, B be a set of neighbourhoods of points N_b for $b \in (b_n)$ as shown in Fig. 11.2. Assume $\mathcal{F}$ is a family of admissible covers of X. The dotted circles in Fig. 11.2 represent neighbourhoods of points in X, where

$$A = N_{a_1} \cup N_{a_2} \cup N_{a_3} \cup N_{a_4},$$
$$B = N_{b_1} \cup N_{b_2} \cup N_{b_3} \cup N_{b_4}.$$

In particular, each $a_i : i \in \{1,2,3,4\}$ in (a_n) is the centre of a neighbourhood N_{a_i}. Similarly, each $b_j : j \in \{1,2,3,4\}$ in (b_n) is the centre of a neighbourhood N_{b_j}. Observe that $A\ \delta\ B$, since

$$B \cap St(A,\mathcal{U}) \neq \varnothing, \text{ where,}$$
$$St(A,\mathcal{U}) = \{B \in \mathcal{U} : B \cap A \neq \varnothing\}.$$

Given a cover $\mathcal{U}$ of X and $A \subset X$, the **star of** A with respect to cover $\mathcal{U}$ is denoted by $St(A,\mathcal{U})$. That is, for each $N_{a_i} \in B$, observe that $N_{a_i} \cap \mathcal{U}$ has a nonempty intersection with some N_{b_j} in B. ∎

Descriptive admissible covers of an EF-proximity space X are useful in the study of the appearance of spherical neighbourhoods of points, since it is often the case that it is helpful to consider both spatial as well as non-spatial proximity of neighbourhoods in such covers. Let $\varepsilon \in (0,\infty)$. The spherical neighbourhood

$$B_\Phi(x,\varepsilon) = \{y \in X : d(\Phi(x),\Phi(y)) < \varepsilon\},$$

where d is a semi-metric and the family $\left\{B_\Phi\left(x,\frac{1}{2^n}\right) : n \in \mathbb{N}\right\}$ is a countable base at $x \in X$. Let

$$\mathcal{U}_n = \bigcup\left\{B_\Phi\left(x,\frac{1}{n}\right) : x \in X\right\}$$

be a cover of X. Then $\{\mathcal{U}_n : n \in \mathbb{N}\}$ is a countable family of covers of X and for cover $\mathcal{U}_n$, the **descriptive star** of x (denoted by $St_\Phi(x,\mathcal{U}_n)$) is defined by

$$St_\Phi(x,\mathcal{U}_n) = \bigcup_\Phi\{U \in \mathcal{U}_n : x \in X\}.$$

Conversely, suppose we are given a countable family $\{\mathcal{U}_n : n \in \mathbb{N}\}$ of covers of X and $\{St_\Phi(x,\mathcal{U}_n) : n \in \mathbb{N}\}$ is a descriptive neighbourhood base at $x \in X$. Then define the semi-metric d_Φ to be

$$(****) \qquad d_\Phi(x,y) = \begin{cases} inf\left\{\frac{1}{2^n}\right\}, & \text{if } y \in St_\Phi(x,\mathcal{U}_n), \\ 1, & \text{otherwise.} \end{cases}$$

A cover is **descriptively admissible**, if and only if, $(a_n) \underset{\Phi}{\approx} (b_n)$ implies $St_\Phi(a_m,\mathcal{U}) \neq \varnothing$ for some $n \in \mathbb{N}$. Let $\mathcal{F}$ denote a family of all admissible covers of X. Then $A\ \delta_\Phi\ B$ in X (A is descriptively near B), if and only if,

$$\text{for each } \mathcal{U} \in \mathcal{F}, B \bigcup_\Phi St(A,\mathcal{U}) \neq \varnothing.$$

Hence, $St(A,\mathcal{U}_n : n \in \mathbb{N})$ is a neighbourhood base at $x \in X$ and the semi-metric d_Φ defined by (****) is compatible with the topology of X. In fact,

$$A\ \delta_\Phi\ B \Leftrightarrow D_\Phi(A,B) =$$
$$inf\,\{d(\Phi(a),\Phi(b)) : \Phi(a) \in \Phi(A), \Phi(b) \in \Phi(B)\} = 0.$$

This line of reasoning yields the following result that is a descriptive form of Efremovič-Švarc proximal metrisation, namely,

Theorem 11.8. *A descriptive EF-proximity space is metrisable, if and only if,*

*(**Φ**.i) Descriptively near sets in admissible covers:*

$$A\ \delta_\Phi\ B \text{ in } X \Leftrightarrow (a_n) \underset{\Phi}{\approx} (b_n), \text{ for } (a_n) \in A, (b_n) \in B.$$

*(**Φ**.ii) The family of descriptive admissible covers has a countable base.*

There is an aspect of descriptive admissible covers that is attractive in the study of microfossil images. This attractiveness stems from the fact that one need only consider the cardinality of the descriptive intersection. Given a cover $\mathcal{U}$ of X, A, B in X, and the star of A with respect to cover $\mathcal{U}$ of X $(St(A,\mathcal{U}))$. Descriptive intersections are easy to determine and the count of the number of members of that are descriptively near each other is computationally simple. A measure of the degree of nearness μ_{δ_Φ} is defined by

$$\mu_{\delta_\Phi}(A,B) = \frac{\left|B \underset{\Phi}{\cap} St(A,\mathcal{U})\right|}{\left|B \underset{\Phi}{\cup} St(A,\mathcal{U})\right|}.$$

Then $A\ \delta_\Phi\ B$, if and only if, $\mu_{\delta_\Phi} > 0$. Before we attempt to apply descriptive admissible covers of a set, we first consider a simple example.

Example 11.2. Sample Descriptive Admissible Cover.
Let (X,δ) be an EF-proximity space. Let A be a set of neighbourhoods of points N_a for $a \in (a_n)$, B be a set of neighbourhoods of points N_b for $b \in (b_n)$ as shown in Fig. 11.3. Assume $\mathcal{F}$ is a family of descriptive admissible covers of X. The dotted circles in Fig. 11.3 represent neighbourhoods of points in X, where

$$A = N_{a_1} \underset{\Phi}{\cup} N_{a_2} \underset{\Phi}{\cup} N_{a_3} \underset{\Phi}{\cup} N_{a_4},$$
$$B = N_{b_1} \underset{\Phi}{\cup} N_{b_2} \underset{\Phi}{\cup} N_{b_3} \underset{\Phi}{\cup} N_{b_4}.$$

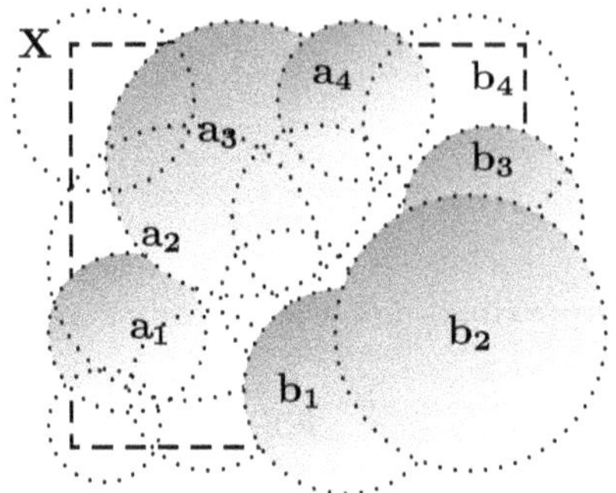

Fig. 11.3 Descriptive admissible cover of **X**

Unlike the spatial nearness of the neighbourhoods of points in Fig. 11.2, notice, in Fig. 11.3, several pairs of neighbourhoods are disjoint but descriptively similar, if we let $\Phi = [\phi_g]$, where $\phi(x)$ is the greylevel intensity of point $x \in X$. For example, N_{a_1} is descriptively close to N_{b_1} in Fig. 11.3. Each $a_i : i \in \{1,2,3,4\}$ in (a_n) is the centre of a neighbourhood N_{a_i}. Similarly, each $b_j : j \in \{1,2,3,4\}$ in (b_n) is the centre of a neighbourhood N_{b_j}. Observe that

$$A \ \delta_\Phi \ B, \text{ since } B \underset{\Phi}{\cap} St(A,\mathcal{U}) \neq \varnothing.$$

That is, for at least one $N_{a_i} \in B$, observe that $N_{a_i} \underset{\Phi}{\cap} \mathcal{U}$ has a descriptive nonempty intersection with some N_{b_j} in B. Then the cover $\mathcal{U}$ shown in Fig. 11.3 is descriptively admissible, since $\Phi(b_n) \in St_\Phi(a_n,\mathcal{U})$ for some $n \in \mathbb{N}$. For example, this is the case with N_{a_1} and N_{b_1} that contain pixels with matching greylevel intensities. Hence, $\mu_{\delta_\Phi} > 0$. ■

Three ostracod age groups are shown in Fig. 11.5, namely,

(**age**.1) Young ostracod labelled Y (there are many young ostracod visible in Fig. 11.5),

(**age**.2) Juvenile ostracod labelled J (there is only one juvenile ostracod visible in Fig. 11.5),

(**age**.3) Adult ostracod labelled A (there are two adult ostracod visible in Fig. 11.5).

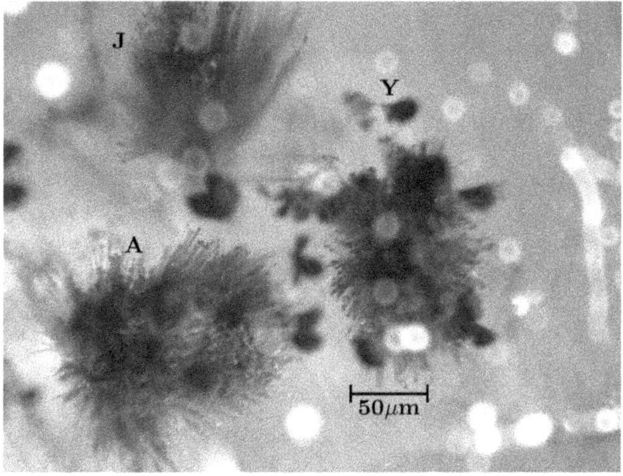

Fig. 11.4 Ostracod microfossils

In the sequel, only the nearness (apartness) of the adult microfossils in Fig. 11.5 is considered.

There is a subtly in the definition of descriptive neighbourhoods that needs to be observed. Let $N_{\Phi,x}$ denote a **descriptive neighbourhood** with centre x and member descriptions defined by $\Phi = [\phi_1, \ldots, \phi_n]$. The only points that are members of $N_{\Phi,x}$ are those y such that $d(\Phi(x), \Phi(y)) < \varepsilon$ for $\varepsilon \in (0, \infty)$, *i.e.*, the descriptions of x and y are sufficiently near. Hence, when a descriptive neighbourhood is displayed, the only points that shown are those points y with descriptions that *sufficiently near* to the description of neighbourhood center x. This observation explains what is shown in two different cases in Fig 11.5 and in Fig. 11.6. This is explained in Examples 11.3 and 11.4.

Example 11.3. Microfossil Intensity Neighbourhoods.
Let X be a set of points in a microfossil image. Assume that the microfossil image contains a descriptive admissible cover $\mathcal{U}$ of X that is a collection of open descriptive neighbourhoods of points. Let $\Phi = [\phi_g]$, where $\phi_g(x)$ equals the greylevel intensity of $x \in X$ and $\varepsilon \in (0, \infty)$. Each neighbourhood $N_{\Phi,x} \in \mathcal{U}$ is defined by

$$N_{\Phi,x} = \{y \in X : d(\phi_g(x)\phi_g(y)) < \varepsilon\}.$$

Fig. 11.5 Microfossils intensity neighbourhoods

For simplicity, only two descriptive neighbourhoods of points are shown in Fig. 11.5, namely, $N_{\Phi,a}, N_{\Phi,b}$. A translucent mask has been superimposed over the points in each of these two neighbourhoods. It is easy to verify that many pairs of pixels from the two neighbourhoods have matching greylevel intensities. Hence, Hence, $\mu_{\delta_\Phi} > 0$. ■

Fig. 11.6 Four microfossils gradient neighbourhoods

Example 11.4. Gradient-based Microfossil Neighbourhoods.
Let X be a set of points in a microfossil image. Assume that the microfossil image contains a descriptive admissible cover $\mathcal{U}$ of X that is a collection of open descriptive neighbourhoods of points. Let $\Phi = [\phi_\Delta]$, where $\phi_\Delta(x)$

equals the gradient direction of $x \in X$ and $\varepsilon \in (0, \infty)$. In this example, we consider only the gradient direction of the pixels along the setae (stiff hair-like or bristle-like structures protruding from the bodies of each adult ostracod in Fig. 11.6). Let $\varepsilon = 0.1$. Four gradient direction-based neighbourhoods are shown in Fig. 11.6, namely,

(Δ.1) N_{Φ,a_1} with centre a_1,
(Δ.2) N_{Φ,a_2} with centre a_2,
(Δ.3) N_{Φ,b_1} with centre b_1,
(Δ.4) N_{Φ,b_2} with centre b_2.

Neighbourhood N_{Φ,a_2} contains the largest number of setae with pixels y such that $d(\phi_\Delta(a_2), \phi_\Delta(y)) < \varepsilon$. Observe that neighbourhood N_{Φ,b_2} contains the next largest number of setae with pixels with gradient direction close to the gradient direction of b_2. It is a straightforward task to verify that the gradient direction of the setae in these two neighbourhoods are sufficiently near, *i.e.*, in many cases, $d(\phi_\Delta(a_2), \phi_\Delta(b_2)) < \varepsilon$. Hence,

$$B \underset{\Phi}{\cup} St(A, \mathcal{U}) \neq \varnothing,$$

implies $A\ \delta_\Phi\ B$ and $\mu_{\delta_\Phi} > 0$. ■

11.6 Problems

(**11.6**.1) Construct a semi-metric space (X, d) in which the 'open' ball

$$B(x, \varepsilon) = \{y \in X : d(x, y) < \varepsilon \text{ for } \varepsilon > 0\}$$

is not open.

(**11.6**.2) In a T_1 space X, let $\{\mathcal{V}_{n+1} : n \in \mathbb{N}\}$ be a countable family of covers of X such that $\mathcal{V}_{n+1}$ is a refinement of $\mathcal{V}_n$ and $\{St(x, \mathcal{V}_n) : n \in \mathbb{N}\}$ is a neighbourhood base at each $x \in X$. Show that d defined by

$$d(x, y) = \begin{cases} inf\left\{\frac{1}{2^n}\right\}, & \text{if } y \in St(x, \mathcal{V}_n), \\ 1, & \text{otherwise,} \end{cases}$$

is a compatible semi-metric on X.

(**11.6**.3) If, in Problem 11.6.2, the covers are open, show that the semi-metric d is upper semi-continuous.

(**11.6**.4) Let (X, δ) be an EF-proximity space and let $(a_n), (b_n)$ be sequences in X. Further, let

$$(a_n) \approx (b_n) \Leftrightarrow \text{for each infinite subset } M \subset \mathbb{N}, \\ \{a_n : n \in \mathbb{N}\}\ \delta\ \{b_n : n \in \mathbb{N}\}.$$

Show that $\approx$ is an equivalence relation.

(**11.6**.5) Prove that in a semi-metric space (X, d), the triangle inequality, for each $x \in X$, is equivalent to

$$\text{for each } x \in X, B(B(x, \frac{1}{2^{n-1}}), \frac{1}{2^{n-1}}) \subset B(x, \frac{1}{2^n}).$$

(**11.6**.6) Prove Corollary 11.1.

(**11.6**.7) Prove Theorem 11.6.

(**11.6**.8) Prove Theorem 11.7.

(**11.6**.9) Given a cover $\mathcal{U}$ of X, nonempty A, B in X, a measure of the degree of spatial nearness μ_δ is defined by

$$\mu_\delta(A, B) = \frac{|B \cap St(A,\mathcal{U})|}{|B \cup St(A,\mathcal{U})|}.$$

Let μ_{δ_Φ} be the measure of the degree of descriptive nearness of A and B as defined in Section 11.5. Assume that $St(A,\mathcal{U})$ and $St_\Phi(A,\mathcal{U})$ are nonempty. Show that $\mu_\delta(A, B) \le \mu_{\delta_\Phi}(A, B)$.

Chapter 12

Function Space Topologies

12.1 Topologies and Convergences on a Set of Functions

This chapter introduces topologies and convergences on a set of functions from one topological space to another, an important topic in topology. There are variations of this topic that take into account proximities and uniformities similar to uniform convergence in real and complex analysis and metric spaces (see Notes and Further Readings for a review). Generally, most texts consider only continuous functions. Here, this restriction is absent.

Let $(X, \tau), (Y, \tau')$ be topological spaces and let $\mathcal{F}$ denote a set of functions on X to Y. Each $f \in \mathcal{F}$ is identified with the graph

$$G(f) = \{(x, f(x)) : x \in X\} \subset X \times Y.$$

There are several ways to define topologies on functions spaces. We present three approaches, next. Other approaches will be given later in the text or in problems.

($\mathbf{Sp}_{fn}$.1) **Set-open topology**.

A **network** α on X is a family of subsets of X such that, for each $x \in U \in \tau$, there is an $A \in \alpha$ with $x \in A \subset U$. Network α is closed (compact), provided each of its members is closed (compact). Without any loss of generality, we assume that α is a closed network on X that contains finite unions and finite intersections of its members and contains all singletons (assuming X is T_1).

The **set-open topology** τ_α on $\mathcal{F}$ is generated by

$$\{[A, U] : A \in \alpha, U \in \tau'\},$$

where $[A, U] = \{f \in \mathcal{F} : f(A) \subset U\}$. In case Y has an L-proximity

δ, then τ_α has been generalised to a **proximal set-open topology** by replacing $[A, U]$ by

$$[A :: U] = \{f \in \mathcal{F} : f(A) \ll_\delta U\}.$$

($\mathbf{Sp}_{fn}$.2) **Graph Topology.**
A typical basic open set in graph topology is

$$W^+ = \{f \in \mathcal{F} : G(f) \subset W\},$$

where W is an open subset of $X \times Y$. Notice that when A is a closed set in X and U is an open set in Y,

$$[A, U] = [(X \times U) \cup (X - A) \times Y)]^+.$$

($\mathbf{Sp}_{fn}$.3) **Convergence.**
Recall that $\mathcal{F}$ is a set of functions on X to Y. Convergence of a net $(f_n) \in \mathcal{F}$ of functions to $f \in \mathcal{F}$ is defined using topology, proximity or uniformity on Y.

12.2 Pointwise Convergence

The idea of *pointwise convergence* has existed since the early days of calculus/mathematical analysis, especially in the study of power series and trigonometric series. Its study in topology began with Tychonoff's definition of a product space, since a set of functions $\mathcal{F}$ on X to Y is a subset of the product Y^X. Tychonoff showed that pointwise convergence topology $\mathcal{P}$ on $\mathcal{F}$ is precisely the subspace topology induced by the product topology on Y^X.

We illustrate the equivalence of the three approaches to defining topologies on function spaces in terms of point-open topology or pointwise convergence topology $\mathcal{P}$.

A net (f_n) in $\mathcal{F}$ **converges pointwise** to $f \in \mathcal{F}$, if and only if, for each $x \in X$, the net $(f_n(x))$ converges to $f(x)$ in Y. That is, for each open set U in Y containing $f(x)$, eventually $f_n(x) \in U$. If X is T_1, this is equivalent to

$$[\{x\}, U] = [(X \times U) \cup ((X - \{x\})) \times Y]^+,$$

that is open in the resulting topology on $\mathcal{F}$.

Thus, if X is T_1, the topology of pointwise convergence is precisely the **closed-open topology**, when the network α on X is the family of all *finite* subsets of X. A **base** for the point-open topology is

$$\bigcap\{[x_j, U_j] : 1 \le j \le m\} = \{f \in \mathcal{F} : f(x_j) \in U_j\},$$

where $x_j \in X$ and U_j is open in Y.

There is a visual way to look at a pointwise topology on $\mathcal{F}$. Consider $X = Y = [0,1]$. Then a typical basic open set in pointwise topology is the set of functions with graphs that intersect finitely many open vertical segments in $X \times Y$.

It has already been noted that $\mathcal{F}$ is a subset of the product

$$Y^X = \prod \{Y = Y_x : x \in X\},$$

and the pointwise topology is merely the subspace topology on $\mathcal{F}$ induced by the product topology on Y^X. Clearly, a pointwise topology depends solely on the topology of Y and the topology on X plays no role. Obviously, $\mathcal{F}$ inherits the separation axioms $T_1, T_2 =$ Hausdorff, T_3, or Tychonoff from Y, in case Y satisfies one of these conditions.

If $\mathcal{F}$ is a set of functions on X to Y, then, for each $x \in X$, write

$$\mathcal{F}(x) = \{f(x) : f \in \mathcal{F}\} \subset Y.$$

If $\operatorname{cl}\mathcal{F}(x)$ is compact, then, by the Tychonoff Product Theorem, $\prod\{\operatorname{cl}\mathcal{F}(x) : x \in X\}$ is compact. Further, if $\mathcal{F}$ is closed in Y^X in the pointwise topology, then it is compact. In many applications, such as in analysis, it is important to know when a function space $\mathcal{F}$ is compact in a finer topology. The usual trick is to put a condition (*e.g.*, equicontinuity or even continuity) on $\mathcal{F}$ so that the topology at hand equals the pointwise topology and use the above result. Here we summarize the results for pointwise topology.

Let $\mathcal{F}$ be a family of functions on (X, τ) to (Y, τ') with pointwise topology.

Theorem 12.1. *If Y is $T_j, j = 1, 2, 3, 3.5$, then so is $\mathcal{F}$.*

Further, if Y is Hausdorff and $\mathcal{F}$ is compact, then $\mathcal{F}$ is closed in Y^X and $\operatorname{cl}(\mathcal{F}(x))$ is compact for each $x \in X$.

Theorem 12.2. *If $\mathcal{F}$ is closed in Y^X and $\operatorname{cl}(\mathcal{F}(x))$ is compact for each $x \in X$, then $\mathcal{F}$ is compact.*

Suppose we have space X, the product $Y = \prod\{Y_\lambda : \lambda \in \Lambda\}$ and maps $f_\lambda : X \to Y_\lambda$. In §8.2, we studied evaluation map $e : X \to Y$ given by $e(x) = (f_\lambda(x))$. In the present case, we have spaces X, Y and the set of functions $\mathcal{F} \subset Y^X = \prod\{Y = Y_x : x \in X\}$. For each $x \in X$, there is an evaluation e_x, a function on $\mathcal{F} \to Y$ given by $e_x(f) = f(x)$. The pointwise topology on $\mathcal{F}$ is the coarsest topology such that each evaluation is continuous and so depends only on Y and the family $\mathcal{F}$.

12.3 Compact Open Topology

In 1945, R.H. Fox discovered compact open topology (a special case of set-open topologies) on function spaces. It generalized uniform convergence on compacta (see Sect. 12.5 for details). This has proven to be useful in homotopy and functional analysis.

Let $(X,\tau),(Y,\tau')$ be topological spaces and let $\mathcal{F}$ denote a set of functions on X to Y. A typical member of an open subbase for a **compact open topology** is

$$[A,U] = \{f \in \mathcal{F} : f(A) \subset U\},$$

where A is a compact set in X and U is an open set in Y. Thus, if X is Hausdorff, the compact open topology is a set open topology, provided the network α on X is the family of all *compact* subsets of X.

Obviously, compact open topology is finer than a pointwise topology. So, if Y is T_1 or T_2, then so is $\mathcal{F}$ with compact open topology. However, if Y is T_3 for $\mathcal{F}$ to be T_3, we need

$$\mathcal{F} \subset C(X,Y) = \text{ family of continuous functions on } X \text{ to } Y.$$

It is sufficient to show that if f is continuous and belongs to $[A,U]$, where A is a compact subset of X and U is an open subset of Y, then $[A,U]$ contains a closed neighbourhood of f. Since f is continuous, $f(A)$ is a compact subset of U. Since X is T_3, there is an open set V in Y such that $f(A) \subset V \subset \mathrm{cl}\, V \subset U$ (see Prob. 12.11.1). The result follows if we show that $[A, \mathrm{cl}\, V]$ is closed. For each $x \in A, [\{x\}, \mathrm{cl}\, V]$ is closed in the pointwise topology and so is

$$[A, \mathrm{cl}\, V] = \bigcap \{[\{x\}, \mathrm{cl}\, V] : x \in A\}.$$

Hence, $[A, \mathrm{cl}\, V]$ is closed in the compact open topology that is finer than pointwise topology.

Next, we give an outline of a proof that if Y is Tychonoff, then $\mathcal{F} \subset C(X,Y)$ is Tychonoff in the compact open topology. It suffices to prove that if $f \in \mathcal{F} \cap [A,U]$, then there exists a continuous function $g : \mathcal{F} \to [0,1]$ such that $g(f) = 0$, $g(h) = 1$ for $h \in \mathcal{F} - [A,U]$ (see Prob. 12.11.2). Since Y is Tychonoff, there is a continuous function $g : Y \to [0,1]$ such that $g(f(A)) = 0$ and $g(U^c) = 1$. For each $h \in \mathcal{F}$, set $g(h) = sup\{h(x) : x \in A\}$.

Theorem 12.3. *Let $\mathcal{F}$ denote a family of functions on (X,τ) to (Y,τ') with compact open topology.*

(12.3(a)) If Y is $T_j, j = 1, 2$, then so is $\mathcal{F}$.
(12.3(b)) If $\mathcal{F} \subset C(X, Y)$, then it inherits T_3(regular) or $T_{3.5}$.

Let $\mathcal{F} = C(X, Y)$. A topology τ for $\mathcal{F}$ is called **jointly continuous** (**admissible**, **conjoining**), if and only if, the map $P : \mathcal{F} \times X :\to Y$ given by $P(f, x) = f(x)$ is continuous. If $\mathcal{F}$ has discrete topology, then P is continuous, *i.e.*, the discrete topology on $\mathcal{F}$ is the finest admissible topology. Generally, there is no coarsest topology on $\mathcal{F}$ that makes P continuous.

Compact open topology on $\mathcal{F}$ is coarser than any jointly continuous topology τ. Let τ a topology for $\mathcal{F} \subset C(X, Y)$ be **jointly continuous on campacta**, *i.e.*, the map $P : \mathcal{F} \times X :\to Y$ is continuous on $\mathcal{F} \times A$, where A is a compact set in X. If U is open in Y, we show that $[A, U]$ is open in τ. Note that $V = (\mathcal{F} \times A) \cap P^{-1}(U)$ is open in $\mathcal{F} \times A$. If $f \in [A, U]$, then $\{f\} \times A$ is compact and is contained in V Hence, there is a τ-neighbourhood N of f and $N \times A \subset V$ (see Prob. 12.11.3). Hence, $[A, U]$ is τ-open. This shows that a compact open topology is coarser than τ, a topology that is jointly continuous on compacta.

The converse holds if X is locally compact Hausdorff (in which case, X is Tychonoff). Suppose $x \in A$ (a compact set in X), U is an open set in Y and $(f, x) \in P^{-1}(U)$. Then x has a compact neighbourhood H in A such that $f(H) \subset U$. So, the compact open neighbourhood $[H, U] \times H$ of (f, x) is contained in $P^{-1}(U)$. This shows that P is jointly continuous on compacta.

Theorem 12.4. *Let $\mathcal{F}$ be a family of functions on (X, τ) to (Y, τ') with compact open topology $\mathcal{K}$.*

(12.4(a)) $\mathcal{K}$ is coarser than any topology on $\mathcal{F}$ that is jointly continuous on compacta.

(12.4(b)) If X is locally compact Hausdorff and $\mathcal{F} \subset C(X, Y)$, then $\mathcal{K}$ is the coarsest topology on $\mathcal{F}$ that is jointly continuous on compacta (cf. Prob. 12.11.5).

In differential equations and functional analysis, we need to know when a set $\mathcal{F}$ of functions on topological spaces X to Y is compact in the compact open topology $\mathcal{K}$. Observe

(i) Theorem 12.2 shows when a set of functions is compact in the pointwise topology $\mathcal{P}$.
(ii) If $\mathcal{L}$ is any jointly continuous topology on $\mathcal{F}$, then $\mathcal{P} \subset \mathcal{K} \subset \mathcal{L}$.

So the trick is to put a condition on $\mathcal{F}$ that will ensure joint continuity and this will make all three topologies equal. Intuitively, *the family* $\mathcal{F}$ *is evenly continuous, if and only if, for each* $x \in X, y \in Y, f \in \mathcal{F}$, *if* $f(x)$ *is near* y, *then* f *maps points near* x *into points near* y.

Formally, $\mathcal{F}$ is said to be **evenly continuous**, if and only if, for each $x \in X, y \in Y$ and each neighbourhood U of y, there are neighbourhoods V of x and W of y such that $f(x) \in W$ implies $f(V) \subset U$. In other words, for each neighbourhood U of y, there is a neighbourhood V of x and a neighbourhood $[\{x\}, W]$ of f in topology $\mathcal{P}$ such that $P([\{x\}, W], V) \subset U$, *i.e.*, P is jointly continuous. Members of a family of evenly continuous functions are obviously continuous.

If $\mathcal{F}$ is an evenly continuous family and Y is T_3 (regular), then $\mathrm{cl}\,\mathcal{F}$ (closure of $\mathcal{F}$ in the pointwise topology $\mathcal{P}$) is also evenly continuous. Let $x \in X, y \in Y$ and let U be a closed neighbourhood of y. Then there is a neighbourhood of V of x and an open set W containing y such that $f(x) \in W$ implies $f(V) \subset U$. If $h \in \mathrm{cl}\,\mathcal{F}$, then there is a net $(f_n) \in \mathcal{F}$ that converges pointwise to h. If $f(x) \in W$, then *eventually* $f_n(x) \in W$. This shows that, for each $v \in V$, *eventually* $f_n(v) \in U$. Since U is closed, $f(V) \subset U$.

Theorem 12.5. *Pointwise closure* $\mathrm{cl}\,\mathcal{F}$ *of an evenly continuous family* $\mathcal{F}$ *from a topological space* X *to a regular space is also evenly continuous and, hence,* $P : \mathrm{cl}\,\mathcal{F} \times X \to Y$ *is jointly continuous.*

The next result is due to Kelley and Morse.

Theorem 12.6. (**Ascoli Theorem**) *Let* X *be a locally compact Hausdorff space,* Y *a regular space and let the family* $\mathcal{F} \subset C(X, Y)$ *have compact open topology* $\mathcal{K}$, *then* $\mathcal{F}$ *is compact, if and only if,*
(12.6(a)) $\mathcal{F}$ *is closed in* $C(X, Y)$.
(12.6(b)) For each $x \in X, \mathrm{cl}\,\mathcal{F}(x)$ *is compact.*
(12.6(c)) $\mathcal{F}$ *is evenly continuous.*

Proof. If $\mathcal{F}$ is compact Hausdorff in the compact open topology $\mathcal{K}$, then it is jointly continuous and equals the coarser pointwise topology (see Prob. 12.11.6). Since X is locally compact, $\mathcal{P} = \mathcal{K}$ is jointly continuous and hence evenly continuous. This proves 12.6(c). Results 12.6(a) and (b) follow trivially. Conversely, 12.6(c) implies $\mathcal{P}$ is jointly continuous and hence equals $\mathcal{K}$. The compactness of $\mathcal{K}$ follows form Theorem 12.2. □

Note: The result in Theorem 12.6 can be extended to X being a k-space, *i.e.*, a space in which a set U is open, if and only if, its intersection with each compact set K is open in K.

12.4 Proximal Convergence

In analysis, we study that the pointwise limit of a sequence of continuous functions need not be continuous, *e.g.*, $f_n : [0,1] \to [0,1]$ defined by $f_n(x) = x^n$. Sequence (f_n) converges pointwise to f, where $f(x) = 0$ for $0 \leq x < 1$ and $f(1) = 1$. To preserve continuity, we need a stronger convergence. There are several such convergences (see Sect. 12.5). S. Leader discovered proximal convergence in 1959. First, we give motivation by looking at pointwise convergence in terms of proximity.

Let $(X,\delta),(Y,\eta)$ be EF-proximity spaces, $\mathcal{F}$ a set of functions on X to Y and a net (f_n) in $\mathcal{F}$ that converges pointwise to $f \in \mathcal{F}$. This is equivalent to *for each* $x \in X$, *each set* F *in* Y, $f(x)\ \underline{\eta}\ F$ *implies eventually* $f_n(x)\ \underline{\eta}\ F$. We get proximal convergence (**prxc**) by replacing x by any subset $A \subset X$.

Precisely (prxc): a net (f_n) in $\mathcal{F}$ **converges proximally** to $f \in \mathcal{F}$ iff, for each $A \subset X$ and $F \subset Y, f(A)\ \underline{\eta}\ F$ implies *eventually* $f_n(A)\ \underline{\eta}\ F$.

Notice that '*for each* $A \subset X$ *and* $F \subset Y, f(A)\ \underline{\eta}\ F$ *implies* eventually $f_n(A)\ \underline{\eta}\ F$' is equivalent to '*for each* $A \subset X$ *and* $U \subset Y, f(A) <<_\eta U$ *implies* eventually $f_n(A) <<_\eta U$'.

Proximal convergence preserves continuity *and* proximal continuity. The proofs are analogous. Suppose a net (f_n) of p-continuous functions **converges proximally** to f. We show that $A\ \delta\ B$ and $f(A)\ \underline{\eta}\ f(B)$ leads to a contradiction. Observe that $f(A)\ \underline{\eta}\ f(B)$ implies that there is an $E \subset Y$ such that $f(A)\ \underline{\eta}\ E$ and $(Y - E)\ \underline{\eta}\ f(B)$. Proximal convergence implies eventually $f_n(A)\ \underline{\eta}\ E$ and $(Y - E)\ \underline{\eta}\ f_n(B)$, which shows that $f_n(A)\ \underline{\eta}\ f_n(B)$. Since f_n is proximally continuous, $A\ \underline{\delta}\ B$ is a contradiction.

Theorem 12.7. *Let* $(X,\delta),(Y,\eta)$ *be EF-proximity spaces,* $\mathcal{F}$ *a set of functions on* X *to* Y. *If a net* (f_n) *of continuous (proximally continuous) functions in* $\mathcal{F}$ *converges proximally to* $f \in \mathcal{F}$, *then* f *is continuous (proximally continuous).*

A generalization of (prxc) substitutes each $A \subset X$ with *each set* A *from a network* α *of nonempty subsets of* X. The resulting *proximal set open topology* contains as special cases all set-open topologies.

12.5 Uniform Convergence

From Sect. 12.4, we know that pointwise convergence need not preserve continuity. To rectify this deficiency, uniform convergence was discovered in 1847-1848 by Stokes and Seidel independently. In an 1841 paper (published in 1894), Weierstrass used the notion of uniform convergence. Modifications of uniform convergence were studied by others.

Let $(X, d), (Y, e)$ denote metric spaces and let $\mathcal{F}$ be a set of functions on X to Y. Then a net (f_n) in $\mathcal{F}$ **converges uniformly** to $f \in \mathcal{F}$, if and only if, for each $\varepsilon > 0$, there exists an $m \in \mathbb{N}$ such that, for all $x \in X$ and $n > m$, $e(f_n(x), f(x)) < \varepsilon$.

Notice in the definition of uniform convergence, m depends on ε only and is independent of $x \in X$. Proof that uniform convergence preserves (uniform) continuity is analogous to the proof studied in analysis. After the discovery of uniform spaces, there was a natural extension of uniform convergence to uniform spaces.

Uniformity on nonempty set Y is a family $\mathcal{E}$ of pseudometrics satisfying some conditions (see Chapter 5.5). Let X be a topological space, $(Y, \mathcal{E})$ a uniform space and $\mathcal{F}$ a set of functions on X to Y. The definition of uniform convergence (uc) arises easily, here.

(**uc**) A net (f_n) in $\mathcal{F}$ **converges uniformly** to $f \in \mathcal{F}$, if and only if, for each $\varepsilon > 0$ and $e \in \mathcal{E}$, there exists an $m \in \mathbb{N}$ such that, for all $x \in X$ and $n > m, e(f_n(x), f(x)) < \varepsilon$.

Suppose each f_n is continuous and the net (f_n) converges uniformly to f. Then the function f is continuous.

Proof. Let $\varepsilon > 0$. To show that f is continuous at $p \in X$, notice that there is an $n \in \mathbb{N}$ such that, for all $x \in X$, $e(f_n(x), f(x)) < \frac{\varepsilon}{3}$. Since f_n is continuous at p, there is a neighbourhood U of p such that $x \in U, e(f_n(x), f_n(p)) < \frac{\varepsilon}{3}$. By the triangle inequality,

$$\begin{aligned} e(f(x), f(p)) &\leq e(f(x), f_n(x)) + e(f_n(x), f_n(p)) + e(f_n(p), f(p)) \\ &< \frac{\varepsilon}{3} + \frac{\varepsilon}{3} + \frac{\varepsilon}{3} = \varepsilon. \end{aligned}$$

Hence, f is continuous at p. □

In case X is a uniform space and f_n is uniformly continuous, then a modification of the previous line of reasoning shows that f is uniformly continuous.

Theorem 12.8. *Uniform convergence preserves (uniform) continuity.*

U. Dini modified uniform convergence to **simple uniform convergence**: (f_n) converges pointwise to f and, for each $\varepsilon > 0$ and $e \in \mathcal{E}$, *frequently*, for all $x \in X$, $e(f_n(x), f(x)) < \varepsilon$. This convergence also preserves (uniform) continuity.

12.6 Pointwise Convergence and Preservation of Continuity

Pointwise convergence need not preserve continuity and uniform convergence does. But uniform convergence is too strong. It is sufficient but not necessary. For example, let f_n, f be functions on $\mathbb{R}$ to $\mathbb{R}$ defined by

$$f_n(x) = \left(x + \frac{1}{n}\right)^2, f(x) = x^2.$$

Here, the sequence (f_n) of continuous functions converges pointwise to the continuous function f but the convergence is not uniform. This is so because $|f_n(n) - f(n)| > 2$.

A search began to find necessary and sufficient conditions for preservation of continuity. To accomplish this, modifications of uniform convergence were studied by Arzela, Dini, Young and others. Much of this interesting classical material is not found in contemporary analysis texts. Recently, there has been some work on preservation of continuity in uniform spaces (see Historical Notes section).

Let $(X, d), (Y, e)$ be metric spaces, $\mathcal{F}$ the set functions on X to Y, and $C(X, Y)$ the subset of continuous functions. Let a sequence $(f_n : n \in \mathbb{N})$ in $C(X, Y)$ converge pointwise to $f \in \mathcal{F}$, then

(**Dini**.1) A necessary and sufficient condition that f is continuous is that for each $x \in X, \varepsilon > 0$, *eventually*, for each $n \in \mathbb{N}$, there is a neighbourhood U of x such that, for each $z \in U_x, e(f_n(z), f(z)) < \varepsilon$.

(**Dini**.2) A necessary and sufficient condition that f is continuous is that for each $x \in X, \varepsilon > 0$, *frequently*, there exists $n \in \mathbb{N}$ and a neighbourhood U of x such that, for each $z \in U_x, e(f_n(z), f(z)) < \varepsilon$.

N. Bouleau generalised (Dini.1) to uniform spaces. Let X be a Hausdorff space, $(Y, \mathcal{E})$ a uniform space, $\mathcal{F}$ the set of functions on X to Y and $C(X, Y)$ the set of continuous functions. A net $(f_n : n \in \mathcal{D})$ in $\mathcal{F}$ Bouleau (**sticking**)

converges to $f \in \mathcal{F}$, if and only if, for each $x \in X$, for each $\varepsilon > 0$ and $e \in \mathcal{E}$, there is an $n_0 \in \mathcal{D}$ such that, for each $n > n_0$, there is a neighbourhood U_x of x such that, for each $z \in U_x$, $e(f_n(z), f(z)) < \varepsilon$.

Suppose each f_n is continuous and Bouleau converges to $f \in \mathcal{F}$. Then, for each $x \in X$, $\varepsilon > 0$ and $e \in \mathcal{E}$, *eventually* there is a neighbourhood U_1 of x such that, for each $z \in U_1$, $e(f_n(z), f(z)) < \frac{\varepsilon}{2}$. Since f_n is continuous, there is a neighbourhood U_2 of x such that $z \in U_2$ implies $e(f_n(z), f(x)) < \frac{\varepsilon}{2}$. This shows that for all $z \in U = U_1 \cap U_2$, $e(f(z), f(x)) < \varepsilon$, *i.e.*, f is continuous. Conversely, suppose each f_n is continuous and pointwise converges to $f \in C(X,Y)$. By pointwise convergence for $\varepsilon > 0$ and $e \in \mathcal{E}$, there is an $n \in \mathcal{D}$ such that $e(f_n(x), f(x)) < \frac{\varepsilon}{3}$. Since both f_n and f are continuous, there exists a neighbourhood U of x such that, for each $z \in U$, $e(f_n(z), f_n(x)) < \frac{\varepsilon}{3}$ and $e(f(z), f(x)) < \frac{\varepsilon}{3}$. Hence, for $z \in U$, $e(f_n(z), f(z)) < \varepsilon$, *i.e.*, f_n Bouleau converges to f. We summarise the results in Theorem 12.9.

Theorem 12.9. *Let X be a Hausdorff space, $(Y, \mathcal{E})$ a uniform space, $\mathcal{F}$ the set of functions on X to Y and $C(X,Y)$ the set of continuous functions. Then*

Bouleau.*1 $C(X,Y)$ is closed in $\mathcal{F}$ with Bouleau topology.*

Bouleau.*2 Bouleau topology and pointwise topology coincide on $C(X,Y)$.*

In other words, the pointwise limit of a net of continuous functions is continuous, if and only if, the convergence is Bouleau. If X is compact (in Bouleau convergence of $(f_n : n \in \mathcal{D})$ to f), for each $n_0 \in \mathcal{D}$, X has a finite open cover $(U_{nk} : 1 \le k \le m)$ with $n_k > n_0$ and, for each $z \in X$, there is an n_k such that $z \in U_{nk}$ and so $e(f_{nk}(z), f(z)) < \varepsilon$. Thus we have the theorem of Arzelà extended by Bartle. In this case, the convergence is called **quasi-uniform**.

Theorem 12.10. *Let X be a compact Hausdorff space, $(Y, \mathcal{E})$ a uniform space. Then the net $(f_n : n \in \mathcal{D})$ of continuous functions on X to Y converges to a continuous function f, if and only if, (f_n) converges pointwise to f and, for each $\varepsilon > 0$ and $e \in \mathcal{E}$, eventually there exists a finite set $\{n_k : 1 \le k \le m\}$ such that, for each $x \in X$, there exists a k with $e(f_{nk}(x), f(x)) < \varepsilon$.*

Next, we consider Dini's result. A net of functions $(f_n : n \in \mathcal{D})$ on a Hausdorff space X to $\mathbb{R}$ is **monotone increasing**, if and only if, for each $n \ge m$ and for all $x \in X$, $f_n(x) \ge f_m(x)$. Monotone decreasing is defined similarly. Suppose $\{f_n : n \in \mathcal{D}\}$ is a monotone increasing net of real-valued continuous functions on a compact Hausdorff space and (f_n)

converges pointwise to a continuous function f. Since f is continuous, the convergence is quasi-uniform by Theorem 12.10. It follows that, for $\varepsilon > 0$, *eventually* there exists a finite set $\{n_k : 1 \le k \le m\}$ such that, for each $x \in X$, there exists k with $f(x) < f_{nk}(x) + \varepsilon$. Hence, eventually $f(x) < f_n(x) + \varepsilon$. This proves Dini's result given in Theorem 12.11. There are many more results on this topic (we give a few as problems).

Theorem 12.11. *Let (f_n) be a net of real-valued monotone continuous functions on a compact Hausdorff space converging pointwise to a function f. Then f is continuous, if and only if, the convergence is uniform.*

12.7 Uniform Convergence on Compacta

Uniform convergence is quite strong and a modification has been used extensively in real and complex analysis. Let X be a topological space, $(Y, \mathcal{E})$ a uniform space and $\mathcal{F}$ a set of functions on X to Y. Let $\mathcal{C}$ denote the family of compact subsets in X.

(ucc) A net (f_n) in $\mathcal{F}$ **converges uniformly on compacta** to $f \in \mathcal{F}$, if and only if, for each $\varepsilon > 0, K \in \mathcal{C}$ and $e \in \mathcal{E}$, there exists an $m \in \mathbb{N}$ such that, for all $x \in K$ and $n > m, e(f_n(x), f(x)) < \varepsilon$.

If each f_n is continuous on members of $\mathcal{C}$, then (ucc) implies f is also continuous on members of $\mathcal{C}$. It is an interesting fact that in Euclidean spaces, a function that is continuous on compacta is continuous. This fact leads to an interesting family of topological spaces. A topological space in which a set U is open, if and only if, its intersection with each compact set K is open in K, is called a **k-space**. In a k-space, a function that is continuous on compacta is continuous. Locally compact and first countable spaces are k-spaces (see Prob. 12.11.4). Euclidean spaces are well known examples of k-spaces. The result in Theorem 12.12 shows that a compact open topology is a nice generalization of the classical topology of uniform convergence on compacta.

Theorem 12.12. *Let X be a topological space, $(Y, \mathcal{E})$ a uniform space and $\mathcal{F} \subset C(X, Y)$ (the set of continuous functions on X to Y). Then on $\mathcal{F}$, compact open topology equals topology of uniform convergence on compacta.*

Proof. Let (f_n) in $\mathcal{F}$ converge uniformly on compacta to $f \in \mathcal{F}$ and let $f \in [K,U]$, where K is compact in X and U is open in Y. Since f is continuous and K is compact, $f(K)$ is compact. So there is a $\varepsilon > 0$ such that $S_e(f(K),\varepsilon) \subset U$. Uniform convergence on compacta implies that, for any $e \in \mathcal{E}$, *eventually*, for all $x \in K$, $e(f_n(x), f(x)) < \frac{\varepsilon}{2}$. By the triangle inequality, it follows that *eventually* $S_e(f_n(K),\varepsilon) \subset U$, *i.e.*, $f_n \in [K,U]$. So (f_n) converges to f in compact open topology.

Next, let (f_n) converge to f in compact open topology and let K be a compact set in X. Then, since f is continuous, $f(K)$ is compact. Let $e \in \mathcal{E}$ and $\varepsilon > 0$. By standard compactness argument and noting Y is regular, $f(K)$ can be covered by finitely many closed sets $\{V_j : 1 \le j \le m\}$ such that the diameter of each V_j is less than $\frac{\varepsilon}{3}$ and the set $\{\text{int } V_j : 1 \le j \le m\}$ cover $f(K)$ (see Prob. 12.11.7). Then each $K_j = K \cap V_j$ is compact in X and $K = \bigcup\{K_j : 1 \le j \le m\}$. Since (f_n) converges to f in compact open topology and $f \in \bigcap\{[K_j, V_j] : 1 \le j \le m\}$, *eventually* $f_n(K_j) \subset V_j$ for each j. Hence, *eventually*, for each $x \in K$, $e(f_n(x), f(x)) < \varepsilon$. □

We conclude this section with *equicontinuity*, one of the most important concepts in mathematics, discovered by two Italian mathematicians, Arzelà and Ascoli. Let X be a topological space, $(Y,\mathcal{E})$ a uniform space and $\mathcal{F}$, a set of functions on X to Y. Then $\mathcal{F}$ is **equicontinuous**, if and only if, for each $x \in X, e \in \mathcal{E}$ and $\varepsilon > 0$, there is a neighbourhood U of x such that, for each $f \in \mathcal{F}$, $f(U) \subset S_e(f(x),\varepsilon)$. Obviously, members of an equicontinous family of functions are continuous. Moreover, pointwise closure $\text{cl}\,\mathcal{F}$ of an equicontinuous family $\mathcal{F}$ is also equicontinuous. The proof is akin to that of Theorem 12.5 (see Prob. 12.11.8).

Morse and Kelley discovered *even continuity* as a generalization of equicontinuity, which is similar to compact open topology being a generalization of uniform convergence on compacta. Suppose $\mathcal{F}$ is equicontinuous at $x \in X$ and let $y \in Y$ and $U = S_e(y,\varepsilon)$, $e \in \mathcal{E}$ a neighbourhood of y. Since $\mathcal{F}$ is equicontinuous at x there is a neighbourhood V of x such that, for each $f \in \mathcal{F}, f(V) \subset S_e(f(x),\frac{\varepsilon}{2})$. Then, if $f(x) \in S_e(y,\frac{\varepsilon}{2})$, $f(V) \subset U$, which shows that $\mathcal{F}$ is evenly continuous. On the other hand, if $\text{cl}(\mathcal{F}(x))$ is compact, then equicontinuity equals even continuity (the proof is akin to previous proofs (see Prob. 12.11.9)). Using the above results, one gets the standard result in function spaces, a generalization of one in analysis due to Arzelà and Ascoli.

Theorem 12.13. *Let X be a locally compact Hausdorff space, Y a uniform space, and let $\mathcal{F} \subset C(X,Y)$ have the topology of uniform convergence on*

compacta. Then $\mathcal{F}$ is compact, if and only if,
(a) $\mathcal{F}$ is closed in $C(X,Y)$.
(b) For each $x \in X, \mathrm{cl}\,\mathcal{F}(x)$ is compact.
(c) $\mathcal{F}$ is equicontinuous.

12.8 Graph Topologies

Let $(X,\tau),(Y,\tau')$ be topological spaces with at least two points and let $\mathcal{F}$ be the set of functions on X to Y. Each $f \in \mathcal{F}$ is identified with the graph

$$G(f) = \{(x, f(x)) : x \in X\} \subset X \times Y.$$

Function $f \in \mathcal{F}$ is called **almost continuous**, if and only if, every open set W in $X \times Y$ that contains $G(f)$ also contains graph $G(g)$ of a continuous function $g \in \mathcal{F}$. An example of an almost continuous non-continuous function is the topologist's sine curve (see Fig. 1.6).

A topological space has the **fixed point property**, if and only if, every continuous self-map has a fixed point, *i.e.*, a point p with $f(p) = p$. It is easy to show that if a Hausdorff space has the fixed point property, then every almost continuous function also has this property. From the need to assign a suitable topology to a family of almost continuous functions, graph topology was born. Let $\mathcal{F}$ be the set of functions on X to Y. A typical basic open set in graph topology Γ is

$$W^+ = \{f \in \mathcal{F} : G(f) \subset W\},$$

where W is an open subset of $X \times Y$. Graph topology is unusual, since its properties depend on those of both spaces X and Y as compared to other topologies studied so far, which depend more on the properties of Y.

Let X and Y be T_1 and let f, g denote two distinct functions. Then there is a point $a \in X$ such that $f(a) \neq g(a)$. Since Y is T_1, there is an open set U in Y such that $f(a) \in U$ and $g(a) \notin U$. Since X is T_1, the set $(X - \{a\}) \times Y$ is open in $X \times Y$. Hence,

$$[(X \times U) \cup (X - \{a\}) \times Y]^+ \text{ contains } G(f), \text{ but not } G(g).$$

So the graph topology Γ is T_1. The converse is proven in a similar fashion (see Prob. 12.11.10). The Hausdorff case is similar and we state the results.

Theorem 12.14.
(a) Graph topology Γ is T_1, if and only if, both X and Y are T_1.
(b) Graph topology Γ is T_2, if and only if, both X is T_1 and Y is T_2.

How does graph topology Γ compare with other function space topologies? Under simple conditions, Γ is finer than both pointwise and compact open topologies. Suppose X is Hausdorff (T_2). A typical open set in the compact open topology is $[K,U] = \{f : f(K) \subset U\}$, where K is compact in X and U is open in Y. Then

$$[(X \times U \cup (X - K) \times Y]^+ = [K,U].$$

It is easy to show that if, in addition, X is compact, then graph topology equals compact open topology on $C(X,Y)$.

Theorem 12.15.

(a) If X is T_1, then graph topology Γ is finer than the pointwise topology.

(b) If X is T_2, then graph topology Γ is finer than the compact open topology and, if X is compact, the two topologies coincide on $C(X,Y)$.

Next, suppose $(X,\mathcal{D}),(Y,\mathcal{E})$ are uniform spaces and that $f \in \mathcal{U}$ (the family of uniformly continuous functions on X to Y). For $\varepsilon > 0, e \in \mathcal{E}$, there exist $\eta > 0, d \in \mathcal{D}$ such that $d(x,z) < \eta$ implies $e(f(x),f(z)) < \varepsilon$. Put

$$U = \bigcup\{S_d(x,\eta) \times S_e(f(x),\varepsilon) : x \in X\},$$

which contains $G(f)$, *i.e.*, $f \in U^+$ (an open neighbourhood in graph topology). If $g \in \mathcal{U} \cap U^+$, then, for any $p \in X$, there is a $q \in X$ and $(p,g(p)) \in S_d(q,\eta) \times S_e(f(q),\varepsilon)$.

So $g(p) \in S_e(f(q),\varepsilon) \subset S_e(f(p),2\varepsilon)$, showing thereby that the topology of uniform convergence is coarser than graph topology on the set of uniformly continuous functions. The converse holds if X is compact. In that case, all continuous functions are uniformly continuous and their graphs are compact subsets of $X \times Y$. Suppose f is a uniformly continuous function in U^+, where U is open in $X \times Y$. Then $G(f)$ is compact and is a subset of an open set Y, there are $\varepsilon > 0, \eta > 0$ such that

$$\bigcup\{S_d(x,\eta) \times S_e(f(x),\varepsilon) : x \in X\} \subset U.$$

This shows that U^+ is open in the topology of uniform convergence. Thus

Theorem 12.16. *Let X and Y be uniform spaces. Then on the set of uniformly continuous functions, graph topology Γ is finer than the topology of uniform convergence. Further, if X is compact, then Γ and $\mathcal{U}$ are equal.*

12.9 Inverse Uniform Convergence for Partial Functions

Suppose X and Y are topological spaces. In this section, maps are not necessarily continuous. A map f from a nonempty subset of X to Y is called a partial map. Partial maps occur as inverse functions in elementary analysis, as solutions of ordinary differential equations, as utility functions in mathematical economics and so on. Yet there is very little literature on topologies on partial functions.

In many applications, X and Y are metric spaces and there is a need to have a *uniform* convergence on a family of partial functions. Since partial maps do not have a common domain, the usual uniform convergence is not available. Notice that in many situations, all maps of a family under consideration have a common range, we study *inverse uniform convergence* **(iuc)** that is complementary to the usual one. This (iuc) does not preserve continuity but preserves (uniform) openness. The usefulness of (iuc) stems from the fact that it can be used when uniform convergence cannot be defined. Moreover, in some situations where both uniform convergence and (iuc) are available, (iuc) satisfies our intuition but not the former.

Suppose (X, d) is a metric space and Y a topological space. A partial map f on X to Y is a subset of $X \times Y$ such that there is a set $\pi(f)$ (domain of f), a nonempty subset of X, such that for each $x \in \pi(f)$, there is a unique point $f(x) \in Y$. Here is an example of a family $\mathcal{F}$ of partial maps. For each $z \in \mathbb{R}, f_z(x) = log(x + z)$. The domain $\pi(f_z) = (-z, \infty)$ is an open subset of $\mathbb{R}$ that varies with z and the family $\mathcal{F} = \{f_z : z \in \mathbb{R}\}$ has a common range $\mathbb{R}$. Looking at the graphs of the family, it is easy to observe that as $z \to 0, f_z \to f_0$. Hence, the problem is to put a topology on the family $\mathcal{F}$ such that the map

$$(****)\ \psi : \mathbb{R} \to \mathcal{F} \text{ given by } \psi(z) = f_z$$

is continuous. It will be seen later that a similar problem *continuity of the solutions with respect to initial conditions* occurs in ordinary differential equations.

Let P denote a family of partial maps on X with a common range Y. In the literature on partial maps, Kuratowski considers continuous maps with *compact* domains, Sell and Abdallah-Brown deal with continuous maps with *open* domains, while Back and Filippov work with continuous maps with *closed* domains. Here, we study partial maps that are not necessarily continuous and whose domains are arbitrary subsets of X.

In this case, for each $y \in Y$ and $f \in P, f^{-1}(y)$ need not be a closed subset of X. Hence, we use the Hausdorff pseudometric d_H on $2^X - \{\varnothing\}$. It is known that d_H is a metric on $CL(X)$, the family of all nonempty closed subsets of X. On P, we now define $\rho = \rho_d$, the metric of **inverse uniform convergence**, assuming that, for any two distinct functions f and g in P, there is a $y \in Y$ such that $d_H(f^{-1}(y), g^{-1}(y)) > 0$. This condition is certainly satisfied when all the functions in P are continuous. For f and g in P, the metric ρ is defined by

$$(\rho): \qquad \rho(f,g) = sup\{d_H(f^{-1}(y), g^{-1}(y)) : y \in Y\}.$$

If, for a net $(f_n), \rho(f_n, f) \to 0$, then we say that (f_n) **converges inverse uniformly** to f. It is easily shown that ρ is an infinite-valued metric. In many cases, the members of P are injective and ρ has a simple form given in the sequel.

There is a map $\lambda : f \to \pi(f)$ from (P, ρ) to $(\{2^X - \varnothing\}, d_H)$. It is easy to see that $\rho(f_n, f) \to 0$ implies $d_H(\pi(f_n), \pi(f)) \to 0$. This shows that λ is continuous. That the converse is not true is shown by the following example: $\{f_z : z \in \mathbb{R}\}$, where

$$f_z(x) = z + log(x) : x \in (0, \infty).$$

The following examples show that uniform convergence and inverse uniform convergence are independent and complement each other. To see this, we assume that Y is a metric space with metric e and it gives rise to μ, the metric of uniform convergence on Y^X, namely,

$$(\mu): \qquad \mu(f,g) = sup\{e(f(x), g(x)) : x \in X\}.$$

(**iuc**.1) Let $F = \{f_z : z \in \mathbb{R}\}$, where,

$$f_z(x) = cos(x + z) : x \in [0, 2\pi].$$

Here the maps have a common range, namely, $[-1, 1]$. If we look at the map $\psi : \mathbb{R} \to F$, given (****), this map is continuous when F is given the metric ρ or μ. So both uniform convergence and (iuc) are equally efficacious. Notice that with ρ on F, ψ is an isometry!

(**iuc**.2) Let $F = \{f_z : z \in \mathbb{R}\}$, where,

$$f_z(x) = (x + z)^3 : x \in \mathbb{R}.$$

Here the maps have a common range, namely, $\mathbb{R}$. In this case, the map ψ defined in (****), is not continuous when F is given the metric μ of uniform convergence but it is an isometry with respect to the metric ρ of (iuc).

(**iuc**.3) Let $F = \{f_z : z \in \mathbb{R}\}$, where,

$$f_z(x) = z + x^{\frac{1}{3}} : x \in \mathbb{R}.$$

Here the maps have a common range, namely, $\mathbb{R}$. In this case, the map ψ is an isometry when F is given the metric μ of uniform convergence but is not continuous with respect to the metric ρ of (iuc).

(**iuc**.4) Let $F = \{f_z : z \in \mathbb{R}\}$, where,

$$f_z(x) = (x+z)^3 \text{ for } x \in (-z, \infty),$$

and

$$f_z(x) = z + x^{\frac{1}{3}} \text{ for } x \in (-\infty, -z).$$

Here the maps have $\mathbb{R}$ as a common domain as well as a common range. In this case, however, the map ψ is not continuous when F is assigned the metric ρ or μ.

Next, we give an example to show that (iuc) does not preserve continuity, namely,

$$f(x) = sin(x^{-1}) \text{ for } x > 0, f(0) = 0,$$
$$f_n(x) = sin(x^{-1}) \text{ for } x \geq (n\pi)^{-1} \text{ for } n \in \mathbb{N}.$$

Here, $X = [0, \infty), Y = [-1, 1]$, and each f_n is a continuous partial map and $\rho(f_n, f) \to 0$. We can extend this example to ordinary maps on X by setting $f_n(x) = 0$, for $x \in [0, (n\pi)^{-1}]$.

Uniform convergence preserves continuity and uniform continuity. Naturally, (iuc) is expected to preserve (uniform) openness. Suppose each f_n is uniformly open and $\rho(f_n, f) \to 0$. We must show that f is uniformly open, *i.e.*, for each $\varepsilon > 0$, there is a $\delta > 0$ such that,

$$\text{for each } x \in \pi(f), S_e(f(x), \delta) \subset f(S_d(x, \varepsilon)).$$

Since f_n inverse uniformly converges to f, eventually $\rho(f_n, f) < \frac{\varepsilon}{3}$. Also, the assumption that f_n is uniformly open implies that there is a $\delta > 0$ such that

$$S_e(f(x), \delta) \subset S_e(f_n(x_n), \delta) \subset f_n(S_d(x_n, \frac{\varepsilon}{3})) \subset f(S_d(x, \varepsilon)).$$

The case of preservation of openness is shown similarly and this yields the result in Theorem 12.17.

Theorem 12.17. *Suppose (f_n) is a net of (uniformly) open maps in P that converges inverse uniformly to $f \in P$. Then f is (uniformly) open.*

It is obvious from the proof of Theorem 12.17 that the result remains true if only a subnet of (f_n) converges inverse uniformly to f. In this case, we say that (f_n) **simply inverse uniformly converges** to f.

Here is an application of (iuc) to ordinary differential equations (ODEs). In elementary analysis, several inverse functions that occur are partial functions. We saw such a function at the beginning of this section. Obviously, inverse trigonometric functions as well as square root functions are partial maps. These functions and others occur naturally as solutions of the ODE

$$y' = f(y),$$

where f is a continuous function in an interval. Next, we give some examples. In each case, $z \in \mathbb{R}$.

(**ODE**.1) $y' = exp(-y), y \in \mathbb{R}$, whose solutions are

$$y = log(x + z), x \in (-z, \infty).$$

Here, the solutions are partial maps whose domains are *open* sets of $\mathbb{R}$. This example was considered earlier in another context.

(**ODE**.2) $y' = [cos(y)]^{-1}, y \in [0, \frac{\pi}{2})$, whose solutions are

$$y = arcsin(x + z), x \in [-z, -z + 1].$$

Here, the solutions are partial maps whose domains are *closed* subsets of $\mathbb{R}$.

(**ODE**.3) $y' = y\sqrt{y^2 - 1}, y \in [1, \infty)$, whose solutions are

$$y = sec(x + z), x \in [-z, -z + \frac{\pi}{2}].$$

In (ODE.3), the solutions are partial maps whose domains are neither open nor closed subsets in $\mathbb{R}$.

Notice that if both f_z and $f_{z'}$ are any two maps from the same family in any of the above examples, then $\rho(f_z, f_{z'}) = |z - z'|$. This shows that although each family consists of partial maps having different character from that of another family, when each is assigned the metric of the (iuc) ρ, the family is isometric to $\mathbb{R}$. Also notice that if we view z as an *initial condition*, then we have a much stronger result than the one mentioned in the theory of ODE, namely, *solutions are continuous functions of the initial conditions.* Also notice that this gives the best solution to the problem posed in the motivation.

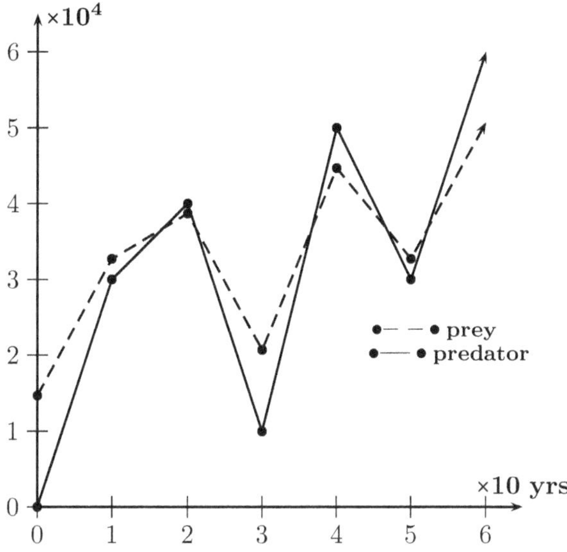

Fig. 12.1 Lotka-Volterra predator-prey model

12.10 Application: Hit and Miss Topologies in Population Dynamics

This section introduces an approach to representing and analysing population dynamics of interacting species using hit and miss topologies. The Volterra principle asserts that an intervention in a prey-predator system that removes both prey and predators in proportion to their population sizes has the effect increasing average prey populations [34]. Evidence of the Volterra principle at work can be seen, for example, in the study of pelt-trading records of the Hudson-Bay Company, where there is a near periodic oscillation in the number of trapped snowshoe hares and lynxes between 1840 and 1930 [165]. Lotka and Volterra independently proposed a mathematical model for prey and predator population dynamics. Let $U(t), V(t)$ denote the number of prey and predators at time t, respectively. The parameters α, β denote the average per capita birthrate of prey and deathrate of predators, respectively. The parameter γ is the fraction of prey caught per predator per unit time. In the absence of intervention (hunting or fishing, *e.g.*), the Lotka-Volterra predator-prey equations are

$$\frac{dU}{dt} = \alpha U - \gamma UV, \qquad \frac{dV}{dt} = e\gamma U - \beta V$$

An explanation of this model is given in detail in [34] and not repeated here. This variation in prey-predicator population sizes is partially represented in Fig. 12.1.

The basic approach in this application of topology is to consider populations in terms of hit and miss sets of species in a hypertopological space. This approach complements the traditional approach in representing the spatial interaction of predator-prey populations in terms of ordinary differential equations and as graphs of predator vs. prey densities from which conditions for population stability of the interaction are predicted [202]. It has been observed by V.A.A. Jensen and A.M. de Roos that the classical ODE model may not accurately represent the interactions among individuals when they move diffusively [53]. The proposed topological approach in representing predator-prey populations permits not only modeling spatial relations (*e.g.*, predators and prey occupying the same patch in what are known as hit sets) but also representations of population members in terms of their features. For example, a consideration of descriptive near or descriptively far sets can be used to model predator-prey populations, taking into account energy gain relative prey features such size (big vs. small) [114].

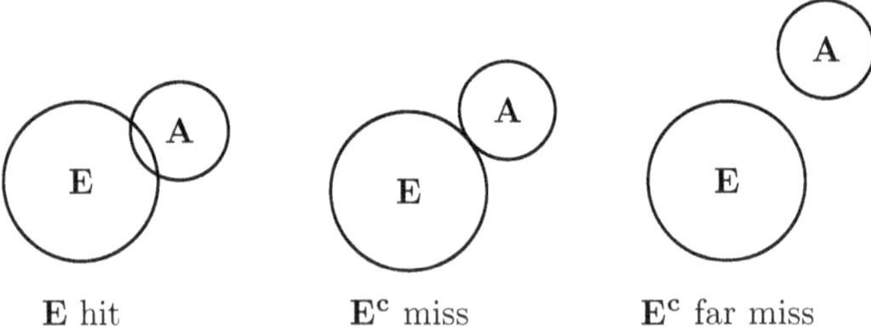

Fig. 12.2 Hit, miss and far miss sets

Let X be a Hausdorff space containing population members, $E \subset X$ and let $CL(X)$ denote a hyperspace (collection of all nonempty closed subsets of X). For the details about hyperspaces, see [105; 26; 50; 158]. The basic approach is to split $CL(X)$ into hit and miss sets.

(**E**.1) **hit**: In a Vietoris topology on $CL(X)$, the hit part is the set $E \subset X$ in

$$E^{-} := \{A \in CL(X) : A \cap E \neq \varnothing, \text{ for hit set } E \subset X\}.$$

(**E**.2) **miss**: A nonempty closed set A misses $E \subset X$, provided $A \cap E = \varnothing$, *i.e.*, $A \in (E^c)^+$, where the miss set E^c in the Vietoris topology on $CL(X)$, is generated by the collection E^+:

$$E^{+} := \{A \in CL(X) : A \subset E\}, \text{ miss set } E^c.$$

(**E**.3) **strongly miss**: A nonempty closed set A strongly misses $E \subset X$, provided $A \cap E = \varnothing$, *i.e.*, $A \in (E^c)^{++}$, where E^c is the *far* set and A is far from E^c in

$$E^{++} := \{A \in CL(X) : A \ll_\delta E, i.e., A \underline{\delta} E^c\}, \text{ far set } E^c.$$

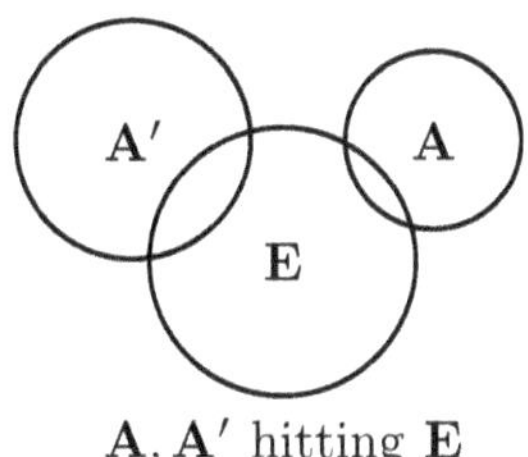

A, A′ hitting **E**

Fig. 12.3 Multiple sets hitting E

Sample hit, miss, and far miss sets are shown in Fig. 12.2. The set E^- contains those closed sets that hit E. Only one closed set is shown hitting E in Fig. 12.2. Assuming that sets A contain a combination of predators and prey that begin migrating into the patch occupied by animals in E, then that situation is represented by Fig. 12.3. Similar situations can be represented for miss and far miss sets by expanding the view of H^+, H^{++} in Fig. 12.2. The join of the finite hit part and miss part is known as *Vietoris topology* [234].

Lemma 12.1. *For a Hausdorff space X and $A \in CL(X)$, $A \in (E^c)^+$, if and only if, $A \in (E^-)^c$.*

Proof. $A \in (E^c)^+$, if and only if, $A \cap E = \varnothing$, if and only if, $A \in (E^-)^c$. □

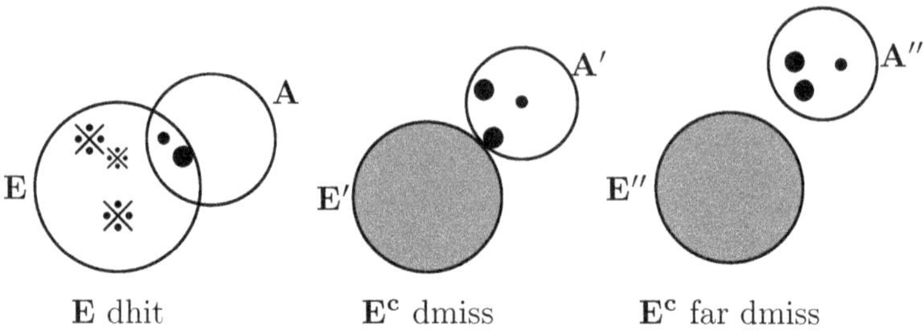

Fig. 12.4 Descriptive hit and miss sets

In keeping with the need to consider the features of members of a population in considering the stability or instability of local ecological systems, a descriptive Vietoris topology is introduced, where one considers descriptive hit, miss and far miss sets.

(E_Φ.1) **descriptive hit**: In a descriptive Vietoris topology on $CL(X)$, the hit part is the set $E \subset X$ that has a nonempty intersection $\underset{\Phi}{\cap}$ with sets A in $CL(X)$, *i.e.*,

$$E_\Phi^- := \left\{ A \in CL(X) : A \underset{\Phi}{\cap} E \neq \emptyset, \text{ for hit set } E \subset X \right\}.$$

There are a number of cases to consider for E_Φ^-. One of these cases is illustrated in Fig. 12.4.

(E_Φ.2) **descriptive miss**: A nonempty closed set A *descriptively* misses $E \subset X$, provided $A \underset{\Phi}{\cap} E = \emptyset$, *i.e.*, $A \in (E^c)_\Phi^+$, where the miss set E^c in the Vietoris topology on $CL(X)$, is generated by the collection E_Φ^+:

$$E_\Phi^+ := \left\{ A \in CL(X) : A \underset{\Phi}{\subset} E \right\}, \text{ miss set } E^c,$$

where

$$A \underset{\Phi}{\subset} E = \{ a \in A : \Phi(a) \in \Phi(E) \}.$$

(E_Φ.3) **descriptive strongly miss**: A nonempty closed set A strongly *descriptively* misses $E \subset X$, provided $A \underset{\Phi}{\cap} E = \emptyset$, *i.e.*, $A \in (E^c)_\Phi^{++}$, where E^c is the *far* set in

$$E_\Phi^{++} := \{ A \in CL(X) : A \ll_{\delta_\Phi} E, i.e., A \underline{\delta}_\Phi E^c \}, \text{ far set } E^c.$$

Theorem 12.18. *For a descriptive Hausdorff space X and $A \in CL(X)$, $A \in (E^c)_\Phi^+$, if and only if, $A \in (E_\Phi^-)^c$.*

Proof. Immediate from the descriptive counterpart of Lemma 12.1. □

Example 12.1. Descriptive Hits and Misses.
Sample descriptive hit, miss, and far miss sets are shown in Fig. 12.4. In the descriptive hit case in Fig. 12.4, E is the dhit set, since E and A have, for example, descriptively similar elements such as • and ● in common. That is, the intersection of E and A contain members that resemble each other (in Fig. 12.4, both sets contain black discs). By contrast, the spatial gap between E' and A' in Fig. 12.4 is close to zero and E' and A' are descriptively different. The uniformly dark shading of E' contrasts with the uniformly white appearance of A' and represents the descriptive difference between E' and A' in Fig. 12.4. The spatial gap between E'' and A'' in Fig. 12.4 is large (A'' *strongly misses* E'') and E'' and A'' are descriptively different. ■

The join of the finite descriptive hit part and descriptive miss part is known as a *descriptive Vietoris topology.* By varying the construction of descriptive hit and miss sets, we obtain models for stable and unstable populations. This approach works well in terms of the patch approach in the study of predator-prey populations. A *patch* is local neighbourhood wherein resides predators or prey or both predators and prey.

It is well-known that space complicates ecological interactions inasmuch as it allows for non-uniform patterns of environment and population density as well as movement from one location to another. Local uniqueness is considered the most fundamental contributor to spatial heterogeneity [126].

In keeping with the need to consider features of predators and prey (population description) as well as the spatial distribution population members in patches or ecological niches, different cases for hit and miss sets are introduced. Let ⋇, ※, ※ represent an indeterminate number of different size prey, and let •,•,● represent an indeterminate number of different size predators. Let X, E, A be sets of animals, where $E \subset X, A \in CL(X)$.

(**case**.1) **(sp)Hit**. Let E(sp)hit denote a spatial hit set, where the intersection of the hit set E and set A is not empty. For example, $E \cap A$ have predator groups •,• in common. This is case (E.1), where $E\ \delta\ A$.

(**case**.2) **(sp+d)Hit**. Let E(sp+d)hit denote a spatial hit set, where the intersection of the hit set E and set A is not empty such that

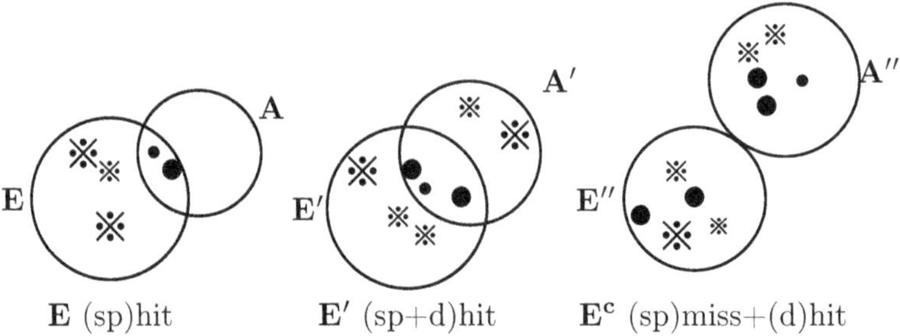

Fig. 12.5 Forms of hit and miss sets

A has members that are descriptively near but not spatially near members of E. For example, in Fig. 12.5, $E \cap A$ have predator groups •,● in common and A has prey ※, ※ that are descriptively but not spatially near members of E. This is case (E.1), where $E\ \delta\ A$ combined with case (E$_\Phi$.1), where $E\ \delta_\Phi\ A$.

(**case**.3) **(sp)Miss+(d)Hit**. Let E^c(sp)miss+(d)hit denote a spatial miss set that is also a descriptive hit set, where the intersection of the hit set E and $A \subset E^c$ is empty but A has members that are descriptively near but not spatially near members of E. For example, in Fig. 12.5, $E \cap A$ is empty but A has prey ※, ※ and predators •,● that are descriptively but not spatially near members of E. This is case (E$_\Phi$.1), where $E\ \delta_\Phi\ A$ combined with case (E.2), where $E\ \underline{\delta}\ E^c$.

Other cases are possible, if you consider (E.3) and (E$_\Phi$.3), but not considered here. To quantify the nearness of hit and miss sets, use

($\boldsymbol{\mu}$.1) **(sp)Hit**. Spatial nearness of hit set measure relative to $CL(X)$:

$$\mu_{(sp)E^-}(X) = \frac{|E^-|}{|CL(X)|}.$$

($\boldsymbol{\mu}$.2) **(d)Hit**. Descriptive nearness of hit set measure relative to $CL(X)$:

$$\mu_{(d)E^-}(X,\Phi) = \frac{|E^-_\Phi|}{|CL(X)|}.$$

($\boldsymbol{\mu}$.3) **(sp+d)Hit** Spatial-descriptive nearness of hit set measure relative to $CL(X)$:

$$\mu_{(sp+d)E^-}(X,\Phi) = \frac{|E^- \cap E^-_\Phi|}{|CL(X)|}.$$

(μ.4) **(sp)Miss**. Measure of degree of spatial miss relative to $CL(X)$:

$$\mu_{(sp)E^+}(X) = \frac{|E^+|}{|CL(X)|}.$$

(μ.5) **(d)Hit**. Measure of degree of descriptive miss relative to $CL(X)$:

$$\mu_{(d)E^+}(X, \Phi) = \frac{|E_\Phi^{++}|}{|CL(X)|}.$$

(μ.6) **(sp+d)Miss**. Measure of degree of spatial-descriptive miss relative to $CL(X)$:

$$\mu_{(sp+d)E^+}(X, \Phi) = \frac{|E^+ \cap E_\Phi^+|}{|CL(X)|}.$$

(μ.7) **(sp+d)Hit-Miss**. Measure of degree of spatial-descriptive hit-miss relative to $CL(X)$:

$$\mu_{(sp+d)h-m}(X, \Phi) = \frac{|E^- \cap E_\Phi^- \cap E^+ \cap E_\Phi^+|}{|CL(X)|}.$$

12.11 Problems

(**12.11**.1) In a T_3 space, show that if $A \subset U$, where A is compact and U is open, then there is an open set V such that $A \subset V \subset \mathrm{cl}\, V \subset U$.

(**12.11**.2) Give a detailed proof of Theorem 12.3(b) for the Tychonoff case.

(**12.11**.3) Let A, B be compact subsets of X, Y, respectively. If W is a neighbourhood of $A \times B$ in $X \times Y$, show that there are neighbourhoods U of A, V of B, such that $U \times V \subset W$.

(**12.11**.4) Show that locally compact and first countable spaces are k-spaces.

(**12.11**.5) Show that the result in Theorem 12.4(b) is true if X is a Hausdorff k-space.

(**12.11**.6) Show that a compact Hausdorff topology is the coarsest Hausdorff topology.

(**12.11**.7) Let $(Y, \mathcal{E})$ be a uniform space, $e \in \mathcal{E}, \varepsilon > 0$ and M a compact subset of Y. Then prove that M can be covered by finitely many closed sets $\{V_j : 1 \le j \le m\}$ such that $\{\mathrm{int}\ V_j : 1 \le j \le m\}$ cover M.

(**12.11**.8) Show that the pointwise closure of an equicontinuous family of functions is equicontinuous.

(**12.11**.9) Let X be a topological space, $(Y, \mathcal{E})$ a uniform space and $\mathcal{F}$ a set of functions on X to Y. $\mathcal{F}$ is evenly continuous at $x \in X$ and $\mathrm{cl}\,\mathcal{F}(x)$ is compact. Show that $\mathcal{F}$ is equicontinuous at x.

(**12.11**.10) **Graph Topology** Γ. Prove

(a) $\{W^+ : W \text{ is open in } X \times Y\}$, where $W^+ = \{f \in \mathcal{F} : G(f) \subset W\}$ is a base for Γ.

(b) Γ is T_2, if and only if, X is T_1 and Y is T_2.

(c) Uniform topology is coarser than Γ on $C(X, Y)$.

(d) If X is T_1, then Γ is finer than the pointwise topology.

(e) A Hausdorff space X is compact, if and only if, Γ and compact open topology coincide on $C(X, Y)$.

(**12.11**.11) **Alexandroff Convergence**. Let (f_n) be a sequence of functions from a topological space X to a metric space (Y, e). Then (f_n) is *Alexandroff convergent* to a function $f : X \to Y$, if and only if, (f_n) converges pointwise to f and, for every $\varepsilon > 0$ and $n_0 \in \mathbb{N}$, there exists a countable open cover $\{G_n\}$ of X and a sequence (n_k) of positive integers greater than n_0 such that, for each $x \in G_k$, $e(f_{nk}(x), f(x)) < \varepsilon$. If each f_n is continuous, show that f is continuous, if and only if, (f_n) Alexandroff converges to f.

(**12.11**.12) **Strong Convergence** [61]. Let (f_n) be a net of functions on a space X to a regular space Y. Then (f_n) **converges strongly**(cs) to f, if and only if, (a) (f_n) converges pointwise to f and (b), for each open cover $\mathcal{G}$ of Y and $N \in D$, there is a finite set $F \subset D$, each member of F greater than N, such that, for each $x \in X$, there exists a $k \in F$ and $G \in \mathcal{F}$ with $\{f_k(x), f(x)\} \subset G$.

(**cs**.1) Show that if each f_n is continuous and converges strongly to f, then f is continuous.

(**cs**.2) If Y is a uniform space, show that strong convergence implies quasi-uniform convergence. If Y is compact, then the two convergences are equal.

(**cs**.3) Show that X is compact, if and only if, strong convergence equals pointwise convergence.

(**12.11**.13) **Duality in Function Spaces** [71]. Let X and Y be regular T_3 spaces. Let $\mathcal{F}$ be a set of functions on X to Y. With each $x \in X$, associate a function $\underline{x} : \mathcal{F} \to Y$ given by $\underline{x}(f) = f(x)$. Thus there is a natural map $v : X \to \underline{X}$ given by $v(x) = \underline{x}$. Prove the following results.

(**duality**.1) $\mathcal{F}$ has the point-open topology, if and only if, $\underline{x} \in \underline{X}$ is continuous. Dually, $\underline{X}$ has the point-open topology, if and only if, each $f \in \mathcal{F}$ is continuous.

(**duality**.2) If $\underline{X}$ has the point-open topology, then the *natural map* $v : X \to \underline{X}$ is continuous.

(**12.11**.14) **Proximal Convergence** [52; 154]. Let $(f_n : n \in D)$ be a net of functions on a uniform space X to a uniform space Y. Let X and Y have induced Efremovič proximities. Prove

(**pc**.1) If each f_n is continuous and the net (f_n) proximally converges to $f : X \to Y$, then the set $\{f_n\}$ is eventually equicontinuous and the convergence is uniform on compacta.

(**pc**.2) If X is compact, then on $C(X, Y)$ proximal convergence equals uniform convergence.

(**pc**.3) If the uniformity on Y is totally bounded, then on $C(X, Y)$ proximal convergence equals uniform convergence [123].

(**12.11**.15) **Hausdorff Convergence** [22; 119; 153]. Let $(X, d), (Y, e)$ be metric spaces and (f_n) a sequence of functions on X to Y. Let $X \times Y$ be assigned the product metric ρ. Let $\rho(f, g)$ denote the Hausdorff distance between $G(f)$ and $G(g)$. If $\rho(f_n, f) \to 0$, then it is called Hausdorff convergence. Prove

(**Hc**.1) Let f_n, f be functions on $\mathbb{R}$ to $\mathbb{R}$, given by

$$f_n(x) = (x + \frac{1}{n})^2, f(x) = x^2.$$

Then $(f_n) \to f$ is both pointwise and Hausdorff convergence but not uniform.

(**Hc**.2) Uniform convergence implies Hausdorff convergence. These two types of convergence are equal, if and only if, X is a UC space, *i.e.* on X, continuous functions are uniformly continuous.

(**Hc**.3) Compare Hausdorff convergence with convergence in graph topology.

Chapter 13

Hyperspace Topologies

13.1 Overview of Hyperspace Topologies

We begin with a brief overview of hyperspace topologies and introduce three forms of hyperspace topologies. During the early part of the 20th century, there were two main studies on the hyperspace[1] of all nonempty closed subsets of a topological space X. Vietoris introduced a topology consisting of two parts:
Vietoris.(a) Lower *finite hit* part, and
Vietoris.(b) Upper *miss* part.
This leads to

> The join of Vietoris.(a) and Vietoris.(b) is known as **Vietoris topology** or finite topology. Replacing Vietoris.(b) (upper miss part) by *far* when X has proximity, gives rise to **proximal topology** on the hyperspace. Note that upper Vietoris topology on hyperspace $X \times Y$ was used in defining graph topology in Ch. 12.

As observed in §1.17, Hausdorff defined a metric on the hyperspace of a metric space and the resulting topology depends on the *metric* rather than on the topology of the base space. Two equivalent metrics on a metrisable space induce equivalent **Hausdorff metric topologies**, if and only if, they are uniformly equivalent. With the discovery of uniform spaces, the Hausdorff metric was generalized to **Hausdorff-Bourbaki uniformity** or H-B uniformity on the hyperspace of uniformisable (Tychonoff) space. In

[1]Recall that the set of all nonempty closed subsets of a space X (denoted $CL(X)$) is called hyperspace.

the 1960s, this led to the celebrated Isbell-Smith problem to find necessary and sufficient conditions that two uniformities on the base space are H-equivalent, *i.e.*, they induce topologically equivalent H-B uniformities on the hyperspace.

We show that the H-B uniform topology has two parts, namely,

(**HB**.1) lower *locally finite hit* part, and
(**HB**.2) upper *proximal miss* part.

Hyperspaces are useful in areas such as unilateral analysis, optimization theory, convex analysis, geometric functional analysis, mathematical economics, and the theory of random sets (see Historical Notes section for details).

13.2 Vietoris Topology

The basics of what is known as Vietoris topology (denoted τ_V) are presented in this section. Let X be a nonempty set with a T_1 topology τ and let $CL(X)$ (respectively, $K(X)$) denote the family of all nonempty closed (respectively, compact) subsets of X. The Vietoris topology τ_V on $CL(X)$ has two parts, namely,

(**Vietoris**.1) lower *hit* part in which a typical subbase element consists of members of $CL(X)$ that hit (intersect) an open set V (denoted by V^-).

(**Vietoris**.2) upper *miss* part in which a typical subbase element consists of members of $CL(X)$ that miss closed set $V^c = X - V$, where V is an open set. This miss part is denoted by

$$V^+ = \{E \in CL(X) : E \subset V\}.$$

Thus, Vietoris topology is a **hit-and-miss topology**. A typical basis element for the Vietoris topology τ_V is $W^+ \cap (\bigcap \{V_j^- : 1 \leq j \leq k\})$, where W and V_j are open sets in X. Sets $\{V_j\}$ are called **hit sets**. The above basis is equivalent to a basis whose typical member is

$$\langle V_1, \ldots, V_k\rangle = \{E \in CL(X) : E \cap V_j \neq \varnothing, E \subset \bigcup(V_j)\},$$

i.e., a typical member consists of closed subsets that intersect each V_j and are subsets of their union (see Problem 1).

Map $i : x \to \{x\}$ from X into $CL(X)$ is a homeomorhphism (see Problem 1(b)), a fact that is called *admissible*. A natural omnibus problem arises, namely, which topological properties such as separation axioms, connectedness, (local) compactness of the base X, are inherited by the hyperspace. Here we discuss a few properties and give some as problems. Some properties such as those specified by separation axioms get weaker from X to $CL(X)$.

Suppose X is compact. Then to show that $CL(X)$ is compact, we use Alexander's Lemma that it is sufficient to consider open covers of subbasic sets. If $\{V_j^+\}$ is an open cover of $CL(X)$, then there must be one V_j containing X, *i.e.*, $V_j = X$ and this certainly contains all closed subsets. In the case where $\{V_j^-\}$ is an open cover of $CL(X)$, then, for each $x \in X$, $\{x\}$ is in some V_j and the compactness of X implies that there is a finite subset V_js covering X and so every closed set intersects one of these. Conversely, suppose $CL(X)$ is compact and $\{V_j\}$ is an open cover of X. Then $\{\langle X, V_j\rangle\}$ is an open cover of $CL(X)$ and compactness gives a finite subcover $\{\langle X, V_j\rangle : 1 \le j \le k\}$. Then $\{V_j : 1 \le j \le k\}$ is a subcover of $\{V_j\}$. Thus we have proved Theorem 13.1.

Theorem 13.1. *Let (X, τ) be a T_1 space, $CL(X)$ the space of all nonempty closed sets with Vietoris topology τ_V. Then X is compact, if and only if, $CL(X)$ is compact.*

Next consider separation axioms. The axioms get weaker from X to $CL(X)$ but most are preserved from X to $K(X)$. As an example, suppose X is T_3 (regular) and A, B are two members of $CL(X)$. Then we may suppose that there is a $p \in A - B$. By regularity, p and B have disjoint open neighbourhoods U, V, respectively. Then $A \in U^-, B \in V^+$ and $\langle X, U\rangle, \langle V\rangle$ are disjoint, showing $CL(X)$ is Hausdorff. Conversely, if X is not regular, there is a point p and a closed set B which do not have disjoint neighbourhoods. Then B and $B \cup \{p\}$ do not have disjoint neighbourhoods.

Theorem 13.2. *Let (X, τ) be a T_1 space, $CL(X)$ the space of all nonempty closed sets with Vietoris topology τ_V. Then X is regular, if and only if, $CL(X)$ is Hausdorff.*

13.3 Proximal Topology

Let (X, τ) be a T_1 space with a compatible L-proximity δ. Then Vietoris topology can be generalized to proximal topology τ_δ on $CL(X)$ by replacing the upper *miss* part by *far*. A typical subbase element of upper proximal topology τ_δ consists of members of $CL(X)$ that are far from a closed set $V^c = X - V$, where V is an open set. This is denoted by

$$V^{++} = \{E \in CL(X) : E \ll V\}.$$

Lower proximal topology is the same the lower Vietoris topology. Proximal topology is a **hit-and-far** topology. Notice that if the proximity is fine (*i.e.*, $A\ \delta_0\ B$, if and only if, closures of A, B intersect), then proximal topology equals the Vietoris topology and X is a UC-space (continuity equals proximal continuity).

As one would expect, an important special case of proximal topology occurs when the proximity is EF (Efremovič). In this case, τ_δ is coarser than τ_V. Suppose V is an open set in the Tychonoff space X with EF-proximity δ. If a closed set $A \in V^{++} \in \tau_\delta$, that is, $A \ll V$, then there is an open set U such that $A \ll U \ll V$. So, $A \in U^+ \in \tau_V$ and $U^+ \subset V^{++}$. This shows that $\tau_\delta \subset \tau_V$. Obviously, τ_δ and τ_V are equal, if and only if, $\delta = \delta_0$. In which case, X is normal by the Urysohn Lemma.

Theorem 13.3. *Let X be a T_1 Tychonoff space with EF-proximity δ. Proximal topology τ_δ is Tychonoff and coarser than Vietoris topology τ_V. Further, $\tau_\delta = \tau_V$, if and only if, δ is fine ($\delta = \delta_0 = \delta_F$). In which case, X is a UC normal T_4 space. Notice that UC spaces are complete in case X is metrisable.*

13.4 Hausdorff Metric (Uniform) Topology

In Section 14 of Chapter 1, we briefly discussed Hausdorff metric topology. Now we look into this topology in greater detail and compare it with Vietoris topology and proximal topology.

Let (X, d) be a metric space and $CL(X)$ the family of all nonempty closed subsets of X. The **Hausdorff metric** d_H on $CL(X)$ is

$$d_H(A,B) = \begin{cases} inf\{\varepsilon > 0 : A \subset S(B,\varepsilon), B \subset S(A,\varepsilon)\}, & \text{if there is } \varepsilon > 0, \\ \infty, & \text{otherwise.} \end{cases}$$

A neighbourhood of $A \in CL(X)$ in the **Hausdorff metric topology** τ_{Hd} can be split into two parts, namely,

($\boldsymbol{\tau_{Hd}}$.1) lower $\{E \in CL(X) : A \subset S(E,\varepsilon)$ for some $\varepsilon > 0\}$.

($\boldsymbol{\tau_{Hd}}$.2) upper

$$\{E \in CL(X) : E \ll_\delta S(A,\varepsilon) \text{ for some } \varepsilon\},$$

where δ is a metric proximity on X. Also, note that $A \ll_\delta S(A,\varepsilon)$.

As explained in ch. 1, the Hausdorff metric topology τ_{Hd} depends on the metric and not on the topology of X alone.

We now see how the lower Hausdorff metric topology can be expressed as a *hit* topology. For $A \in CL(X)$ and $\varepsilon > 0$, there are maximal $\frac{\varepsilon}{2}$-discrete subsets $\{A_\varepsilon\}$ of A by Zorn's Lemma. In other words, for each $A_\varepsilon \subset A$, the family of open balls $\{S(x, \frac{\varepsilon}{2}) : x \in A\}$ is pairwise disjoint and is maximal in satisfying this property. Moreover, the family $\{S(x, \frac{\varepsilon}{2}) : x \in A_\varepsilon\}$ is discrete, *i.e.*, each point of X has a neighbourhood that intersects at most one member of the family. If $E \in CL(X)$ intersects (hits) each member of $\{S(x, \frac{\varepsilon}{2}) : x \in A_\varepsilon\}$, then $A \subset S(E, \varepsilon)$. So the lower Hausdorff topology can be described as a *discrete hit* and far topology.

Theorem 13.4. *Let (X, d) be a metric space. Then the Hausdorff metric topology τ_{Hd} is a* **discrete-hit-and-proximal topology** *and is finer than the proximal topology.*

Each A_ε is finite, if and only if, (X, d) is totally bounded. In that case, the Hausdorff metric topology τ_{Hd} equals the proximal topology. Hence, we have the following result.

Theorem 13.5. *Let (X, d) be a metric space. Then, on $CL(X)$, $\tau_\delta \subset \tau_V$ and $\tau_\delta \subset \tau_{Hd}$. Moreover, $\tau_\delta = \tau_V$, if and only if, X is UC and $\tau_\delta = \tau_{Hd}$, if and only if, X is totally bounded. Hence, $\tau_V = \tau_{Hd}$, if and only if, X is compact.*

13.5 Application: Local Near Sets in Hawking Chronologies

This section considers near sets in Hawking chronologies that provide a framework for the theory of general relativity. A general mathematical foundation for the topological approach to general relativity proposed by S.W. Hawking, A.R. King and P.J. McCarthy in 1976 [84] is summarised in [183] and reported in [147; 148]. Using C.J. Mozzochi's results on symmetric generalized uniformity, M.S. Gagrat and S.A. Naimpally characterized developable spaces as those which have compatible upper semi-continuous semi-metrics [68]. This result was used by R.Z. Domiaty and O. Laback in a study of semi-metric spaces in general relativity [56]. This approach gives the approach to general relativity proposed by Hawking,

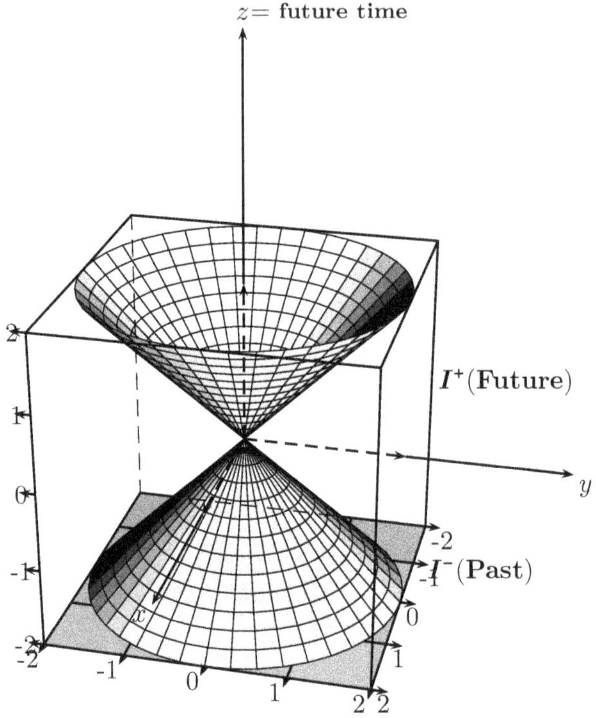

Fig. 13.1 Local chronological past and future

King and McCarthy a more general mathematical foundation [84] (see, also, [158], [150]).

A **Minkowski space** M is a 4-dimensional pseudo-Euclidean space proposed by H. Minkowski as a model of space-time in relativity theory [144]. Corresponding to each event, there is a point in M with 4 coordinates in R^3 and a fourth coordinate ct, where c is the velocity of light and t is the time of the event.

A partial representation of a chronological future in Minkowski space is shown in Fig. 13.1, where the cone above the horizontal plane (denoted by I^+) represents a collection of timelike paths projecting upward from a point in M. For example, future paths would be followed by a free particle undergoing a finite number of collisions [84]. In Fig. 13.1, a sample future path starts at $(0,0)$ in $\mathbb{R}^2$, a future time cone projecting upward along the time axis. Also shown in Fig. 13.1 is a chronological past cone projecting downward from the horizontal plane (denoted by I^-), a pasttime cone that contains timelike paths projecting downward along the time axis from a point in M. In keeping with R. Domiaty's suggestion [56], it is helpful to replace Hawking's consideration of the entire Minkowski space with a subspace X in M. Accordingly, X is called a local Minkowski space. The motivation for limiting our considerations to local Minkowski spaces is the fact that our definite geometrical knowledge of space-time is of a local nature. This approach in the study of chronologies in Minkowski space also befits the fact that timelike curves are assumed to be of finite extent [84].

Space-time is assumed to be a connected, Hausdorff, paracompact, C^∞ real 4-dimensional manifold M without a boundary, with a C^∞ Lorentz metric and associated pseudo-Riemannian connection [84]. The local space X is assumed to be time-orientable throughout, *i.e.*, M admits a nonvanishing time-orientable vector field.

Let $\mathcal{P}(X)$ be the set of all subsets in X, $A, B \subset X$. For timeline vector fields of X, the local **chronological future** I^+ of A *relative to* B is defined by

$$\begin{aligned} I^+(A,B) = \{b \in B : b &\text{ reachable from } A \\ &\text{by a future-directed timelike curve in } B \text{ of finite extent}\}. \end{aligned}$$

In terms of the geometric model of $I^+(A,B)$ in Fig. 13.1, a local, finite future-directed timelike curve would start at some point lower in the cone above the horizontal plane and reach upward to some higher point in the cone. A future point $b \in B$ has features such as reachable (from A), unreachable (from A), soonest (reachable via the shortest timelike curve from a point $a \in A$). The local **causal future** $J^+(A,B)$ is defined by

$$\begin{aligned} J^+(A,B) = \bigcup\{A \cap B : b &\text{ reachable from } A \\ &\text{by a future-directed causal curve in } B \text{ of finite extent}\}. \end{aligned}$$

The **future horismos** $E^+(A,B)$ of A relative to B is defined by

$$E^+(A,B) = J^+(A,B) - I^+(A,B).$$

Let $A, B, A', B' \subset X$ in chronological futures $I^+(A,B), I^+(A',B')$. Future sets $I^+(A,B), I^+(A',B')$ are near, provided the pair of local chronological futures have points in common, *i.e.*,

$$I^+(A,B)\ \delta\ I^+(A',B') \Leftrightarrow I^+(A,B)\ \cap\ I^+(A',B') \neq \varnothing.$$

Chronological futures with no points in common are far, *i.e.*,

$$I^+(A,B)\ \underline{\delta}\ I^+(A',B') \Leftrightarrow I^+(A,B)\ \cap\ I^+(A',B') = \varnothing.$$

The assumption made here is that δ is an L-proximity on nonempty chronology sets in X, *i.e.*, δ satisfies (P.0)-(P.4) given in Section 3.1. Let δ_Φ be a descriptive proximity defined relative to the features of points in future chronology sets.

Example 13.1. Near Future Chronology Sets.
Let $A, B, B' \subset X$, $a \in A$. Future chronology sets $I^+(\{a\}, B)$, $I^+(\{a'\}, B')$ are descriptively near, provided the pair of local chronological futures have points in common, *i.e.*,

$$I^+(\{a\}, B)\ \delta_\Phi\ I^+(\{a\}, B') \Leftrightarrow I^+(\{a\}, B) \underset{\Phi}{\cap} I^+(\{a'\}, B') \neq \varnothing.$$

When such future chronology sets are descriptively near, both sets will have future points with common features such as reachable, provided the intersection of $I^+(\{a\}, B), I^+(\{a'\}, B')$ is not empty. ∎

In terms of the geometric model of $I^-(A, B)$ in Fig. 13.1, a local, finite past-directed timelike curve would start at some point higher in the cone below the horizontal plane and reach downward to some lower point in the cone. A past point $b \in B$ has features such as reachable (from A), unreachable (from A), earliest (reachable via the shortest timelike curve from a point $a \in A$). The local **chronological past** I^- of A *relative to* B is defined by

$$\begin{aligned} I^-(A, B) = \{b \in B : b \text{ reachable from } A \\ \text{by a past-directed timelike curve in } B \text{ of finite extent}\}. \end{aligned}$$

A **complete chronology** $T^\pm(A, B)$ of local events in X is defined by

$$T^\pm(A, B) = \bigcup_{A, B \subset X} I^+(A, B) \cup I^-(A, B).$$

For $A, B \subset X$, a **restricted chronology** $\mathbb{T}^\pm(A, B)$ of local events in X is defined by

$$\mathbb{T}^\pm(A, B) = I^+(A, B) \cup I^-(A, B).$$

Example 13.2. Descriptive Hausdorff Chronology Space.
Let (X, δ_Φ) be a descriptive proximity space for local chronologies in X. From Corollary 36 in Chapter 5, descriptively distinct points belong to disjoint descriptive neighbourhoods. Then, from Theorem 37, the descriptive proximity space X is a Hausdorff space. ∎

13.6 Problems

(**13.6**.1) **Vietoris topology**: (X, τ) is a T_1 space, $CL(X)$ is the space of all nonempty closed sets with Vietoris topology τ_V. Prove the following statements [142].

(a) $\langle V_1, \ldots, V_k \rangle$ is a typical basis element for τ_V.

(b) Map $i : x \to \{x\}$ on X into $CL(X)$. is a homeomorphism. If X is Hausdorff, then $i(X)$ is closed in $CL(X)$.

(c) $CL(X)$ is connected, if and only if, X is connected.

(d) $K(X)$ is locally compact, if and only if, X is locally compact and $K(X)$ is open in $CL(X)$. X is locally compact does not imply $CL(X)$ is locally compact.

(e) $CL(X)$ is separable, if and only if, X is separable. $K(X)$ is second countable, if and only if, X is second countable.

(f) The following statements are equivalent.

(13f.i) X is normal.

(13f.ii) $CL(X)$ is Tychonoff.

(13f.iii) $CL(X)$ is regular.

(**13.6**.2) **Proximal topology**: Let (X, τ) be a T_1 space with a compatible L-proximity δ. Let $CL(X)$ have proximal topology τ_δ. Consider results analogous to those in Problem 13.6.1.

(**13.6**.3) **Hausdorff metric topology**: Let (X, d) be a metric space and τ_{Hd} Hausdorff metric topology on $CL(X)$. Prove the following assertions [23].

(a) Two compatible metrics on X induce equal Hausdorff metric topologies on $CL(X)$, if and only if, the metrics are uniformly equivalent.

(b) The following statements are equivalent.

(i) (X, d) is totally bounded.

(ii) $(CL(X), d_H)$ is totally bounded.

(iii) $(CL(X), d_H)$ is second countable.

(c) (X, d) is complete, if and only if, $(CL(X), d_H)$ is complete.

(d) Map $i : x \to \{x\}$ on X into $CL(X)$ is an isometry.

(e) Vietoris, Proximal and Hausdorff metric topologies coincide on $K(X)$.

Chapter 14

Selected Topics: Uniformity and Metrisation

This chapter considers a number of selected topics that serve to strengthen the overall view of topology. In addition to two additional forms of uniform structures (uniformities), namely, entourage uniformity and covering uniformity, this chapter introduces Tietze's extension theorem as well as further results concerning topological metrisation.

14.1 Entourage Uniformity

There are several equivalent ways to define uniformity. For example, uniformity is defined in terms of families of pseudometrics in Section §6.1. In this chapter, the focus is on entourage uniformity and covering uniformity.

Recall that **uniformity** on a nonempty set X is a family $\mathcal{D}$ of pseudometrics satisfying the following properties.

($\mathcal{D}$.1) $d, d' \in \mathcal{D} \Rightarrow max\{d, d'\} \in \mathcal{D}$.
($\mathcal{D}$.2) If a pseudometric $e \in \mathcal{D}$ satisfies the condition

> for all $\varepsilon > 0$, exists $d \in \mathcal{D}, \delta > 0 : x, y \in X, d(x, y) < \delta \Rightarrow e(x, y) < \varepsilon$,

then $e \in \mathcal{D}$.

With each pseudometric d on X and $\varepsilon > 0$, associate an **entourage** $V_{d,\varepsilon}$ defined by

$$V_{d,\varepsilon} = \{(x, y) \in X \times X : d(x, y) < \varepsilon\}.$$

Set $B = \{V_{d,\varepsilon} : d \in D \text{ and } \varepsilon > 0\}$.

Members of B satisfy the following properties.

($\boldsymbol{V}$.1) Each entourage $V_{d,\varepsilon}$ contains the diagonal $\Delta = \{(x,x) : x \in X\}$.

($\boldsymbol{V}$.2) $V_{d,\varepsilon} = V_{d,\varepsilon}^{-1}$ is symmetric, where

$$V_{d,\varepsilon}^{-1} = \{(x,y) : (y,x) \in V_{d,\varepsilon}\}.$$

($\boldsymbol{V}$.3) $e = max\{d, d'\} \in \mathcal{D}$ implies $V_{e,\varepsilon} \subset V_{d,\varepsilon} \cap V_{d',\varepsilon}$.

($\boldsymbol{V}$.4) $V_{d,\varepsilon} \circ V_{d,\varepsilon} \subset V_{d,2\varepsilon}$ (triangle inequality), where

$$U \circ V = \{(x,z) : \text{ for some } y \in X, (x,y) \in V, (y,z) \in U\}.$$

It is obvious that B is a filter base in $X \times X$ and the filter generated by B is called an **entourage uniformity** or just a **uniformity** on X.

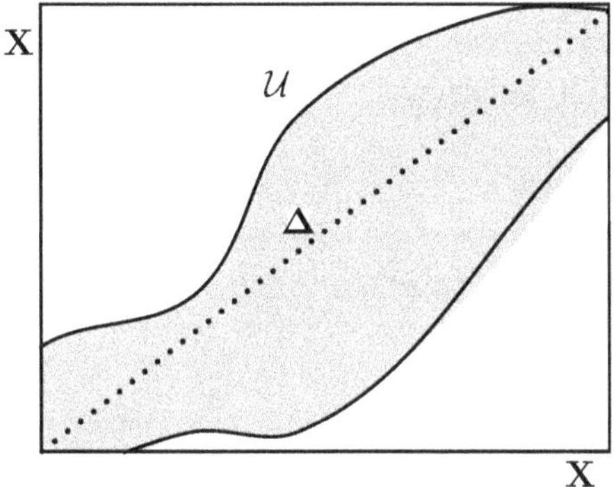

Fig. 14.1 Sample entourage $\mathcal{U}$

We now give a formal definition of uniformity. A **uniformity** $\mathcal{U}$ on a nonempty set X is a family of entourages (subsets of $X \times X$) satisfying the following properties.

($\mathcal{V}$.1) Each $U \in \mathcal{U}$ contains the diagonal $\Delta = \{(x,x) : x \in X\}$.

($\mathcal{V}$.2) Each $U \in \mathcal{U}$ contains a symmetric $V = V^{-1} \in \mathcal{U}$.

($\mathcal{V}$.3) $U, V \in \mathcal{U}$ implies $(U \cap V) \in \mathcal{U}$.

($\mathcal{V}$.4) For each $U \in \mathcal{U}$, there is $V \in \mathcal{U}$ such that $V \circ V \subset U$ (triangle inequality).

($\mathcal{V}$.5) If $U \in \mathcal{U}$ and $U \subset V$, then $V \in \mathcal{U}$.
Generally, an additional axiom is introduced to make the induced topology Hausdorff, namely,

($\mathcal{V}$.6) If $x \neq y$, then there is a $U \in \mathcal{U}$ such that $(x, y) \notin U$.

A sample entourage of the uniformity $\mathcal{U}$ defined on X is shown in Fig. 14.1.

While working with uniformity, it is often convenient to consider an *open* symmetric base and occasionally to consider a *closed* symmetric base. In a uniformity, 'distance' is eliminated and, in general, there is no longer 'first countability'. Also, if there is a countable uniformity base, then the uniformity is metrisable as we saw in Chapter 6. Obviously, every metric space (X, d) induces a uniformity $\mathcal{U}$ that is the filter generated by the countable family of entourages $\{V_{d,\varepsilon}\}$, where $\varepsilon > 0$ takes rational values or $\left\{\frac{1}{n} : n \in \mathbb{N}\right\}$.

It is a simple exercise to show that every uniformity $\mathcal{U}$ induces the Efremovič proximity $\delta = \delta_{\mathcal{U}}$ and, in turn, a Tychonoff (completely regular) topology such that

$$A \; \delta \; B \Leftrightarrow \text{ for each } U \in \mathcal{U}, U(A) \cap B \neq \varnothing,$$

where

$$U(A) = \{x \in X : (a, x) \in U, \text{ for some } a \in A\}.$$

If uniformity $\mathcal{U}$ induces an Efremovič proximity δ (topology τ), then $\mathcal{U}$ and δ (respectively, τ) are called **compatible**. Thus, the structures are related in the following order.

$$\text{uniformity} \longrightarrow \text{proximity} \longrightarrow \text{topology}.$$

Thus, Efremovič proximity lies between Tychonoff topology and uniformity.

The intersection to two uniformities need not be a uniformity but the union of any family of uniformities is a base for a uniformity. Given a Tychonoff space X, the union of all compatible uniformities generates a compatible uniformity called **fine uniformity** or **universal uniformity** on X. The coarsest compatible uniformity on X exists, if and only if, the corresponding topological space is locally compact Hausdorff.

Uniform continuity and uniform convergence (studied in Chaper 1) can be extended to uniform spaces. If $(X,\mathcal{U}), (Y,\mathcal{V})$ are uniform spaces, then a function $f : X \to Y$ is **uniformly continuous**, if and only if, any one of the following properties is satisfied.

($\boldsymbol{UC}$.1) For each $V \in \mathcal{V}$, there is a $U \in \mathcal{U}$ such that

$$(x,y) \in U \text{ implies } (f(x), f(y)) \in V.$$

($\boldsymbol{UC}$.2) For each $V \in \mathcal{V}$,

$$(f \times f)^{-1}(V) = \{(x,y) \in X \times X : (f(x), f(y)) \in V\} \in \mathcal{U}.$$

($\boldsymbol{UC}$.3) For each $V \in \mathcal{V}$, there is $U \in \mathcal{U}$ such that for all $x \in X, f(U(x)) \subset V(f(x))$.

Every uniformly continuous function is (proximally) continuous with respect to the associated (proximities) topologies.

Let X be a topological space, $(Y,\mathcal{V})$ a uniform space. A net (f_n) of functions from X to Y **converges uniformly** to function g, if and only if, for each $V \in \mathcal{V}$ and for all $x \in X$, *eventually* $f_n(x) \in V(g(x))$. Obviously, uniform convergence is stronger than proximal and pointwise convergences as well as preserving continuity and uniform continuity, provided X is a uniform space.

A net $(x_n : n \in D)$ in a uniform space $(X,\mathcal{U})$ is **Cauchy**, if and only if, for each $V \in \mathcal{U}$, *eventually*, for all $m, n \in D, (x_m, x_n) \in V$. In terms of filters, $\mathcal{F}$ is Cauchy, if and only if, for each $V \in \mathcal{F}$,

(**Cauchy**.1) There is an $x \in X$ such that $V(x) \in \mathcal{F}$, or
(**Cauchy**.2) there is an $F \in \mathcal{F}$ such that $F \times F \in V$.

A uniform space $(X,\mathcal{U})$ is **complete**, if and only if, each Cauchy net (filter) converges. A uniform space $(X,\mathcal{U})$ is **totally bounded** (**precompact**), if and only if, for every $V \in \mathcal{U}$, there is a finite subset F of X such that $V(F) = X$.

14.2 Covering Uniformity

Another avatar of uniformity on a nonempty set X is in the form of covers. If d is a metric on X, then the open balls $\{S(x,\varepsilon) : x \in X\}$ with radius $\varepsilon > 0$, provide a *uniform* cover $\mathcal{U}$ of X of the same size ε.

If $\mathcal{U}$ is an entourage uniformity or base, the family $\{U(x) : x \in X\}$ is a 'uniform' cover of X and properties satisfied by the family of such covers can be used as a motivation for defining covering uniformity. A covering uniformity has one advantage over an entourage uniformity of a nonempty set X, namely, the cardinality of its covers. Before we give a precise definition of a covering uniformity, we first introduce some terminology.

(**term**.1) A cover $\mathcal{A}$ is a **refinement** of a cover $\mathcal{B}$, if and only if, each member of $\mathcal{A}$ is contained in some member of $\mathcal{B}$ and is written $\mathcal{A} < \mathcal{B}$.

(**term**.2) **Star** of $E \subset X$ with respect to $\mathcal{A}$ is denoted by

$$St(E, \mathcal{A}) = \bigcup \{P \in \mathcal{A} : P \cap E \neq \emptyset\}.$$

(**term**.3) A cover $\mathcal{A}$ is a **star refinement** of $\mathcal{B}$, if and only if, for each $A \in \mathcal{A}$, there is a $B \in \mathcal{B}$ such that $St(A, \mathcal{A}) \subset B$ (denoted by $\mathcal{A} *< \mathcal{B}$).

A family μ of covers of a nonempty set X is called a **covering uniformity**, if and only if, μ satisfies the following requirements.

(**cμ**.1) If $\mathcal{A} \in \mu$ and $\mathcal{B} \in \mu$, then there is a $\mathcal{C} \in \mu$ that refines both $\mathcal{A}$ and $\mathcal{B}$.

(**cμ**.2) If $\mathcal{B} \in \mu$, there is an $\mathcal{A} \in \mu$ such that $\mathcal{A} *< \mathcal{B}$.

(**cμ**.3) If $\mathcal{B}$ is a cover of X and there is an $\mathcal{A} \in \mu$, such that $\mathcal{A} < \mathcal{B}$, then $\mathcal{B} \in \mu$.

The transition from covering uniformity μ to entourage uniformity $\mathcal{V}$ is given as follows.

> A **base** for an entourage uniformity $\mathcal{V}$ is a family of sets of the form $\bigcup \{E \times E : E \in \mathcal{A}\}$, where $\mathcal{A} \in \mu$.

14.3 Topological Metrisation Theorems

In Chapter 11, we studied some metrisation theorems for topological, proximity and uniform spaces, mainly via semi-metrics. Here, we give outlines of two well-known results in topological metrisation, namely,

(**tm**.1) Special topological metrisation case due to Urysohn and Tychonoff for *separable* metric spaces, and

(**tm**.2) General topological metrisation case due to Nagata-Smirnov and Bing.

The topological metrisation problem is summarised as follows. Given a topological space (X, τ), find necessary and sufficient *topological* conditions that the topology τ is induced by a metric on X.

For motivation, notice that a metric space (X, d) is first countable, *i.e.*, each point $x \in X$ has a countable open neighbourhood base $\{S(x, \frac{1}{n}) : n \in \mathbb{N}\}$, where

$$S(x, \frac{1}{n}) = \left\{y \in X : d(x, y) < \frac{1}{n}\right\}.$$

So a *countability* condition is bound to come up, somewhere. Observe that there are topological spaces that are first countable but not metrisable. Hence, we need a stronger condition for metrisability. Usually, metrisability is shown in one of two ways for a topological space (X, τ).

(**met**.1) Explicit construction of a metric compatible with the topology τ.
(**met**.2) Embedding the topological space in a known metric space such as a countable product of $\mathbb{R}$ or the closed unit interval $I = [0, 1]$.

In Chapter 11, we studied several metrisation results using (Met.1). First, one constructs a semi-metric and then adds an extra condition to get the triangle inequality (*cf.* Theorems 11.1, 11.4, and Corollary 11.1). The Alexandroff-Urysohn metrisation Theorem 11.2 is universally recognized as the first metrisation theorem. However, opinions differ on its status among metrisation results. Some consider Theorem 11.2 unsatisfactory, since it is a uniform not a topological metrisation result.

We have shown that uniformity can be characterized in terms of covers. So a simple *topological* characterization is the existence of a countable base for a covering uniformity.

Suppose X it T_1 and $\{\mathcal{U}_n : n \in \mathbb{N}\}$ is a countable family of open covers of X with $\{St(x, \mathcal{U}_n) : n \in \mathbb{N}\}$ a neighbourhood base at each $x \in X$. Such a family of covers is called a **development**. The collection $\{\mathcal{U}_n\}$ is a **regular development**, if and only if, for each $n \in \mathbb{N}$, G, G' are two intersecting sets in $\mathcal{U}_{n+1}$, then the union $(G \cup G') \subset H$ for some $H \in \mathcal{U}_n$. The Alexandroff-Urysohn Theorem in its topological avatar is as follows.

Theorem 14.1. Alexandroff-Urysohn.
A T_1 topological space X is metrisable, if and only if, X has a regular development.

We now turn to Urysohn's trick to use method (met.2). In a T_2 normal (T_4) space X, there are continuous functions on X to I separating two disjoint closed sets (Urysohn's Lemma) (see Lemma 5.1 in §5.2). A suitable candidate in which X is embedded, is the Hilbert cube[1], which is a countable product of the closed unit interval I. This led Urysohn to consider the special case of a normal T_4 space X with a countable open base $\mathcal{B}$.

Let $\mathcal{M}$ denote the countable family of pairs

$$\mathcal{M} = \{(U,V) : U, V \in \mathcal{B} \text{ with } \operatorname{cl} U \subset V\}.$$

Since X is normal, then, for each $\{U,V\}$, there is a continuous function $f_{(U,V)}$ on X to I taking $\operatorname{cl} U$ to $\{0\}$ and $(X - V)$ to $\{1\}$. The countable family $\{f_{(U,V)} : (U,V) \in \mathcal{M}\}$ separates points from closed sets and, hence, X is embedded into the Hilbert cube.

Theorem 14.2. Urysohn
A T_1 normal (T_4) space with a countable base is metrisable.

Obviously this condition is not necessary, since there are metric spaces without a countable base. This result was improved by Tychonoff by showing that a T_1 regular space X with a countable base $\mathcal{B}$ is normal.

To see, suppose that A, B are disjoint closed subsets of X. For each $a \in A$, there is an open set U_a in B such that $a \in U_a$ and $(\operatorname{cl} U_a) \cap B = \varnothing$. Hence, there is a countable open cover $\{U_n\}$ of A with $(\operatorname{cl} U_n) \cap B = \varnothing$ for each $n \in \mathbb{N}$. Similarly, there is a countable open cover $\{V_n\}$ of B with $(\operatorname{cl} V_n) \cap A = \varnothing$ for each $n \in \mathbb{N}$. Then set

$$U'_n = U_n - \bigcup\{\operatorname{cl} V_m : m < n\} \text{ and } V'_n = V_n - \bigcup\{\operatorname{cl} U_m : m < n\}.$$

Then $[\bigcup\{U'_n : n \in \mathbb{N}\}]$ and $[\bigcup\{V'_n : n \in \mathbb{N}\}]$ are disjoint open neighbourhoods of A, B, respectively. This shows that X is normal and we obtain the following result.

Theorem 14.3. Tychonoff
A T_1 regular (T_3) space with a countable base is normal and hence metrisable.

To improve the result in Theorem 14.3, it is necessary to weaken the *countable base* condition with a condition that is satisfied by all metric spaces. This is what was done independently by Nagata, Smirnov and Bing. The story begins with paracompactness, a nice generalization of compactness discovered by J. Dieudonné.

---

[1]Recall that the Hilbert cube is the product of the intervals $[0,1] \times [0, \frac{1}{2}] \times \cdots \times [0, \frac{1}{n}] \times \cdots, n \in \mathbb{N}$. This is homeomorphic to the countable product of $[0,1]$.

A family of subsets $\mathcal{U}$ of a topological space X is called **locally finite** (**discrete**), if and only if, each point $x \in X$ has a neighbourhood that intersects at most finitely many sets (respectively one set) in $\mathcal{U}$. A countable union of locally finite (discrete) families is called **σ-locally finite** (**discrete**). A topological space is **paracompact**, if and only if, it is Hausdorff and every open cover has a locally finite refinement.

Hausdorff paracompact spaces are normal, which takes us close to metrisability. Dieudonné proved that every separable metric space is paracompact and this result was extended to all metric spaces by A.H. Stone. Stone's paper provided arguments that led to a necessary condition for metrisation. E. Michael showed that a regular T_3 topological space is paracompact, if and only if, every open cover has a σ-locally finite open refinement. With further manipulations, this leads to the fact that every metric space has a σ-locally finite open base. J. Nagata and Y. Smirnov showed that a regular T_3 topological space is metrisable, provided it has a σ-locally finite open base.

Since a regular T_3 space is normal T_4, we begin with a T_4 space with a σ-locally finite open base $\{\mathcal{B}_n : n \in \mathbb{N}\}$. Fix $(m, n) \in \mathbb{N} \times \mathbb{N}$. For each $B \in \mathcal{B}_n$, set

$$B_0 = \bigcup \{\mathrm{cl}\, E \subset B : E \in \mathcal{B}_n\}.$$

Since $\mathcal{B}_n$ is locally finite, B_0 is closed. Urysohn's Lemma gives a continuous function $f_B : X \to I$ with $f_B(B_n) = 1, f_B(X - B) = 0$. Define a pseudometric

$$d_{(m,n)}(x, y) = min\{1, \left(\sum |f_B(x) - f_B(y)|\right) : B \in \mathcal{B}_n\}\}.$$

Rearrange the countable set $\left\{d_{(m,n)} : (m, n) \in \mathbb{N} \times \mathbb{N}\right\}$ as $\{d_k : k \in \mathbb{N}\}$. Next, define ρ on $X \times X \to [0, \infty)$ by

$$\rho(x, y) = \sum \left\{\frac{1}{2^k} d_k(x, y) : 1 \le k < \infty\right\}.$$

Then it can be shown that ρ is a compatible metric on X.

Theorem 14.4. (Nagata-Smirnov) *A topological space is metrisable, if and only if, it is regular T_3 and has a σ-locally finite open base.*

Theorem 14.5. (Bing) *A topological space is metrisable, if and only if, it is regular T_3 and has a σ-discrete open base.*

Proof. The proof is similar to the proof of Theorem 14.4. □

14.4 Tietze's Extension Theorem

In Section §10.2 (Theorem 10.2 (Taimanov)), we studied continuous extensions of functions from a *dense* subspace to the whole space. There is a similar problem concerning the extension from *any* subspace.

The function $f(x) = \frac{1}{x}$ on $(0,1) \to \mathbb{R}$ cannot be extended to a continuous function on $\mathbb{R} \to \mathbb{R}$. Hence, the subset must be closed. This extension result was first proved by H. Lebesgue for the plane, by H. Tietze for metric spaces, and finally P. Urysohn showed that *a topological space is normal* T_4, *if and only if, every continuous* f *on a closed subset* $A \to [-1,1]$ (or $\mathbb{R}$) has a continuous extension to $g : X \to [-1,1](\text{or}\mathbb{R})$. That is, for each $a \in A, f(a) = g(a)$. However, this result is generally known as Tietze's Theorem.

Let X be normal T_4, A a closed subset of X, and $f : A \to [-1,1]$ a continuous function. Then $C_1 = \{x \in A : f(x) \leq -\frac{1}{3}\}$ is disjoint from $D_1 = \{x \in A : f(x) \geq \frac{1}{3}\}$. By Urysohn's Lemma, there is a continuous function $g_1 : X \to [-\frac{1}{3}, \frac{1}{3}]$ such that $g_1(C_1) = -\frac{1}{3}$ and $g_1(D_1) = \frac{1}{3}$. For all $x \in A, |f(x) - g_1(x)| \leq \frac{2}{3}$. The function $f_1 = f - g_1 : A \to [-\frac{2}{3}, \frac{2}{3}]$ is continuous. Inductively, we get for each $n \in \mathbb{N}$,

$$g_{n+1} : X \to \left[-\left(\frac{1}{3}\right)\left(\frac{2}{3}\right)^n, \left(\frac{1}{3}\right)\left(\frac{2}{3}\right)^n\right], \text{ as well as a continuous function,}$$

$$f_{n+1} = f_n - g_{n+1} : A \to \left[-\left(\frac{2}{3}\right)^n, \left(\frac{2}{3}\right)^n\right].$$

Define the function $g : X \to [-1,1]$ by $g = \sum g_n$, which is continuous due to uniform convergence by the well-known Weierstrass test. Also, g is an extension of f. Since $\mathbb{R}$ is homeomorphic to $(-1,1)$, the result can be extended to real-valued continuous functions.

Theorem 14.6. *(Tietze-Urysohn). A Hausdorff space* X *is normal* T_4, *if and only if, every continuous function on any closed subset* A *of* X *to* $I(\mathbb{R})$ *has a continuous extension to* $I(\mathbb{R})$.

14.5 Application: Local Patterns

This section introduces a mathematical basis for pattern recognition provided by spatially near sets as well as descriptively near sets in identi-

Fig. 14.2 Barcode patterns

Fig. 14.3 Partial kangaroo tiling pattern

fying local patterns in metric, proximity, and topological spaces. This study of patterns stems from recent work on near sets [183; 179] as well as [163; 43; 221; 222; 179].

14.5.1 *Near Set Approach to Pattern Recognition*

In general, a **pattern** is a set containing a repetition of some form in the parts of the set. The repetition that defines a pattern can be either spatial or descriptive. In either case, a repetition in the arrangement of the parts of a set determines a pattern. For example, in Fig. 14.2, the various arrangements of the vertical lines determine barcode patterns. Again, for example, in the kangaroo tiling in Fig. 14.3, the repetition of the various shapes produces a tiling pattern.

In terms of a near set approach to **pattern recognition** (PR), we have several choices to consider.

(**choice**.1) **Spatial**. The spatial approach to PR entails the following steps:
  (a) Identify a nonempty set X.
  (b) Identify a spatial proximity space (X, δ). The choices can be found in [163].
  (c) Identify spatially near sets.
  (d) Recognise a spatial pattern such that $A \ \delta \ B$ for a particular $A \in \mathcal{P}$ and for $B \in \mathcal{P}(X)$. That is, look for a collection of subsets $\mathcal{B} \in \mathcal{P}^2(X)$ such that each $B \in \mathcal{B}$ is spatially near some A. A good source of visual spatial patterns is given by M.C. Escher [60] (see, also, [94]).

(**choice**.2) **Descriptive**. The descriptive approach to PR entails the following steps:
  (a) Identify a nonempty set X.
  (b) Identify Φ, a set of probe functions used to represent the features of objects in X.
  (c) Identify a descriptive proximity space (X, δ_Φ). Sample choices for a descriptive near set approach to PR are given in this section (see, also, [179; 185]).
  (d) Identify descriptively near sets.
  (e) Recognise a descriptive pattern such that $A \ \delta_\Phi \ B$ for a particular $A \in \mathcal{P}$ and for $B \in \mathcal{P}$. That is, look for a collection of subsets $\mathcal{B} \in \mathcal{P}^2(X)$ such that each $B \in \mathcal{B}$ is descriptively near some A. A good source of visual descriptive patterns is also given by M.C. Escher [60].

(**choice**.3) **Spatial-Descriptive**. This approach to PR results from a combination of the spatial and descriptive near set approaches to PR.

Let $\mathfrak{I}$ be a nonempty set, X a subset of $\mathfrak{I}$, $\mathfrak{P} \in \mathcal{P}(X) \times \mathcal{P}^2(X)$. A pattern is typically represented by $\{A, \mathcal{B}\}$, where $A \in \mathcal{P}(X), \mathcal{B} \in \mathcal{P}^2(X)$ such that A is either spatially or descriptively near one or more subsets B in $\mathcal{B}$. The collection $\mathfrak{P}$ is a **local pattern**, if and only if, $\mathfrak{P}$ is defined in terms of the proximity of the parts of $X \subset \mathfrak{I}$, *i.e.*, X is a proper subset of $\mathfrak{I}$.

Example 14.1. Roo Pattern.
Let A be the set of grey kangaroo heads in the image in Fig. 14.3 and let $\mathcal{B}$ be the collection dark kangaroo shapes in the same image. The heads in A are spatially near B in $\mathcal{B}$, since the kangaroos in B overlap the head

shapes in A, *i.e.*, $A \ \delta \ B$. Hence, $\mathfrak{P} = \{A, \mathcal{B}\}$ is an example of a local spatial pattern.

In terms of a descriptive roo pattern, let Φ be a set of probe functions representing greylevel intensity of picture elements. Then A is descriptively near at least one B in $\mathcal{B}$, since the eyes of the roos in A descriptively match the eyes of each of the dark roos in $\mathcal{B}$. In other words, $A \ \delta_{\Phi} \ B$. Hence, $\mathfrak{P}_{\Phi} = \{A, \mathcal{B}\}$ is an example of a local descriptive pattern. ■

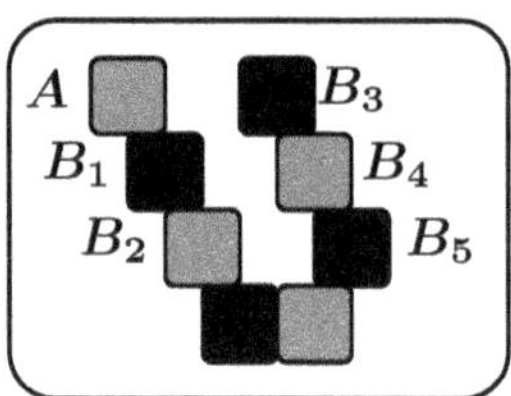

Fig. 14.4 Local metric weave pattern

14.5.2 *Local Metric Patterns*

This section introduces local metric patterns based on the nearness of sets. Two forms of metric patterns are given, namely, spatial metric patterns and descriptive metric patterns. The term *metric* is used here, since the patterns in the section are defined relative to a metric.

Let (X, δ) be an EF-space, δ a *spatial metric proximity*, $\mathcal{B}$ a collection of subsets in $\mathcal{P}^2(X)$, $B \in \mathcal{B}$, and A in $\mathcal{P}(X)$. Recall that $A \ \delta \ B$, provided $D(A, B) = 0$, *i.e.*, the intersection of A and B is not empty. After selecting a metric $d : X \to \mathbb{R}$, the distance D between sets A, B is defined by

$$D(A, B) = inf\{d(a, b) : a \in A, b \in B\}.$$

The collection $\mathfrak{P} = \{A, \mathcal{B}\}$ is a **local spatial metric pattern** on X, if and only if, $A \ \delta \ B$ for at least one $B \in \mathcal{B}$. We illustrate spatial metric patterns in terms of cells in a weave.

From Grünbaum and Shephard [79], a *fabric* consists of two layers of congruent strands in the same plane E so that the strands of different layers are nonparallel and 'weave' over and under each other. A *strand* is a doubly open strip of constant width so that the set of points of the plane lie strictly between two parallel straight lines [79]. This form of fabric is restricted to two-fold weaves that are periodic [94]. Let X denote a nonempty set of cells in the strands of a fabric. The places where the strands cross are taken to be square cells (without boundary) tessellating the plane. The points not on the boundary of any strand are arranged into *cells* (in each cell, one strand is uppermost). The parallel strands of each layer in a weave are perpendicular to those strands in the other layer, making the cells square [220]. For simplicity, assume that a *cell* in a fabric is a set of points from that part of a strand that overlaps another strand. In other words, a cell is that part of a fabric that is visible. A *weave* is a collection of cells $\mathcal{P}(X)$ in a fabric. A *local weave* is a set of cells $A \in \mathcal{P}(X)$. In terms of weave cells, $A\ \delta\ B$, provided $D(\text{cl}\, A, \text{cl}\, B) = 0$, *i.e.*, only adjacent cells are spatially near.

Example 14.2. Spatial Metric Weave Pattern.
Let X be a set of weave cells endowed with a metric proximity δ. Each weave cell is represented by a shaded box ▪ in Fig. 14.4 that is part of a larger weave (not shown). Let $A, \mathcal{B} = \{B_1, B_2\}$ be as shown in Fig. 14.4. Observe that $A\ \delta\ B_1$, since $D(A, B_1) = 0$, but $A\ \underline{\delta}\ B_2$, since $D(A, B_2) \neq 0$. Hence, $\mathfrak{P} = \{A, \mathcal{B}\}$ is an example of a spatial metric pattern. ■

Choose Φ, a set of probe functions $\phi : X \to \mathbb{R}$ representing features of members of X. Let δ_Φ be a *descriptive metric proximity* defined in terms of $\Phi(A)$ (set of descriptions of members of A) and $\Phi(B)$ (set of descriptions of members of B). The set A is descriptively near B (denoted $A\ \delta_\Phi\ B$), provided, for a metric d, we have

$$D(\Phi(A), \Phi(B)) = 0 \text{ for some } B \in \mathcal{B}, \text{ where}$$
$$D(\Phi(A), \Phi(B)) = inf\{d(\Phi(a), \Phi(b) : \text{cell } a \in A, \text{cell } b \in B\}.$$

That is, the description of one or more cells in A matches the description of one or more cells in $B \in \mathcal{B}$. The collection $\mathfrak{P}_\Phi = \{A, \mathcal{B}\}$ is a **local descriptive metric pattern** on X, if and only if, $A\ \delta_\Phi\ B$ for at least one $B \in \mathcal{B}$. We illustrate descriptive metric patterns in terms of weaving patterns in a fabric.

The interest here is in discerning cells as well as collections of cells that are, in some sense, near (proximate) or remote in a descriptive EF weave space. A *descriptive EF weave space* is a descriptive EF space in which the members are weave cells. In the context of proximity spaces, a weaving pattern is defined in terms of nonempty sets of points endowed with a proximity relation with well-known properties. Specifically, a *proximal weaving pattern* is a set containing a repetition in a collection of cells of some form in a weave. The repetition that defines a weaving pattern can be either spatial or descriptive. In either case, a repetition in the arrangement of the points in a weave determines a weaving pattern.

Example 14.3. Local Descriptive Metric Pattern in a Weave. Let (X, δ_Φ) be a descriptive EF space represented by the weave in Fig. 14.4. Let A be represented by the grey box ■ in Fig. 14.4 and let $\mathcal{B} \doteq \{B_1, B_2, B_3, B_4, B_5\}$ in the same weave. Choose Φ be a set of probe functions that represent greylevel intensity and colour of weave cells in X. Observe that $A\ \delta_\Phi\ B_2$ and $A\ \delta_\Phi\ B_4$. Hence, $\mathfrak{P}_\Phi = \{A, \mathcal{B}\}$ is an example of a local descriptive metric weave pattern. $\mathfrak{P}_\Phi$ is an example of a **motif**, a repeated descriptive metric pattern in a weave. Notice that the spatial metric pattern $\mathfrak{P}$ on X contains only A and the adjacent cell B_1, since $A\ \delta\ B_1$ but $A\ \underline{\delta}\ E$ for all other cells E in X. ■

Fig. 14.5 Checkerboard pattern

Let (X, δ_Φ) be a descriptive EF space, $x, y \in X$. The descriptive proximity δ_Φ is **descriptively separated**, provided $\Phi(x) = \Phi(y)$ implies $x = y$. Otherwise, δ_Φ is not descriptively separated.

Example 14.4. Let (X, δ_Φ) be a descriptive EF space represented by the checkerboard pattern in Fig. 14.5. B, W are neighbouring black and white cells, W, B' neighbouring white, black cells in the checkerboard. Let $x \in B, y \in B'$. Then $\Phi(x) = \Phi(y)$ but $x \neq y$. Hence, δ_Φ is not descriptively separated.

Lemma 14.1. *Let (X, δ_Φ) be a descriptive EF space in which every member has a different description. Then δ_Φ is descriptively separated.*

Proof. Let (X, δ_Φ) be a descriptive EF space such distinct points $x, y \in X$ have different descriptions. That is, $x \neq y$ implies $\Phi(x) \neq \Phi(y)$. Hence, δ_Φ is descriptively separated. □

Theorem 14.7. *Let (X, δ_Φ) be a descriptive EF weave space in which every cell has a different description. Then δ_Φ is separated.*

Proof. Immediate from Lemma 14.1. □

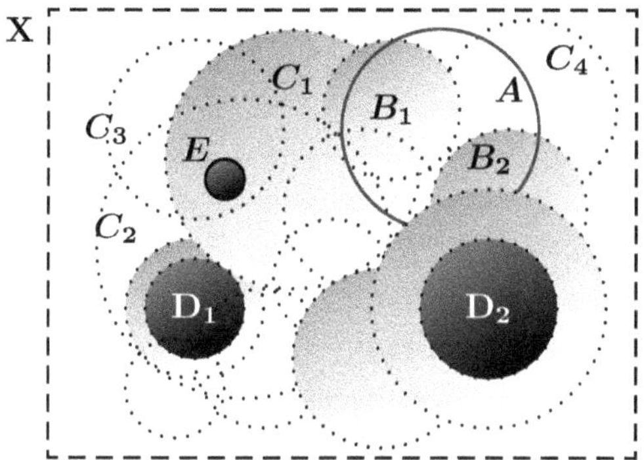

Fig. 14.6 Sample local topological patterns

14.5.3 *Local Topological Patterns*

Let (X, τ) be a local topological space, A a subset of X, $B \in \tau$, δ a spatial EF-proximity relation on X. Recall that $A\ \delta\ B$, if and only if, $A \cap B \neq \emptyset$. The collection $\mathfrak{P} = \{A, \tau\}$ is a local **spatial topological pattern**, if and only if, $A\ \delta\ B$ for at least one B in τ.

Example 14.5. Spatial Topological Pattern.
The dotted circles in Fig. 14.6 represent open subsets in a topology τ on X. Let $A \in \mathcal{P}(X)$ overlaps $B_1, B_2 \in \tau$ as well as other sets in the upper right hand corner of Fig. 14.6. Observe that $A\ \delta\ B_1$ and $A\ \delta\ B_2$. Hence, $\mathfrak{P} = \{A, \tau\}$ is an example of a local topological pattern that is a repetition of near sets such as $A\ \delta\ B_1, A\ \delta\ B_2$. ∎

Let (X, τ) be a local topological space, $A \in \mathcal{P}(X)$, $B \in \tau$, δ_Φ a descriptive EF-proximity relation on X. Choose Φ to be a set of probe functions $\phi : X \to \mathbb{R}$ representing features of points in X. Recall that $A\ \delta_\Phi\ B$, if and only if, $A \underset{\Phi}{\cap} B \neq \emptyset$. The collection $\mathfrak{P}_\Phi = \{A, \tau\}$ is a local **descriptive topological pattern**, if and only if, $A \underset{\Phi}{\cap} B$ for at least one B in τ.

Example 14.6. Descriptive Topological Pattern.
From Example 14.5, $A \in \mathcal{P}(X)$ is represented by the circle in the upper right hand corner in Fig. 14.6. Choose Φ to be a set of probe functions $\phi : X \to \mathbb{R}$ representing the greylevel intensity and the colour of points in X. Observe, for example, $A\ \delta_\Phi\ C_1$ and $A\ \delta_\Phi\ C_2$ for $C_1, C_2 \in \tau$, since A, C_1, C_2 have white pixels. Hence, $\mathfrak{P}_\Phi = \{A, \tau\}$ is an example of a local

descriptive topological pattern that is a repetition of descriptively near sets. ∎

Recall that A refines B (denoted $A < B$), provided $A \subseteq B$. Let (X, τ) be a local topological space. The collection $\mathfrak{P}_{<} = \{A, \tau\}$ is a local **spatial refinement topological pattern** on X, if and only if, for at least one $B \in \tau$, $A < B$.

Example 14.7. Local Spatial Refinement Topological Pattern. From Example 14.5, let the topological space (X, τ) be represented in Fig. 14.6 such that $C_1, C_2, C_3 \in \tau$. Choose $E \in \mathcal{P}(X)$ in Fig. 14.6. Since $E < C_1, E < C_2, E < C_3$, then $\mathfrak{P}_{<} = \{A, \tau\}$ is a local spatial refinement topological pattern. ∎

A nonempty set A descriptively refines B (denoted $A \underset{\Phi}{<} B$), provided $\Phi(A) \subseteq \Phi(B)$, *i.e.*, the description of A is a subset of the description of B. Let (X, τ) be a local topological space. The collection $\mathfrak{P}_{\underset{\Phi}{<}} = \{A, \tau\}$ is a local **descriptive refinement topological pattern** on X, if and only if, for at least one $B \in \tau$, $A \underset{\Phi}{<} B$.

Example 14.8. Local Descriptive Refinement Topological Pattern. From Example 14.5, let the topological space (X, τ) be represented in Fig. 14.6 such that $D_1, D_2 \in \tau$. Choose $E \in \mathcal{P}(X)$ in Fig. 14.6. Also, choose Φ to be a set of probe functions that represent greylevel intensity and colour of the points in X. Since $E \underset{\Phi}{<} D_1, E \underset{\Phi}{<} D_2$, then $\mathfrak{P}_{\underset{\Phi}{<}} = \{A, \tau\}$ is a local descriptive refinement topological pattern. ∎

14.5.4 *Local Chronology Patterns*

This section briefly illustrates two forms of local chronology patterns arising out the discussion in Chapter 13.

Example 14.9. Local Chronology Refinement Pattern. Let X be a subset of Minkowski space M, $\mathcal{C}$ a cover of X, I^+ a chronology in X in Fig. 14.7 and let A, B be subsets in M and $\mathfrak{P} = \{I^+(A, B), \mathcal{C}\}$. The collection $\mathfrak{P}$ is a local **chronology refinement pattern** on M, if and only if, $I^+(A, B) < \mathcal{C}$ (one future is part of another future). ∎

Example 14.10. Local Chronology Star Pattern. Let X be a subset of Minkowski space M, $\mathcal{C}$ a cover of X, I^+ a chronology in X in Fig. 14.7 and let A, B be subsets in M and $\mathfrak{P} = \{\{I^+(A, B)\}, \mathcal{C}\}$.

The collection $\mathfrak{P}$ is a local **chronology star pattern** on M, if and only if, $St(I^+(A,B),\mathcal{C}) \neq \emptyset$ (one future is near another future). ∎

14.5.5 *Local Star Patterns*

Let $\mathcal{C}$ be a collection of subsets in $\mathcal{P}^2(X)$, A a subset of $\mathcal{P}(X)$. The *star* of A with respect to a cover $\mathcal{C}$ (denoted $\mathrm{St}(A,\mathcal{C})$) is defined to be

$$\mathrm{St}(A,\mathcal{C}) := \bigcup_{B \in \mathcal{C}} \{A \cap B \neq \emptyset\},$$

i.e., $\mathrm{St}(A,\mathcal{C})$ is the union of subsets $B \in \mathcal{C}$ that are near A. The collection $\mathfrak{P} = \{A, \mathrm{St}(A,\mathcal{C})\}$ is a *local star pattern* on X, if and only if, $\mathrm{St}(A,\mathcal{C})$ is nonempty, *i.e.*, $A\ \delta\ B$ for at least one B in $\mathcal{C}$.

Example 14.11. Local Star Pattern.
In Fig. 14.6, let the set X be endowed with an EF proximity δ, $A \in \mathcal{P}(X)$, $\mathcal{B} = \{B_1, B_2, \ldots, B_n\}$ be a cover of X (only partially shown). $\mathrm{St}(A,\mathcal{B}) = \{B_1, B_2\}$, since $A \cap B_1 \neq \emptyset$ and $A \cap B_2 \neq \emptyset$. Hence, $\mathfrak{P} = \{A, \mathrm{St}(A,\mathcal{B})\}$ is a local star pattern. Again, for example, the star of a_n with respect to the cover $\mathcal{U}$ of X in Fig. 11.6 defines the micropalaeontology spatial star pattern $\mathfrak{P} = \{a_n, \mathrm{St}(a_n,\mathcal{U})\}$. ∎

The *descriptive star* of $A \in \mathcal{P}(X)$ with respect to a cover $\mathcal{C}$ (denoted $\mathrm{St}_\Phi(A,\mathcal{C})$) is defined to be

$$\mathrm{St}_\Phi(A,\mathcal{C}) := \bigcup_{B \in \mathcal{C}} \left\{A \underset{\Phi}{\cap} B \neq \emptyset\right\},$$

i.e., $\mathrm{St}_\Phi(A,\mathcal{C})$ is the union of subsets $B \in \mathcal{C}$ that are descriptively near A. The collection $\mathfrak{P}_\Phi = \{A, \mathrm{St}_\Phi(A,\mathcal{C})\}$ is a *local descriptive star pattern* on X, if and only if, $\mathrm{St}_\Phi(A,\mathcal{C})$ is nonempty, *i.e.*, $A\ \delta_\Phi\ B$ for at least one B in $\mathcal{C}$.

Example 14.12. Local Descriptive Star Pattern.
In Fig. 14.6, let the set X be endowed with an EF proximity δ_Φ, $A \in \mathcal{P}(X)$, and let $\mathcal{C} = \{C_1, C_2, C_3, C_4, \ldots, C_n\}$ be a cover of X(only partially shown). Choose Φ to be a set of probe functions for greylevel intensity, colour, and shape features for points in X. Then $\mathrm{St}_\Phi(A,\mathcal{C}) = \{C_2, C_3, C_4\}$, since, for example, $A \underset{\Phi}{\cap} C_3 \neq \emptyset$ (A, C_3 each contain white points). Hence, $\mathfrak{P}_\Phi = \{A, \mathrm{St}_\Phi(A,\mathcal{C})\}$ is a local descriptive star pattern. Assume that (X,δ) is an EF space. Then notice that $\mathfrak{P} = \{A, \mathrm{St}(A,\mathcal{C})\}$ is a local spatial star pattern, since $A \cap C_4 \neq \emptyset$ (A, C_4 contain common points). Again, for example, the star of N_{ϕ,a_1} with respect to the cover $\mathcal{U}$ of X in Fig. 11.6 defines a

micropalaeontology descriptive star pattern $\mathfrak{P}_{\Phi} = \{N_{\phi,a_1}, \mathrm{St}_{\Phi}(N_{\phi,a_1}, \mathcal{U})\}$, since $N_{\phi,a_1} \underset{\Phi}{\cap} N_{\phi,b_2} \neq \varnothing$. ■

14.5.6 *Local Star Refinement Patterns*

A cover $\mathcal{C}$ is a *star refinement* of a cover $\mathcal{D}$ (denoted by $\mathcal{C} *< \mathcal{D}$), provided

$$\{\mathrm{St}(A,\mathcal{C}) : A \in \mathcal{C}\} \text{ refines } \mathcal{D}.$$

The collection $\mathfrak{P} = \{\mathrm{St}(A,\mathcal{C}), \mathcal{D}\}$ is a *local star refinement pattern* on X, if and only if, $\mathrm{St}(A,\mathcal{C}) \subseteq \mathcal{D}$.

Example 14.13. Local Star Refinement Pattern.
Let (X,δ) be the EF space with cover $\mathcal{B}$ from Example 14.11 and let $\mathcal{C} = \{C_1, \ldots, C_n\}$ be a cover of X(only partially shown), $A \in \mathcal{P}(X)$. $\mathfrak{P} = \{\mathrm{St}(A,\mathcal{B}), \mathcal{C}\}$ is a star refinement pattern, since $\mathrm{St}(A,\mathcal{B}) \subseteq \mathcal{C}$. ■

A cover $\mathcal{C}$ is a *descriptive star refinement* of a cover $\mathcal{D}$ (denoted by $\mathcal{C} \underset{\Phi}{*<} \mathcal{D}$), provided

$$\{\mathrm{St}_{\Phi}(A,\mathcal{C}) : A \in \mathcal{C}\} \text{ descriptively refines } \mathcal{D}.$$

That is, the descriptions of the members of $\mathrm{St}_{\Phi}(A,\mathcal{C})$ are a subset of the descriptions of the members of $\mathcal{D}$ (denoted $\mathrm{St}(A,\mathcal{C}) \underset{\Phi}{\subseteq} \mathcal{D}$). The collection $\mathfrak{P}_{\underset{\Phi}{*<}} = \{\mathrm{St}(A,\mathcal{C}), \mathcal{D}\}$ is a *local descriptive star refinement pattern* on X, if and only if, $\mathrm{St}(A,\mathcal{C}) \underset{\Phi}{\subseteq} \mathcal{D}$.

Example 14.14. Local Descriptive Star Refinement Pattern.
Let (X,δ) be the EF space from Example 14.11, $E \in \mathcal{P}(X)$, and let $\mathcal{C} = \{C_1, \ldots, C_n\}$ and $\mathcal{D} = \{D_1, \ldots, D_n\}$ be covers of X (only partially shown). $\mathfrak{P}_{\underset{\Phi}{*<}} = \{\mathrm{St}(E,\mathcal{C}), \mathcal{D}\}$ is a star refinement pattern, since $\mathrm{St}(E,\mathcal{C}) \underset{\Phi}{\subseteq} \mathcal{D}$.

From Example 7.1, the pair of descriptive point clusters $\mathcal{C}_{\Phi(x)}, \sigma_{\Phi(y)}$ defines a local descriptive star refinement pattern, since it is reasonable to expect that the cover of a forged cursive character will be rather similar descriptively to the original cursive character. It is evident that this is the case in Example 7.1. Hence, the descriptive star refinement pattern $\mathfrak{P}_{\underset{\Phi}{*<}}$ defined by

$$\mathfrak{P}_{\underset{\Phi}{*<}} = \left\{\mathrm{St}(A, \sigma_{\Phi(y)}), \mathcal{C}_{\Phi(x)}\right\}$$

is an example of a point cluster-based forgery pattern. ■

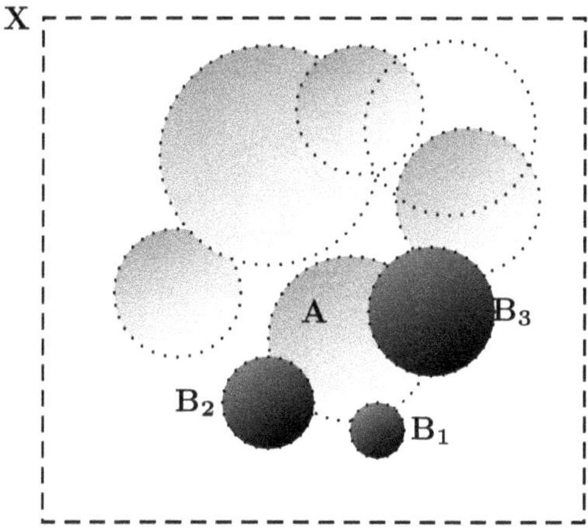

Fig. 14.7 Sample local proximity patterns

14.5.7 *Local Proximity Patterns*

Let X be endowed with an EF-proximity δ, $A \subset X$, $\mathcal{B}$ a collections of subsets in X. Let $(\mathfrak{I}, \delta)$ be an EF-space, X a subset of $\mathfrak{I}$, A a subset of X, $\mathcal{B}$ a collection of subsets in X. The collection $\mathfrak{P} = \{A, \mathcal{B}\}$ is a local **proximity pattern** on X, if and only if, $St(A, \mathcal{B})$ is nonempty, *i.e.*, $A\ \delta\ B$ for at least one B in $\mathcal{B}$.

Example 14.15. Local Proximity Pattern.
Assume that X in Fig. 14.7 is a subset in an EF-space $\mathfrak{I}$ (not shown). A set A is represented by a grey circle ● and the collection $\mathcal{B} = \{B_1, B_2, B_3\}$ is represented by ● in Fig. 14.7. Observe that $A\ \delta\ B_1$, $A\ \delta\ B_2$ and $A\ \delta\ B_3$, *i.e.*, $St(A, \mathcal{B}) = \{B_1, B_2, B_3\}$. Hence, the collection $\mathfrak{P} = \{A, \mathcal{B}\}$ is an example of a local proximity pattern. ■

Let $(X, \delta_\Phi), (Y, \delta_\Phi)$ be descriptive EF spaces, $A \in \mathcal{P}(Y)$, $\mathcal{B}$ a collection of subsets in $\mathcal{P}^2(X)$, Φ a set of probe functions representing features of members of X and Y. The collection $\mathfrak{P}_\Phi = \{A, \mathcal{B}\}$ is a local **descriptive proximity pattern** on $X \cup Y$, if and only if, $A\ \delta_\Phi\ B$ for at least one $B \in \mathcal{B}$.

Example 14.16. Local Descriptive Proximity Pattern.
Let $(X, \delta_\Phi), (Y, \delta_\Phi)$ be descriptive EF spaces in Fig. 14.8, where X, Y are subsets in a kangaroo tiling. Let C_1 be a subset of an open cover $\mathcal{C}$ of X. Let A be an open subset of Y. Choose Φ to be a set of probe functions

Fig. 14.8 Descriptive proximity kangaroo patterns

representing shape and colour features of members of X and Y. The open sets C_1, A are represented by dotted circles. The set A is descriptively near C_1, since there is a shape in A that matches a shape in C_1. In effect, the descriptive star of A with respect to cover $\mathcal{C}$ contains at least one open set in $\mathcal{C}$, namely, C_1. Hence, the collection $\mathfrak{P}_\Phi = \{A, \mathcal{C}\}$ is an example of a local descriptive proximity pattern. A repetition of a pattern in a tiling is called a **motif**. The pattern identified in this example is an example of a motif, since it is repeated in the tiling in Fig. 14.8, especially in the larger tiling in Fig. 14.3. Other examples of local descriptive proximity patterns can be found in the tilings reported in [43; 221]. ■

From the foregoing observations about local patterns, we obtain the following results.

Theorem 14.8. *A Local pattern in an EF space is a collection of near sets.*

Proof. There are two cases to consider.

(1) *Spatial pattern.* Let (X, δ) be a local spatial EF-space, A a subset of X, $\mathcal{C}$ a collection of subsets in X. And let pattern $\mathfrak{P} = \{A, \mathcal{C}\}$. Assume, wlog, $|St(A, \mathcal{C})| > 1$. Let $B \in St(A, \mathcal{C})$, *i.e.*, $A \cap B \neq \emptyset$. Then $A\ \delta\ B$. Hence, A, B are spatially near in the pattern.
(2) *Descriptive pattern.* Let (X, δ_Φ) be a local descriptive EF-space, $A \in \mathcal{P}(X)$, $\mathcal{C} \in \mathcal{P}^2(X)$. Let pattern $\mathfrak{P}_\Phi = \{A, \mathcal{C}\}$. Assume $St_\Phi(A, \mathcal{C})$ contains B, *i.e.*, $A \underset{\Phi}{\cap} B \neq \emptyset$. Then $A\ \delta_\Phi\ B$. Hence, A, B are descriptively near in $\mathfrak{P}_\Phi$. □

The preceding results lead to the notion of near sets in an EF space. This notion is first defined in terms of metric proximity and spatial nearness. In an EF weave space (X, δ), nonempty sets $A, B \in \mathcal{P}(X)$ are *spatially near sets*, provided $D(A, B) = 0$. In a descriptive EF space (X, δ_Φ), nonempty sets $A, B \in \mathcal{P}(X)$ are *descriptively near sets*, provided $D(\Phi(A), \Phi(B)) = 0$.

Corollary 14.1. *A local pattern in a descriptive EF space is a collection of near sets.*

Proof. A space X can be endowed with a descriptive EF proximity δ_Φ. Then the result is immediate from Theorem 14.8. □

Lemma 14.2. *Let $(X, \delta), (X, \delta_\Phi)$ be spatial and descriptive EF-spaces, respectively, with nonempty sets $A, B \in \mathcal{P}^2(X)$, $A \cap B \neq \emptyset$. Then $A \cap B \subseteq A \underset{\Phi}{\cap} B$.*

Proof. Let $A, B \in \mathcal{P}^2(X)$ and assume $A \cap B \neq \emptyset$. If $x \in A \cap B$, then, by definition, $\Phi(x) \in \Phi(A)$ and $\Phi(x) \in \Phi(B)$. Assume, wlog, $x \in A \setminus A \cup B, y \in B \setminus A \cup B$ such that $\Phi(x) = \Phi(y)$. Then $x, y \in A \underset{\Phi}{\cap} B$. Hence, $A \cap B \subseteq A \underset{\Phi}{\cap} B$. □

Given two nonempty sets A and B, if $A \subseteq B$, then A is *coarser* that B (A is *strictly coarser* than B, provided $A \subset B$) and B is termed *finer* (or *strictly finer*) than A.

Theorem 14.9. *Let (X, δ), (X, δ_Φ) be spatial and descriptive EF-spaces, respectively, with nonempty sets of cells A, $B \in \mathcal{P}(X)$. Then $St(A, \mathcal{C})$ is coarser than $St_\Phi(A, \mathcal{C})$.*

Proof. Immediate from Lemma 14.2. □

Theorem 14.10. *A star pattern in a spatial EF space is coarser than a star pattern in a descriptive EF space.*

Proof. Let $(X,\delta),(X,\delta_\Phi)$ be spatial and descriptive EF-spaces, respectively, $A \in \mathcal{P}(X)$, $\mathcal{C} \in \mathcal{P}^2(X)$. From Theorem 14.9, $\mathrm{St}(A,\mathcal{C}) \subseteq \mathrm{St}_\Phi(A,\mathcal{C})$. Hence, $\mathfrak{P} = (A,\mathrm{St}(A,\mathcal{C}))$ is coarser than $\mathfrak{P}_\Phi = (A,\mathrm{St}_\Phi(A,\mathcal{C}))$. □

Corollary 14.2. *A spatial star refinement pattern is coarser than a descriptive star refinement pattern.*

14.6 Problems

(**14.6**.1) Show that every uniformity $\mathcal{U}$ induces Efremovič proximity $\delta = \delta_{\mathcal{U}}$, where

$$A\ \delta\ B \Leftrightarrow \text{ for each } U \in \mathcal{U}, U(A) \cap B \neq \emptyset.$$

In turn, $\mathcal{U}$ induces a Tychonoff (completely regular) topology $\tau(\mathcal{U})$.

(**14.6**.2) Show that a uniformity has (a) an open base and (b) a closed base.

(**14.6**.3) The union of any family of uniformities is a base for a uniformity.

(**14.6**.4) A coarsest compatible uniformity exists (and is totally bounded) on a topological space X, if and only if, X is locally compact Hausdorff.

(**14.6**.5) Every uniformly continuous function is (proximally) continuous with respect to associated (proximities) topologies.

(**14.6**.6) Show that uniform convergence implies proximal convergence and preserves (uniform, proximal) continuity.

(**14.6**.7) Show that the following hold.

(i) The two ways to define a Cauchy filter in a uniform space are equivalent.

(ii) A uniform space $(X,\mathcal{U})$ is totally bounded, if and only if, every ultrafilter in X is Cauchy.

(iii) A uniform space $(X,\mathcal{U})$ is compact, if and only if, $\mathcal{U}$ is complete and totally bounded.

(**14.6**.8) For a covering uniformity, define analogues of various concepts and results discussed in this chapter.

(**14.6**.9) If μ is a covering uniformity on X, show that a **base** for the related entourage uniformity $\mathcal{U}$ is a family of sets of the form

$$\bigcup\{E \times E : E \in \mathcal{A}\}, \text{ where } \mathcal{A} \in \mu.$$

In the following problems, assume X is a regular T_3 space [93].

(**14.6**.10) Show that X is metrisable, if and only if, for each $x \in X$, there is a countable collection $\{V_n(x) : n \in \mathbb{N}\}$ of open neighbourhoods of x such that the following hold.

(i) $\{V_n(x) : n \in \mathbb{N}\}$ is a local base for x.

(ii) For all $n \in \mathbb{N}$, there exists $k \in \mathbb{N}$ such that if $V_k(x) \cap V_k(y) = \varnothing$, then $V_k(y) \subset V_n(x)$.

(**14.6**.11) Show X is metrisable, if and only if, there is a sequence $\{G_n : n \in \mathbb{N}\}$ of closure-preserving open collections in X such that, for all $x \in X$,

$$\{St(x, G_n) : n \in \mathbb{N} \text{ and } St(x, G_n) \neq \varnothing\} \text{ is a local base for } x.$$

(**14.6**.12) Show X is metrisable, if and only if, there is a sequence $\{G_n : n \in \mathbb{N}\}$ of open covers of X such that, for all $x \in X$, $\left\{St^2(x, G_n) : n \in \mathbb{N}\right\}$ is a local base for x.

(**14.6**.13) Show X is metrisable, if and only if, there is a sequence $\{G_n : n \in \mathbb{N}\}$ of open covers of X such that, for all $x \in X$ and every neighbourhood R of x, there is a neighbourhood V of x and $n \in \mathbb{N}$ such that $St(V, G_n) \subset R$.

(**14.6**.14) Show X is metrisable, if and only if, there is a sequence $\{G_n : n \in \mathbb{N}\}$ of open covers of X such that, given disjoint closed sets H and K in X with K compact, there exists $n \in \mathbb{N}$ such that $St(K, G_n) \cap H = \varnothing$.

(**14.6**.15) Show X is metrisable, if and only if, there is a sequence $\{F_n : n \in \mathbb{N}\}$ of locally finite closed covers of X such that, for all $x \in X$ and every neighbourhood R of x, there exists $n \in \mathbb{N}$ such that $St(x, F_n) \subset R$.

(**14.6**.16) Show X is metrisable, if and only if, there is a sequence $\{F_n : n \in \mathbb{N}\}$ of closure-preserving closed covers of X such that, for all $x \in X$ and every neighbourhood R of x, there exists $n \in \mathbb{N}$ such that $St(x, F_n) \subset R$.

(**14.6**.17) Is the following result true?

An Efremovič space (X, δ) is normal T_4, if and only if, every proximally continuous function on any closed subset A of X to $I(\mathbb{R})$ has a proximally continuous extension to $I(\mathbb{R})$.

(**14.6**.18) Give a detailed proof of Theorem 14.7.

(**14.6**.19) Prove Corollary 14.2.

Notes and Further Readings

Fréchet's Class, Écart and Voisinage

In his thesis published in 1906, M. Fréchet [64], n^0 49, p. 30 introduced a set of elements called a class, and concepts called *voisinage* and *écart*. A *class* E is a set of elements with several properties, namely, (1) if $p \in E$, then there is a countable sequence of elements $p_1, p_2, \ldots$ with p as its limit, (2) if $p \in E$ is the limit of sequence $\{p_n\}, n \in \mathbb{N}$ and $n_1, n_2, \ldots$ is an infinite sequence of positive integers such that $n_1 < n_2 < \ldots$, then p is also the limit of the sequence $p_{n_1}, p_{n_2}, \ldots$, and (3) if $p \in E$, then p is the limit of the sequence $p, p, \ldots$ whose elements all coincide with p. R.L. Moore observed that an element $p \in E$ is said to be a *limiting element* of a sub-class M of E, if p is the limit of some infinite sequence of distinct elements belonging to M [146].

Given a class V and a pair of elements $a, b \in V$, $[a, b]$ is called the *voisinage* of a and b that is non-negative, symmetric in a and b and satisfies two additional conditions, namely, (i) $[a, b] = 0$ if, and only if a = b and (ii) and there exists a positive function $f(e)$ such that $lim_{e \to 0} f(e) = 0$ and if $[a, b] < e$ and $[b, c] < e$, then $[a, c] < f(e)$. The écart (denoted by (a,b)) of two elements a, b of a class E differs from a voisinage [a,b] only in that condition (ii) is replaced by the triangle inequality, *i.e.*, if $a, b, c \in E$, then

$$(a, c) \leq (a, b) + (b, c).$$

Fréchet observed that *Ainsi, l'écart est un voisinage...* (Thus, the écart is a voisinage [64]). Eleven years later, E.W. Chittenden proved the equivalence of an écart and a voisinage [41]. ■

Metric Spaces

M. Fréchet [64], introduced metric spaces. The nearness of sets is formalized with the **gap functional** $D_\rho(A, B)$ [123]. R. Lowen [137] observes that perhaps the most appealing structure (initially considered by F. Hausdorff [82]) is the distance between points and sets. The study of metric spaces depends on the limit point of a set, which can be described in terms of nearness of a point to a set (this was suggested by F. Riesz [199]). For a biography of F. Riesz, see [157]. Put $\rho(x, y)$ equal to the distance between points $x, y \in X$. For a point $x \in X$ and a set B, define a Hausdorff lower distance [82; 141] $\rho : X \times X \to \mathcal{P}(\mathbb{R})$ defined by

$$\rho(x, B) = inf\{\rho(x, b), b \in B\}.$$

In every metric space (X, ρ), there is a **closure operator** [116] induced by the metric ρ, where x is a closure point of set $B \subset X$ if, and only if x is near B or $D_\rho(x, B) = 0$.

In the setting of a metric space (X, ρ), we have studied (a) topology: a point is near a set and (b) proximity: nearness of pairs of sets. It is natural to consider when families of sets are near. This was done by V.M. Ivanova and A.A. Ivanov [99] for finite families and by Herrlich [88] for sets of arbitrary cardinality.

Characterisations of compactness in metric spaces stem from the ramifications of an important result from E.L. Lindelöf [131] concerning an m-dimensional, Euclidean space $\mathbb{R}^m$. ∎

Hausdorff Metric Topology
For a metric space (X, ρ), the set of all non-empty, closed subsets of X (denoted $CL(X)$) is called a **hyperspace**. The Hausdorff metric topology is defined on the hyperspace $CL(X)$. The Hausdorff metric topology was first discovered by D. Pompeiu [192; 193] and by F. Hausdorff in 1927. This topology has been used in a number of areas such as function space topologies, approximation theory, and optimisation (see, *e.g.*, [23]). ∎

Finite Complexes
F. Hausdorff[82] wrote: From an alphabet, i.e., a finite set of letters, we may construct a countable assemblage of finite complexes [= ordered sets] of letters, i.e., words among which, of course, meaningless words such as abracadabra occur. If in addition to the letters, other elements are used, such as punctuation marks, typespacings, numerals, notes, etc., we see that the assemblage of all books, catalogs, symphonies and operas is also countable, and would remain countable even if we were to employ a countable set of symbols (but for each complex only a finite number).

On the other hand, if in the case of a finite number of symbols we restrict the complexes to a maximum number of elements, agreeing for example, to rule out words of more than one hundred letters and books of more than one million words, these assemblages become finite; and if we assume with Giordano Bruno an infinite number of heavenly bodies, with speaking, writing, and musical inhabitants, it follows as a mathematical certainty that on an infinite number of these heavenly bodies there will be produced the same opera with the same libretto, the same names of the composer, the author of the text, the members of the orchestra and the singers. However, this whimsical passage from his 1914 book appears in a greatly reduced passage in [83]. For more reflections from and about Hausdorff, see [29]. ∎

Closure
The basic facts about closure of a set were first pointed out by M. Fréchet [64] and elaborated by B. Knaster and C. Kuratowski [112]. ∎

Gap functional
It was [82] who suggested defining the distance between two sets and to treat sets as elements of a metric space. ∎

Near Sets
A proximity relation between the subsets of a set leads to a study of spatially near sets. Many different proximity relations are possible. A complete overview of proximity and spatially near sets is given by S.A. Naimpally [163] (see, also, [235; 236; 207]). Recent work on the nearness of sets and nearness spaces by S. Tiwari has led to the study of approach merotopological spaces [226] (see, also, [107; 110; 109; 108]).

A descriptive proximity relation between the subsets of a set leads to a study of descriptively near sets. Let A and B be subsets of X and let Φ be a set of probe fuctions $\phi : X \to \mathbb{R}$ that represent features of $x \in X$. A feature vector $\Phi(x)$ is a vector of feature values $\phi(x)$ that provide a description of x. Then A is descriptively near B, provided there are $a \in A, b \in B$ such that $d(\Phi(a), \Phi(b)) = 0$. A non-spatial view of near sets appears informally in [200] (see, also, [157]) and formally in [176; 175] (elaborated in [189]) based on the nearness of disjoint sets that resemble each other. An overview of descriptively near sets is given in [183].

> On dit encore que la fonction $f(x)$ est, dans le voisinage d'une valueur particulière attributée à la variable x, fonction continue de cette variable, toutes les fois qu'elle est continue entre deux limites de x, meme très rapprochées, qui renferment la valeur dont il s'agit.
>
> –A.-L. Cauchy, 1821

It has been observed by J.R. Isbell that the notions *near* and *far* in a uniform space are important. Sets A, B are far (**uniformly distal**), provided the $\{A, B\}$ is a discrete collection. A nonempty set U is *uniform neighbourhood* of a set A, provided the complement of U is far from U [98]. ■

Continuity

In 1821, A.-L. Cauchy [40], ch. II, p. 43 refers to neighbourhoods and nearness (closeness) in explaining continuity. He defines the continuity of a function f at x in terms of neighbourhoods of values of x. Intuitive definition: A function f is continuous at a point c, if and only if, x is near c implies $f(x)$ is near $f(c)$. ■

Function Spaces and Convergence of a Family of Functions

The concept of an *arbitrary function* was almost unknown at the beginning of the 19th century. Even so, the notion of *pointwise convergence* existed during the early beginnings of calculus, especially in the study of power series and trigonometric series. The idea of putting a topological structure or convergence on a family of functions began with G.F.B. Riemann and a systematic study began towards the end of the 19th century (see, *e.g.*, [198]). Uniform convergence of series was discovered independently by G.G. Stokes in 1847 and P.L. von Seidel in 1848. In a paper in 1841 but published in 1894, K. Weierstrass [242] used the notion of uniform convergence. Uniform convergence and its various ramifications were studied further by C. Arzelà, U. Dini, H. Hankel, P. du Bois-Reymond, and others (for the details, see, *e.g.*, [92; 80]). ■

Topological Spaces

There are alternate ways of beginning a study of topological spaces (see, *e.g.*, [4; 141; 120; 32; 132]). A lot of experimentation was made to define compactness (see, *e.g.*, [5], [6]) and it was found that the most suitable definition in abstract topological spaces is (**Equiv**.2) (Heine-Borel Theorem) [115; 231]. The naming of this theorem varies, *e.g.*, P. Dugac called it the Dirichlet-Heine-Weierstrass-Borel-Schoenflies-Lebesgue theorem [57]. R. Cooke points out Borel studied analytic continuation along a curve from one point to another. This was done by covering the curve with overlapping disks in which the Taylor series converges and found it necessary to prove that only a finite number of disks are required to get from one point to another (Borel's part of the theorem), Math Forum: `http://mathforum.org/`, 2004. Like most good theorems, the conclusion of the Heine-Borel theorem has become a definition [106].

Recall that a subset $A \subset X$ for a topological space X is termed **dense** if, and only if, any point $x \in X$ belongs to A or x is a limit point of A, *i.e.*, $X = \overline{A}$ [213]. ∎

L- and EF-Proximities

The conditions for an L-proximity were first studied by S. Leader [124] for the non-symmetric case and by M.W. Lodato [133; 134; 135] for the symmetric case.

Observe that up to the 1970s, *proximity* meant EF-proximity, since this is the one that was studied intensively [163]. But in view of later developments, there is a need to distinguish between various proximities. A **basic proximity** or **Čech-proximity** [229] δ is one that satisfies (P.0)-(P.3). ∎

Symmetric Proximity

The symmetric or R_0 axiom ((∗) a point a is near another point b implies b is near a) arose during a consideration of symmetric proximities compatible with a topology[2]. This axiom was discovered by A.S. Davis [47].

Earlier separation axioms had been discovered and were called *Trennungsaxiome* (German) by P. Alexandroff and H. Hopf [5]. Hence, these axioms are known as $T_n, n = 0, 1, 2, 3, 4, 5$.

While working in abstract spaces, it is necessary frequently to have this property of unique limits for nets and filters (see Ch. 6), which generalises sequences. Hausdorff observed this and used this axiom in his work. The corresponding space with *pairs of distinct points belong to disjoint neighbourhoods* [141] is now named after him and is called the T_2 or Hausdorff space.

It has also been observed that every T_1 space is a T_0 space [59]. Putting these observations together, $T_2 \Rightarrow T_1 \Rightarrow T_0$ [243].

L. Vietoris [233] defined a stronger axiom than T_2. A T_1 space is T_3 or **regular** if, and only if, a point a does not belong to a closed set B implies a and B have disjoint neighbourhoods, or *each neighbourhood of a point contains a closed neighbourhood.*

One of the most beautiful, intricate and surprising results in topology was proved by P. Urysohn in 1925 [230] by showing that, in a normal or T_4 space, neighbourhood separation of disjoint closed sets equals functional separation. This result (known as the Urysohn Lemma) is sometimes viewed as the first non-trivial fact of point set topology. ∎

Uniformity

There is a structure finer than proximity that is called a **uniformity**. This was discovered by A. Weil [244]. There are several ways to define a uniformity on a nonempty set X.

---

[2]See §1.20 for an example of an R_0 space, *i.e.*, the (∗) is satisfied due to the partition of an image into equivalence classes. Notice that the space formed by the partition with two or more classes in the example in §1.20 is not T_0, it is possible to have two distinct points a, b in separate classes such that $a \underline{\delta} b$ (a is far from b).

(**Uniformity**.1) family of pseudometrics [74; 244], or
(**Uniformity**.2) family of covers [227; 98; 244], or
(**Uniformity**.3) set of entourages[3] that is a family of subsets of $X \times X$ [33; 106; 244].

All of the approaches to defining a uniformity are in A. Weil's work [244] with further developments by others. This book begins with metric spaces to motivate (**Uniformity**.1) using pseudometrics as in [74]. ■

Filters and Ultrafilters

We need a generalisation of a sequence. This was done in two ways: with structures called nets by E.H. Moore and H.L. Smith [145] and with structures called filters by H. Cartan [38; 37], elaborated in [18; 20] (see, also, [32]).

An **ultrafilter** $\mathcal{L}$ is a maximal filter [37].

If X contains at least two elements, there are at least two distinct ultrafilters on X. Hence, the ordered set of filters on X has no greatest member [33].

It is obvious that both filters and nets are adequate generalisations of sequences and fit like a glove in topological spaces. Their equivalence has been shown by Bartle [18; 20]. ■

Compactness

Compactness is viewed as one of the most valuable concepts in analysis and topology [19]. Nomenclature for compactness differs among mathematicians, *e.g.*, Bourbaki [32] added a Hausdorff axiom to the definition for a compact space and Soviet mathematicians called such spaces bicompact, *e.g.*, [5], [11]. For more on compactness, see [232]. There is considerable literature on minimal Hausdorff spaces (see, *e.g.*, [27; 216; 100; 103]). There are several forms of local compactness in the literature on topological spaces that are equivalent in the important case of Hausdorff spaces, *e.g.*, [33].

The Heine-Borel theorem is now used as the definition of compactness in topology, functional analysis and elsewhere. Other formulations of compactness (*e.g.*, (a) sequential compactness and (b) Bolzano-Weierstrass property) that are equivalent to it in metric spaces have proved useful as counterexamples. There is a vast literature on this (*e.g.*, [115]).

A generalisation *paracompactness* was discovered by J. Dieudonné [54] that has proven to be important in topology and analysis.

A topological space X is called **paracompact** if, and only if, each open cover of X has an open, locally finite refinement that is a cover of X. Usually, the Hausdorff axiom is added in defining a paracompact space (*e.g.*, [32]).

An important result was proved by A.H. Stone: *every metric space is paracompact* [214]. For more information about this, see, *e.g.*, [115; 106].

J.L. Kelley [106] observes that the Tychonoff product theorem on the product of compact spaces is the most useful theorem on compactness and is *probably the most important single theorem of general topology.* ■

Grills, Clusters and Bunches

Grills, clusters and bunches that arise naturally in proximity spaces (see, *e.g.* the 1973 paper by W.J. Thron [223] and, more recently, by others [204; 138; 182; 186]), are covered in this chapter. Grills were introduced by G. Choquet in 1947 [42], clusters by S. Leader in 1962 [122; 123] and bunches by Leader's student M.W. Lodato [133; 134; 135].

---

[3] French: neighbourhoods or *surroundings*.

W.J. Thron observed that grills are duals of filters and one of their important properties is that they are unions of ultrafilters [223]. In addition, observe that in a discrete proximity space, clusters coincide with ultrafilters [163].

A **clan** in an L-proximity space (X, δ) is a grill in which any two members are near [223]. F. Riesz [199] considered certain grills which, in our present terminology, are maximal clans in a proximity space [223]. We use the notation **δ-clan**, if it is necessary to specify proximity.

A bunch in an L-space (X, δ) is called a **cluster**, if and only if, it satisfies property (**) [122; 123]. In the original definition of a cluster, S. Leader [122] used the analogues of the ultrafilter properties (ultrafilter.1)-(ultrafilter.3), replacing intersection $\cap$ with the relation *near*.

Theorem 14.11. [*122*]
(*a*) *Let (X, δ) be an EF-proximity space. If $\mathcal{L}$ is an ultrafilter in X, then $\delta(\mathcal{L})$ is a cluster. Conversely,*
(*b*) *if $\mathcal{L}$ is an ultrafilter in a cluster σ, then $\sigma = \delta(\mathcal{L})$.*
(*c*) *If $A \in \sigma$ is a cluster, then there is an ultrafilter $\mathcal{L}$ in X such that $A \in \mathcal{L} \subset \delta(\mathcal{L}) = \sigma$.*

Corollary 14.3. *If $A\ \delta\ B$ in an EF-proximity space (X, δ), then there is a cluster σ that contains both A and B.* [*122*]

Corollary 14.4. *In an EF-proximity space (X, δ), σ is a cluster, if and only if, it is a maximal bunch. Hence, in an EF-proximity space, every bunch is contained in a unique cluster.*[*66*]

Theorem 14.12. *An EF-proximity space (X, δ) is compact, if and only if, every cluster in the space is a point cluster.* [*122*]

H. Wallman uses the family wX of *all* closed ultrafilters of X and assigns a topology on wX (known as Wallman topology) that makes wX compact. X is embedded in wX via the map that takes a point $x \in X$ to the closed ultrafilter $\mathcal{L}_x$ containing $\{x\}$ (for details, see [237], [106]). Thus, Wallman gets just one T_1 compactification of any T_1 space.

To generalise Wallman compactification, we begin with an arbitrary compatible L-proximity δ on X and follow Wallman step-by-step replacing wX by X^*, the space of all maximal bunches in the L-proximity space (X, δ) and mapping x to the maximal bunch σ_x, the family of all subsets of X near x. The resulting **proximal Wallman compactification** of (X, δ) [69] contains as special cases infinitely many T_1 compactifications as well as the Smirnov compatification, which includes all Hausdorff compactifications.

Theorem 14.13. [*69*]
Let (X, δ) be a T_1 L-proximity space and let X^* be the space of all maximal bunches with the topology τ^* generated by the closed base $\{\sigma \in X^* : E \in \sigma\}$ and fine L-proximity δ_0.
(*a*) *(X^*, τ^*) is a compact in T_1 space.*
(*b*) *Given the map $\phi : X \to X^*$ given by $\phi(x) = \sigma_x$, the cluster point x is a proximal isomorphism on X to $\phi(X)$.*
(*c*) *$\phi(X)$ is dense in X^*.*
(*d*) *$A\ \delta\ B$, if and only if, $cl\ \phi(A) \cap cl\ \phi(B) \neq \varnothing$ in X^*.*

Frink generalised Wallman compactification by choosing a subfamily of all closed sets and this was put in proximal Wallman form in [65; 67].

$(\mathbb{R}^*, \delta_0)$ is proximally isomorphic to the proximal Wallman compactification of (X, δ) in Corollary 9.5 (see [139] for a generalisation).

By Corollary 9.5, it follows that every EF-proximity δ is induced by the fine EF-proximity δ_0 on the compact Hausdorff space X^* of all clusters with Wallman topology. This simple and beautiful result is a culmination of the work of many mathematicians (for details, see [163; 158]). ■

Proximal Continuity

The proximal continuity of a function is considered in, *e.g.*, [163], [97]. Proximal continuity of f is also sufficient when the spaces involved satisfy stronger conditions. A.D. Taimanov [218] first proved a special case in which $Y = \lambda Y$ is compact Hausdorff.

Theorem 14.14. Generalized Taimanov Theorem [218; 66]
Let X be dense in a T_1 space αX and let δ_1 be the L-proximity on X defined by $A\ \delta_1\ B$ in X if, and only if, their closures in αX intersect. Let (Y, δ_2) be an EF-proximity space and let Y^* be its Smirnov compactification. Then a function $f : X \to Y$ has a continuous extension $\overline{f} : \alpha X \to Y^*$ if, and only if, f is proximally continuous.

Theorem 14.15. Taimanov Theorem
Let X be dense in a T_1 space αX and let Y be compact Hausdorff. Then $f : X \to Y$ has a continuous extension $\overline{f} : \alpha X \to Y$ if, and only if, for every pair of disjoint closed sets $A, B \subset Y$, closures of $f^{-1}(A)$ and $f^{-1}(B)$ in αX are disjoint.

Proof. For a proof without proximity, see [59]. □

■

Metrisation

A space X is metrisable if, and only if the uniformity or proximity or topology is induced by a metric. For an interesting paper on the Hausdorff metric topology, measurability of set-valued functions and separable metrisable spaces, see [17]. For an overview of metrisability, see P.J. Collins and A.W. Roscoe [45].

An important special case occurs when the family $\{\mathcal{U}_n : n \in \mathbb{N}\}$ in Theorem 11.1 consists of open covers of X. In that case, the space X is called **developable**. It has been shown that a T_1 space X is developable if, and only if the semi-metric **(***)** is upper semi-continuous [67].

Using the well-known Weierstrass M-test, it follows that

$$d(x, y) = \Sigma \left\{ \frac{1}{2^n} d_n(x, y) : n \in \mathbb{N} \right\}$$

is a pseudometric on X. It is easily verified that d is a actually a compatible metric on X. Thus, we obtain the **Alexandroff-Urysohn uniform metrisation** [6] result in Theorem 11.2.
The **Chittenden metrisation** result is given in [41]
Similarly, we have the **Niemytzki metrisation** result in [164]
The **Arkhangle'skiĭ metrisation** result is given in [12]

Theorem 14.16. A.H. Stone metrisation [215]
A T_1 space X is metrisable if, and only if, it has a countable family of open covers

$\{\mathcal{U}_n : n \in \mathbb{N}\}$ *such that each* $x \in X$ *and neighbourhood* U *of* x*, there a neighbourhood* V *of* x *and an* $n \in \mathbb{N}$ *such that* $St(V, \mathcal{U}_n) \subset U$*.*

Topologies and Convergences on a Set of Functions

Generally, most texts consider only continuous functions (see, *e.g.*, J.L. Kelley [106] and N. Bourbaki [33]).

The **set-open topology** τ_α on $\mathcal{F}$ is generated by

$$\{[A, U] : A \in \alpha, U \in \tau'\},$$

where $[A, U] = \{f \in \mathcal{F} : f(A) \subset U\}$ [63; 106; 140]. In case Y has an L-proximity δ, then τ_α has been generalised to a **proximal set-open topology** [123; 51] by replacing $[A, U]$ by

$$[A :: U] = \{f \in \mathcal{F} : f(A) <<_\delta U\}.$$

A typical basic open set in a graph topology is

$$W^+ = \{f \in \mathcal{F} : G(f) \subset W\},$$

where W is an open subset of $X \times Y$ [153].

Tychonoff [228] showed that pointwise convergence topology $\mathcal{P}$ on $\mathcal{F}$ is precisely the subspace topology induced by the product topology on Y^X.

In 1945, R.H. Fox [63] discovered compact open topology (a special case of set-open topologies) on function spaces. It generalized uniform convergence on compacta (see Sect. 12.5 for details), *cf.* [9; 10].

Theorem 14.17. (Kelley, Ascoli, Morse [106; 15]) *Let* X *be a locally compact Hausdorff space,* Y *a regular space and let the family* $\mathcal{F} \subset C(X, Y)$ *have compact open topology* $\mathcal{K}$*, then* $\mathcal{F}$ *is compact if, and only if,*
(12.6(a)) $\mathcal{F}$ *is closed in* $C(X, Y)$*.*
(12.6(b)) For each $x \in X$*,* $cl\mathcal{F}(x)$ *is compact.*
(12.6(c)) $\mathcal{F}$ *is evenly continuous.*

To preserve continuity, we need a stronger convergence. There are several such convergences (see Sect. 12.5). S. Leader discovered proximal convergence in 1959 [123].

A generalization of (prxc) substitutes each $A \subset X$ with *each set* A *from a network* α *of nonempty subsets of* X [51]. The resulting *proximal set open topology* contains as special cases all set-open topologies. A. Di Concilio [49] has used them in the study of homeomorphism groups.

From Sect. 12.4, we know that pointwise convergence need not preserve continuity. To rectify this deficiency, uniform convergence was discovered in 1847-1848 by Stokes and Seidel independently. In an 1841 paper (published in 1894), Weierstrass used the notion of uniform convergence. Modifications of uniform convergence were studied by others, *e.g.*, [39].

Proof that uniform convergence preserves (uniform) continuity is analogous to the proof studied in analysis, *cf.*, [8]. U. Dini [55] modified uniform convergence to **simple uniform convergence**: (f_n) converges pointwise to f and, for each $\varepsilon > 0$ and $e \in \mathcal{E}$, *frequently*, for all $x \in X$, $e(f_n(x), f(x)) < \varepsilon$.

A search began to find necessary and sufficient conditions for preservation of continuity. To accomplish this, modifications of uniform convergence were studied by Arzela, Dini, Young and others, *cf.* [39]. Much of this interesting classical material is not found contemporary analysis texts and we have *Hobson's Choice (taking the one offered or nothing* [151]) for this material [92]. Recently, there has been some work on preservation of continuity in uniform spaces [61]. There are two results of Dini in the monumental book by Hobson [92], now restated in the current language in Section 12.6.

N. Bouleau [31] generalised (Dini.1) to uniform spaces.

If X is compact (in Bouleau convergence of $(f_n : n \in \mathcal{D})$ to f), for each $n_0 \in \mathcal{D}$, X has a finite open cover $(U_{nk} : 1 \le k \le m)$ with $n_k > n_0$ and, for each $z \in X$, there is an n_k such that $z \in U_{nk}$ and so $e(f_{nk}(z), f_n(z)) < \varepsilon$. Thus we have the theorem of Arzelà [14; 9] extended by Bartle [19]. In this case, the convergence is called **quasi-uniform**.

Morse and Kelley [106] discovered *even continuity* as a generalization of equicontinuity, which is similar to compact open topology being a generalization of uniform convergence on compacta.

A typical basic open set in graph topology Γ is

$$W^+ = \{f \in \mathcal{F} : G(f) \subset W\},$$

where W is an open subset of $X \times Y$ [153; 159; 194; 195].

The problem is to put a topology on the family of partial maps $\mathcal{F}$ such that the map

$$(****)\ \psi : \mathbb{R} \to \mathcal{F} \text{ given by } \psi(z) = f_z$$

is continuous. A similar problem *continuity of the solutions with respect to initial conditions* occurs in ordinary differential equations [206].

In the literature on partial maps, Kuratowski [119] considers continuous maps with *compact* domains, Sell [206] and Abdallah-Brown [1] deal with continuous maps with *open* domains, while Back [16] and Filippov [62] work with continuous maps with *closed* domains. ■

Hyperspace Topologies

During the early part of the 20th century, there were two main studies on the hyperspace. Vietoris [234] introduced a topology on a T_1 space consisting of two parts:
Vietoris.(a) Lower *finite hit* part, and
Vietoris.(b) Upper *miss* part.
This leads to

> The join of Vietoris.(a) and Vietoris.(b) is known as **Vietoris topology** or Vietoris finite topology [142]. Replacing Vietoris.(a) (upper miss part) by *far* when X has proximity, gives rise to **proximal topology** on the hyperspace [24].

As observed in §1.17, Hausdorff [141] defined a metric on the hyperspace of a metric space and the resulting topology depends on the *metric* rather than on the topology of the base space. Two equivalent metrics on a metrisable space induce equivalent **Hausdorff metric topologies** if, and only if, they are uniformly equivalent. With the discovery of uniform spaces, the Hausdorff metric was generalized to **Hausdorff-Bourbaki uniformity** or H-B uniformity on the hyperspace of uniformisable (Tychonoff) space. In the 1960s, this led to the celebrated Isbell-Smith problem to find necessary and sufficient conditions that two uniformities on the base space are H-equivalent, *i.e.*, they induce topologically equivalent H-B uniformities on the hyperspace [98; 209; 239; 240]. For a brief account of this topic, see [163].

In 1966, while working on a problem in convex analysis, Wijsman [246] introduced a convergence for the closed subsets of a metric space (X, d), namely,

$$A_n \to A \Leftrightarrow, \text{ for each } x \in X, d(x, A_n) \to d(x, A).$$

Wisjman topologies are the building blocks of many other topologies, *e.g.*,

(a) Metric proximal topology is the *sup* of all Wisjman topologies induced by uniformly equivalent metrics, and
(b) Vietoris topology is the *sup* of all Wisjman topologies induced by topologically equivalent metrics [24].

Attempts to topologise Wisjman topology led to the discovery of

(a) Ball topology [23], and
(b) Proximal Ball topology [72].

The lower part of the Wisjman topology coincides with the lower Vietoris topology. It is the upper part that is difficult to handle. Later, it was shown that the upper Wisjman topology can be expressed as a *far-miss* topology [155].

Hyperspaces are useful in areas such as unilateral analysis, optimization theory, convex analysis, geometric functional analysis, mathematical economics, and the theory of random sets. Standard references include a seminal paper by Michael [142] on Vietoris and H-B topologies. An extensive account is given by Beer [23] on the literature up to 1993.

Uniformity and Metrisation

There is a vast literature on uniform spaces (see, *e.g.*, [98; 89; 101] and [32]). Chapter 14 presents two forms of uniform structures (uniformities), namely, entourage uniformity [32; 98; 244] and covering uniformity [98],[106; 50].

A **covering** $\mathcal{U}$ of a metric space X is uniform, provided there is a $\varepsilon > 0$ such that for each point $x \in X$, the ε-neighbourhood N_x of x is contained in some member of $\mathcal{U}$. The number ε is called a Lebesgue number for $\mathcal{U}$ [98].

R.E. Hodel [93] gives an excellent survey of metrisation results.

To improve the result in Theorem 14.3, it is necessary to weaken the *countable base* condition with a condition that is satisfied by all metric spaces. This is what was done independently by Nagata, Smirnov and Bing. The story begins with paracompactness, a nice generalization of compactness discovered by J. Dieudonné [54].

Dieudonné proved that every separable metric space is paracompact and this result was extended to all metric spaces by A.H. Stone [214]. Stone's paper provided arguments that led to a necessary condition for metrisation. E. Michael [143] showed that a regular T_3 topological space is paracompact, if and only if, every open cover has a σ-locally finite open refinement. With further manipulations, this leads to the fact that every metric space has a σ-locally finite open base. J. Nagata [152] and Y. Smirnov [208] showed that a regular T_3 topological space is metrisable, provided it has a σ-locally finite open base.

Theorem 14.18. (Nagata-Smirnov [152; 208]) *A topological space is metrisable, if and only if, it is regular T_3 and has a σ-locally finite open base.*

Theorem 14.19. (Bing [28]) *A topological space is metrisable, if and only if, it is regular T_3 and has a σ-discrete open base.*

The function $f(x) = \frac{1}{x}$ on $(0, 1) \to \mathbb{R}$ cannot be extended to a continuous function on $\mathbb{R} \to \mathbb{R}$. Hence, the subset must be closed. This extension result was first proved by H. Lebesgue for the plane [125], by H. Tietze [224] for metric spaces, and finally P. Urysohn [230] showed that *a topological space is normal* T_4*, if and only if, every continuous* f *on an closed subset* $A \to [-1,1]$ (or $\mathbb{R}$) has a continuous extension to X. That is, for each $a \in A, f(a) = g(a)$ [230]. However, this result is generally known as Tietze's Theorem. ■

A good introduction to metric and pseudometric spaces is given by [106], [213], and [217]. One of the most complete, formal presentations of metric spaces is given by C. Kuratowski [118].

Bibliography

[1] Abdallah, A. and Brown, R. (1980). A compact-open topology for partial maps with open domains, *J. London Math. Soc.* **21**, 2, pp. 480–486.

[2] Alexander, J. (1939a). On the concept of a topological space, *Proc. Nat. Acad. of Sciences* **25**, pp. 52–54.

[3] Alexander, J. (1939b). Ordered sets, complexes and the problem of compactifications, *Proc. Nat. Acad. of Sciences* **25**, pp. 296–298.

[4] Alexandroff, P. (1961). *Elementary Concepts of Topology* (Dover, NY).

[5] Alexandroff, P. and Hopf, H. (1935). *Topologie* (Springer-Verlag, Berlin), xiii + 636pp.

[6] Alexandroff, P. and Urysohn, P. (1929). Mémoire sur les espaces topologique compactes, *Verh. Akad. Wettensch. Amsterdam* **14**, pp. 1–96.

[7] Allison, J., Brayer, M., Migayrou, F. and Spiller, N. (2006). *Future City: Experiment and Utopia in Architecture* (Thames & Hudson, London).

[8] Apostol, T. (1957). *Mathematical Analysis* (Addison-Wesley Pub. Co., London), xii + 559pp.

[9] Arens, R. (1946). A topology for spaces of transformations, *Ann. Math.* **47**, pp. 480–495.

[10] Arens, R. and Dugundji, J. (1951). Topologies for function spaces, *Pacific J. Math.* **1**, pp. 5–31.

[11] Arkhangel'skiĭ, A. (1965). Bicompact sets and the topology of spaces, *Trans. Moscow Math. Soc.* **126**, 2, pp. 239–241.

[12] Arkhangel'skiĭ, A. (1966). Mappings and spaces, *Russian Math. Surveys* **21**, pp. 87–114.

[13] Armstrong, H. and Brasier, M. (2005). *Microfossils, 2nd Ed.* (Blackwell Publishing, Malden, MA, U.S.A.).

[14] Arzelà, C. (1895). Sulle funzioni di linee, *Mem. Accad. Sci. Ist. Bologna Cl. Sci. Fis.* **5**, 5, pp. 55–74.

[15] Ascoli (1884). Le curve limiti di una varietà data di curve, *Atti della R. Accad. Dei Lincei Memorie della Cl. Sci. Fis. Mat. Nat.* **18**, 3, pp. 421–586.

[16] Back, K. (1986). Concepts of similarity for utility functions, *J. Math. Economics* **15**, pp. 129–142.

[17] Barbati, A., Beer, G. and Hess, C. (1994). The Hausdorff metric topology, the Attouch-Wets topology, and the measurability of set-valued functions, *J. Convex Analysis* **1**, 1, pp. 107–119.

[18] Bartle, R. (1955a). Nets and filters in topology, *Amer. Math. Monthly* **62**, 8, pp. 551–557.

[19] Bartle, R. (1955b). On compactness in functional analysis, *Trans. Amer. Math. Society* **79**, 1, pp. 35–57.

[20] Bartle, R. (1963). A correction for 'nets and filters in topology', *Amer. Math. Monthly* **70**, 1, pp. 52–53, http://www.jstor.org/stable/2312784.

[21] Bartle, R. (1976). *The Elements of Real Analysis, 2nd Ed.* (John Wiley & Sons, New York).

[22] Beer, G. (1983). On uniform convergence of continuous functions are topological convergence of sets, *Can. Math. Bull.* **26**, pp. 418–424.

[23] Beer, G. (1993). *Topologies on Closed and Closed Convex Sets* (Kluwer Academic Publishers, The Netherlands).

[24] Beer, G., Lechicki, A., Levi, S. and Naimpally, S. (1992). Distance functionals and suprema of hyperspace topologies, *Ann. Mat. Pura. Appl.* **162**, pp. 367–381.

[25] Beer, G. and Levi, S. (2009). Strong uniform continuity, *J. Math. Anal. & Appl.* **350**, 2, pp. 568–589.

[26] Beer, G. and Lucchetti, R. (1993). Weak topologies for the closed subsets of a metrizable space, *Trans. Am. Math. Soc.* **335**, 2, pp. 805–822.

[27] Berri, M. (1963). Minimal topological spaces, *Trans. Amer. Math. Soc.* **108**, pp. 97–105.

[28] Bing, R. (1951). Metrization of topological spaces, *Can. J. Math.* **3**, pp. 175–186.

[29] Blumberg, H. (1920). Hausdorff's grundzüge der mengenlehre, *Bull. Amer. Math. Soc.* **27**, pp. 116–129.

[30] Boardman, R. (1987). *Fossil Invertebrates* (Blackwell Scientific Publications, Oxford, UK).

[31] Bouleau, N. (2006). On the coarsest topology preserving continuity, *Rapport de recherche du CERAMICS* **317**, pp. 1–16.

[32] Bourbaki, N. (1966a). *Elements of Mathematics. General Topology, Part 1* (Hermann & Addison-Wesley, Paris & Reading, MA, U.S.A.), i-vii, 437 pp.

[33] Bourbaki, N. (1966b). *Elements of Mathematics. General Topology, Part 2* (Hermann & Addison-Wesley, Paris & Reading, MA, U.S.A.), i-iv, 363 pp.

[34] Britton, N. (2003). *Essential Mathematical Biology* (Springer, Berlin).

[35] Campbell, P. (1978). The origin of "Zorn's lemma", *Historia Math.* **5**, pp. 77–89.

[36] Cantor, G. (1883). Über unendliche, lineare punktmannichfaltigkeiten, *Mathmatische Annalen* **21**, 4, pp. 545–591.

[37] Cartan, H. (1937a). Filtres et ultafiltres, *Compt. Rend.* **205**, pp. 777–779.

[38] Cartan, H. (1937b). Théorie des filtres, *Compt. Rend.* **205**, pp. 595–598.

[39] Caserta, A., Maio, G. D. and Holá, L. (2010). Arzela's theorem and strong uniform convergence on bornologies, *J. Math. Anal. Appl.* **371**, pp. 384–392.

[40] Cauchy, A. (1821). *Cours d'Analyse, I.^{r}e Partie. Analyse Algébrique* (Gauthier-Villars, Paris).
[41] Chittenden, E. (1917). On the equivalence of ecart and voisinage, *Trans. Amer. Math. Soc.* **18**, 2, pp. 161–166.
[42] Choquet, G. (1947). Sur let notions de filtre et de grille, *Comptes Rendus Acad. Sci. Paris* **224**, pp. 171–173.
[43] Chung, P., Fernandez, M., Li, Y., Mara, M., Morgan, F., Plata, I., Shah, N., Vieira, L. and Wiker, E. (2012). Isoperimetric pentagonal tilings, *Amer. Math. Soc. Notices* **59**, 5, pp. 632–640.
[44] Collins, A., Zomorodian, A., Carlsson, G. and Guibas, L. (2004). A barcode shape descriptor for curve point cloud data, *Computers and Graphics* **28**, 6, pp. 881–894.
[45] Collins, P. and Roscoe, A. (1984). Criteria for metrisability, *Proc. Amer. Math. Soc.* **90**, 4, pp. 631–640.
[46] Darwin, C. (1859). *On The Origin of Species by Means of Natural Selection* (John Murray, Albemarle Street, London).
[47] Davis, A. (1961). Indexed systems of neighborhoods for general topological spaces, *Amer. Math. Monthly* **68**, pp. 886–893.
[48] Dawkins, R. (1986). *Blind Watchmaker* (W.W. Norton, NY).
[49] Di Concilio, A. (2002). Topologizing homeomorphism groups of rim-compact spaces, in *17th summer Conf. Top. & Appl.* (University of Auckland).
[50] Di Concilio, A. (2009). Proximity: A powerful tool in extension theory, functions spaces, hyperspaces, boolean algebras and point-free geometry, in F. Mynard and E. Pearl (eds.), *Beyond Topology, AMS Contemporary Mathematics 486* (Amer. Math. Soc.), pp. 89–114.
[51] Di Concilio, A. and Naimpally, S. (2000). Proximal set-open topologies, *Boll. Unione Mat. Italy* **8**, 1-B, pp. 178–191.
[52] Di Concilio, A. and Naimpally, S. (2006). Proximal convergence, *Monatsh. Math.* **103**, pp. 93–102.
[53] Dieckmann, U., Law, R. and J.A.J. Metz, E. (2000). *The Geometry of Ecological Interactions. Simplifying Spatial Complexity* (Cambridge University Press, Cambridge, UK).
[54] Dieudonné, J. (1944). Une généralization des espaces compacts, *Math. Pure Appl.* **23**, pp. 65–76.
[55] Dini, U. (1878). *Fondamenti per la teorica delle funzioni di variabili realc* (T. Nistri, Piza), german trans., Verlag Harri Deutsch, Frankfurt, 1977.
[56] Domiaty, R. (1980). Life without t_2, *Lecture Notes in Physics, Differential Geometric Methods in Mathematical Physics* **139**, pp. 251–258.
[57] Dugac, P. (1989). Sur la correspondence de borel et le theorem de dirichlet-heine-weierstrass-borel-schoenflies-lebesgue, *Arch. Internat. Hist. Sci.* **39**, 122, pp. 69–110.
[58] Ellison, S. (1951). Microfossils as environment indicators in marine shales, *J Sedimentary Petrology* **21**, 4, pp. 214–225.
[59] Engelking, R. (1989). *General Topology, Revised & completed edition* (Heldermann Verlag, Berlin).

[60] Escher, M. (2001). *M.C. Escher. The Graphic Work* (Taschen GmbH, London, U.K.).
[61] Ewert, J. (1997). On strong form of Arzela convergence, *Int. J. Math. Math. Sci.* **20**, pp. 417–421.
[62] Filippov, V. (1993). The topological structure of solution spaces of ordinary differential equations, *Russian Math. Surveys* **48**, pp. 101–154.
[63] Fox, R. (1997). On topologies for function spaces, *Bull. Amer. Math. Soc.* **51**, 6, pp. 492–432.
[64] Fréchet, M. (1906). Sur quelques points du calcul fonctionnel, *Rend. Circ. Mat. Palermo* **22**, pp. 1–74.
[65] Frink, O. (1964). Compactification and semi-normal space, *Amer. J. Math.* **86**, pp. 602–607.
[66] Gagrat, M. and Naimpally, S. (1971). Proximity approach to extension problems, *Fund. Math.* **71**, pp. 63–76.
[67] Gagrat, M. and Naimpally, S. (1972). Proximity approach to topological problems, in J. Novák (ed.), *General Topology and its Relation to Modern Analysis and Algebra (Toposym 3)* (Academia Pub. House, Czechoslavak Academy of Sciences, Praha), pp. 141–142.
[68] Gagrat, M. and Naimpally, S. (1973a). Proximity approach to semi-metric and developable spaces, *Pacific Journal of Mathematics* **44**, 1, pp. 93–105.
[69] Gagrat, M. and Naimpally, S. (1973b). Wallman compactifications and wallman realcompactifications, *J. Aust. Math. Soc.* **15**, pp. 417–427.
[70] Garg, K. and Naimpally, S. (1971). On some pretopologies associated with a topology, *Proc. Third Prague. Top. Symp.* , pp. 145–146.
[71] G.DiMaio, Meccariello, E. and Naimpally, S. (2006). Duality in function spaces, *Mediterr. J. Math.* **3**, pp. 189–204.
[72] G.DiMaio and Naimpally, S. (1990). Comparison of hypertopologies, *Rend. Istit. Mat. Univ. Trieste* **22**, pp. 140–161.
[73] Ghaemi, N. (2003). Exploring topological psychology, *MCS J of Sci. Math.* **2**, pp. 1–9.
[74] Gillman, L. and Jerison, M. (1960). *Rings of Continuous Functions* (D. Van Nostrand, Princeton, NJ).
[75] Gonzalez, R. and Woods, R. (2008). *Digital Image Processing, 3rd Ed.* (NJ, U.S.A., Pearson Prentice Hall).
[76] Gonzalez, R., Woods, R. and Eddins, S. (2010). *Digital Image Processing Using Matlab®* (NY, U.S.A., Tata McGraw-Hill).
[77] Grimeisen, G. (1960). Gefilterte summation von filtern und iterierte grenzprozesse. i, *Math. Annalen* **141**, 4, pp. 318–342.
[78] Grimeisen, G. (1961). Gefilterte summation von filtern und iterierte grenzprozesse. ii, *Math. Annalen* **144**, 5, pp. 386–417.
[79] Grünbaum, B. and Shephard, G. (1980). Satins and twills: An introduction to the geometry of fabrics, *Math. Mag.* **53**, 3, pp. 139–161.
[80] Hardy, G. (1918). Sir George Stokes and the concept of uniform convergence, *Proc. Cambridge Philos. Soc.* **19**, pp. 148–156.
[81] Hassanien, A., Abraham, A., Peters, J., Schaefer, G. and Henry, C. (2009). Rough sets and near sets in medical imaging: A review, *IEEE Trans.*

Info. Tech. in Biomedicine **13**, 6, pp. 955–968, digital object identifier: 10.1109/TITB.2009.2017017.

[82] Hausdorff, F. (1914). *Grundzüge der Mengenlehre* (Veit and Company, Leipzig), viii + 476 pp.

[83] Hausdorff, F. (1957). *Set Theory* (AMS Chelsea Publishing, Providence, RI), mengenlehre, 1937, trans. by J.R. Aumann, *et al.*

[84] Hawking, S., King, A. and McCarthy, P. (1976). A new topology for curved space-time which incorporates the causal differential and conformal structures, *J. Math. Physics* **17**, pp. 174–181, `http://dx.doi.org/10.1063/1.522874`.

[85] Henrikson, J. (1999). Completeness and total boundedness of the Hausdorff metric, *The MIT Undergraduate Journal of Mathematics* **v1**, pp. 69–79.

[86] Henry, C. (2010). *Near Sets: Theory and Application*, Ph.D. thesis, Manitoba,Canada, supervisor: J.F. Peters.

[87] Henry, C. and Peters, J. F. (2009). Near set evaluation and recognition (near) system, Tech. rep., Computational Intelligence Laboratory, University of Manitoba, uM CI Laboratory Technical Report No. TR-2009-015.

[88] Herrlich, H. (1974a). A concept of nearness, *Gen. Top. & Appl.* **4**, pp. 191–212.

[89] Herrlich, H. (1974b). Topological structures, *Proc. Int. Congress of Mathematicians* , pp. 1–4See Math. Centrum Amsterdam 52, 1974, 59-122.

[90] Herstein, I. (1964). *Topics in Algebra* (Blaisdell Pub. Co., London), viii + 342pp.

[91] Hewitt, E. (1946). On two problems of Urysohn, *Ann. of Math.* **47**, 2, pp. 503–509.

[92] Hobson, E. (1957). *The Theory of Functions of a Real Variable and the Theory of Fourier Series* (Dover Publications, New York).

[93] Hodel, R. (2003). Classical metrisation theorems, in V. Hart, Nagata (ed.), *Encyclopedia of General Topology* (Elsevier), pp. 239–241.

[94] Hoskins, J. and Thomas, R. (1991). The patterns of the isonemal two-colour two-way two-fold fabrics, *Bull. Austral. Math. Soc.* **48**, pp. 33–43.

[95] Hrycay, R. (1970). Generalized connected functions, *Fund. Math.* **68**, pp. 13–17.

[96] Hu, M. (1962). Visual pattern recognition by moment invariants, *IRE Trans. on Info. Theory* **IT-8**, pp. 179–187.

[97] Hunsaker, W. and Naimpally, S. (1974). Hausdorff compactifications as epireflections, *Canad. Math. Bull.* **17**, 5, pp. 675–677.

[98] Isbell, J. (1964). *Uniform Spaces* (Amer. Math. Soc. Surveys 12, Providence, RI).

[99] Ivanova, V. and Ivanov, A. (1959). Contiguity spaces and bicompact extensions of topological spaces (russian), *Dokl. Akad. Nauk SSSR* **127**, pp. 20–22.

[100] James, I. (1976). Characterizations of minimal Hausdorff spaces, *Proc. Amer. Math. Society* **61**, 1, pp. 145–148.

[101] James, I. (1987). *Topological and Uniform Spaces* (Springer Undergraduate Texts in Mathematics, Berlin), 163pp.

[102] Jones, D. (1958). Displacement of microfossils, *J Sedimentary Petrology* **28**, 4, pp. 453–467.

[103] Joseph, J. (1977). On minimal hausdorff spaces, *J. Austral. Math. Society* **23**, A, pp. 476–480.

[104] Keats, J. (1819). Ode on a Grecian Urn, No. 626 in A. Quiller-Couch, The Oxford Book of English Verse: 1250-1900.

[105] Kelley, J. (1942). Hyperspaces of a continuum, *Trans. Amer. Math. Society* **52**, 1, pp. 22–36.

[106] Kelley, J. (1955). *General Topology* (Springer-Verlag, Berlin), xiv + 298 pp.

[107] Khare, M. and Singh, R. (2008). Complete ξ-grills and (L,n)-merotopies, *Fuzzy Sets and Systems* **159**, 5, pp. 620–628.

[108] Khare, M. and Tiwari, S. (2010a). Approach merotopological spaces and their completion, *J. Math. Math. Sci.* **2010**, Article ID 409804, pp. 1–16.

[109] Khare, M. and Tiwari, S. (2010b). Grill determined l-approach merotopological spaces, *Fund. Inform.* **99**, 1, pp. 1–12.

[110] Khare, M. and Tiwari, S. (2012). α^*-uniformities and their order structure, *Afrika Math.* , p. 21pp10.100/s13370-012-0065-y.

[111] Klee, V. and Utz, W. (1954). Some remarks on continuous transformations, *Fund. Math.* **5**, pp. 182–184.

[112] Knaster, B. and Kuratowski, C. (1921). Sur les ensembles connexes, *Fundamenta Mathematicae* **2**, pp. 206–255.

[113] Ko, A., Ridsdale, A., Smith, M., Mostaço-Guidolin, L., Hewko, M., Pegararo, A., Kohlenberg, E., Schattka, B., Shiomi, M., Stolow, A. and Sowa, M. (2010). Multimodal nonlinear optical imaging of atherosclerotic plaque development in myocardial infaraction-prone rabbits, *J. of Biomedical Optics* **15**, 2, pp. 0205011–02050113.

[114] Krebs, J. and Davies, N. (1993). *An Introduction to Behavioural Ecology* (Blackwell Publishing, Oxford, UK).

[115] Kunen, K. and J.E. Vaughan, E. (1984). *Handbook of Set Theoretic Topology* (North Holland, Amsterdam), v + 1273pp.

[116] Kuratowski, C. (1922a). Sur l'operation $\bar{A}$ opération de l'analysis situs, *Fundamenta Mathematicae* **3**, pp. 182–199.

[117] Kuratowski, C. (1922b). Une méthode d'elimination des nombres transfinis des raisonnement mathématiques, *Fundamenta Mathematicae* **3**, pp. 76–108.

[118] Kuratowski, C. (1958). *Topologie I* (Panstwowe Wydawnictwo Naukowe, Warsaw), xiii + 494pp.

[119] Kuratowski, K. (1955). Sur l'espaces des fonctions partielles, *Ann. Di Mat. Pura ed Appl.* **40**, 4, pp. 61–67.

[120] Kuratowski, K. (1962,1972). *Introduction to Set Theory and Topology, 2nd Ed.* (Pergamon Press, Oxford, UK), 349pp.

[121] Le, T., Langohr, I., Locker, M., Sturek, M. and Cheng, J.-X. (2007). Label-free molecular imaging of atherosclerotic lesions using multimodal nonlinear optical microscopy, *Biophysical J.* **12**, 5, pp. 1–23, `coi:10.1117/1.2795437`.

[122] Leader, S. (1959a). On clusters in proximity spaces, *Fundamenta Mathematicae* **47**, pp. 205–213.

[123] Leader, S. (1959b). On completion of proximity spaces by local clusters, *Fundamenta Mathematicae* **48**, pp. 201–216.

[124] Leader, S. (1967). Metrization of proximity spaces, *Proc. Amer. Math. Soc.* **18**, pp. 1084–1088.

[125] Lebesgue, H. (1907). Sur le probleme de Dirichlet, *Rend. del Circ. Mat. di Palermo* **24**, pp. 371–402.

[126] Levin, S. (1976). Population dynamic models in heterogeneous environments, *An. Rev. of Ecology and Systematics* **7**, pp. 287–310.

[127] Lewin, K. (1936). *Principles of Topological Psychology* (McGraw-Hill, NY).

[128] Liao, S. (1993). *Image Analysis By Moments*, Ph.D. thesis, Dept. Elec. Comp. Engg., supervisor: M. Pawlak.

[129] Libby, P. (2006). Atherosclerosis: disease biology affecting the coronary vasculature, *Am. J. Cardiol.* **98**, pp. 3Q–9Q.

[130] Lindeberg, T. (1998). Edge detection and ridge detection with automatic scale selection, *Int. J. of Comp. Vision* **30**, 2, pp. 117–154.

[131] Lindelöf, E. (1903). Sur quelques points de la théorie des ensembles, *Comptes Rendus de l'Académie des Sciences (CRAS) Paris* **137**, pp. 697–700.

[132] Liu, Y.-M. and Luo, M.-K. (1997). *Fuzzy Topology* (World Scientific, Singapore), x + 353pp.

[133] Lodato, M. (1962). *On topologically induced generalized proximity relations, Ph.D. thesis* (Rutgers University).

[134] Lodato, M. (1964). On topologically induced generalized proximity relations i, *Proc. Amer. Math. Soc.* **15**, pp. 417–422.

[135] Lodato, M. (1966). On topologically induced generalized proximity relations ii, *Pacific J. Math.* **17**, pp. 131–135.

[136] Losick, V., Morris, L., Fox, D. and Spradling, A. (2011). Drosophila cell niches: A decade of discovery suggests a unified view of stem cell regulation, *Developmental Cell* **21**, 1, pp. 159–171.

[137] Lowen, R. (1997). *Approach Spaces. The Missing Link in the Topology-Uniformity-Metric Triad* (Oxford University Press, Oxford, UK).

[138] Mandal, D. and Mukherjee, M. (2012). On a type of generalized closed sets, *Bol. Soc. Paran. Mat.* **30**, 1, pp. 65–76.

[139] Marjanovič, M. (1966). Topologies on collections of closed subsets, *Publ. Inst. Math. (Beograd)* **20**, pp. 125–130.

[140] McCoy, R. and Ntantu, I. (1988). *Topological properties of spaces of continuous functions* (Springer).

[141] Mehrtens, H. (1980). *Felix Hausdorff: ein Mathematiker seiner Zeit* (Universität Bonn, Bonn).

[142] Michael, E. (1951). Topologies of spaces of subsets, *Trans. Amer. Math. Soc.* **71**, 1, pp. 152–182.

[143] Michael, E. (1953). A note on paracompact spaces, *Proc. Amer. Math. Soc.* **4**, pp. 831–838.

[144] Minkowski, H. (1909). Raum und zeit, *Phys. Z. Sowjetunion* **10**, pp. 104–115.

[145] Moore, E. and Smith, H. (1922). A general theory of limits, *Amer. J. of Math.* **44**, 2, pp. 102–121.

[146] Moore, R. (1919). On the most general class l of Frechet in which the Heine-Borel-Lebesgue theorem holds true, *Proc. National Acad. of Sciences of the U.S.A.* **5**, 6, pp. 206–210.

[147] Mozzochi, C. (1971a). Symmetric generalized topological structures I, *Pub. Math. Debrecent* **18**, pp. 239–259.

[148] Mozzochi, C. (1971b). Symmetric generalized topological structures II, *Pub. Math. Debrecent* **19**, pp. 129–143.

[149] Mozzochi, C., Gagrat, M. and Naimpally, S. (1976). *Symmetric Generalized Topological Structures* (Exposition Press, Hicksville, NY).

[150] Mozzochi, C. and Naimpally, S. (2009). *Uniformity and proximity* (Allahabad Mathematical Society Lecture Note Series, 2, The Allahabad Math. Soc., Allahabad), xii+153 pp., ISBN 978-81-908159-1-8.

[151] Murray, J., Bradley, H., Craigie, W. and Onions, C. (1933). *The Oxford English Dictionary* (Oxford University Press, Oxford, UK).

[152] Nagata, J. (1950). On a necessary and sufficient condition of metrizability, *J. Inst. Polytech. Osaka City University* **1**, pp. 93–100.

[153] Naimpally, S. (1966). Graph topology for function spaces, *Trans. Amer. Math. Soc.* **123**, 1, pp. 267–273.

[154] Naimpally, S. (1987). Topological convergence and uniform convergence, *Czech. Math. J.* **37**, 4, pp. 608–612.

[155] Naimpally, S. (2002). All hypertopologies are hit-and-miss, *App. Gen. Topology* **3**, pp. 197–199.

[156] Naimpally, S. (2003). Proximity and hyperspace topologies, *Rostok. Math. Kolloq.* , pp. 99–110.

[157] Naimpally, S. (2009a). Near and far. A centennial tribute to Frigyes Riesz, *Siberian Electronic Mathematical Reports* **2**, pp. 144–153.

[158] Naimpally, S. (2009b). *Proximity approach to problems in topology and analysis* (Oldenbourg Verlag, Munich, Germany), 73 pp., ISBN 978-3-486-58917-7.

[159] Naimpally, S. and Pareek, C. (1970). Graph topologies for function spaces ii, *Comment. Math. Prace. Mat.* **13**, pp. 221–231.

[160] Naimpally, S. and Pareek, C. (1991). Generalized metric spaces via annihilators, *Questions and Answers in General Topology* **9**, pp. 203–226.

[161] Naimpally, S., Piotrowski, Z. and Wingler, E. (2006). Plasticity in metric spaces, *J. of Math. Analyis and Applications* **313**, 1, pp. 38–48.

[162] Naimpally, S. and Tikoo, M. (1979). On t_1-compactifications, *Pacific J. of Math.* **84**, 1, pp. 183–190.

[163] Naimpally, S. and Warrack, B. (1970). *Proximity Spaces* (Cambridge University Press, Cambridge,UK), x+128 pp., ISBN 978-0-521-09183-1.

[164] Niemytzki, V. (1927). On the 'third axiom of metric space', *Trans. Amer. Math. Soc.* **29**, pp. 507–513.

[165] Odum, E. and Barrett, G. (1953). *Fundamentals of Ecology, 5th Ed.* (Thomson Brooks-Cole, Australia).

[166] Ohlstein, B., Kai, T., Decotto, E. and Spradling, A. (2004). The stem cell niche: theme and variations, *Curr. Op. Cell Biol.* **16**, 6, pp. 693–699.

[167] Pal, S. and Peters, J. (2010). *Rough Fuzzy Image Analysis. Foundations and Methodologies* (Chapman & Hall/CRC Press Mathematical & Computational Imaging Sciences, London, UK).

[168] Parisi, L. (1960). Symbiotic architecture. Prehending digitality, *Theory, Culture & Society* **29**, 2-3, pp. 346–376.

[169] Pawlak, M. (2006). *Image Analysis by Moments: Reconstruction and Computational Aspects* (Wydawnictwo Politechniki, Wroclaw, Poland).

[170] Pawlak, Z. (1981a). Classification of objects by means of attributes, *Polish Academy of Sciences* **429**.

[171] Pawlak, Z. (1981b). Rough sets, *International J. Comp. Inform. Science* **11**, pp. 341–356.

[172] Pawlak, Z. and Skowron, A. (2007a). Rough sets and boolean reasoning, *Information Sciences* **177**, pp. 41–73.

[173] Pawlak, Z. and Skowron, A. (2007b). Rough sets: Some extensions, *Information Sciences* **177**, pp. 28–40.

[174] Pawlak, Z. and Skowron, A. (2007c). Rudiments of rough sets, *Information Sciences* **177**, pp. 3–27.

[175] Peters, J. (2007a). Near sets. General theory about nearness of objects, *Applied Mathematical Sciences* **1**, 53, pp. 2609–2029.

[176] Peters, J. (2007b). Near sets. special theory about nearness of objects, *Fundam. Inf.* **75**, 1-4, pp. 407–433.

[177] Peters, J. (2011). Sufficiently near sets of neighbourhoods, *Lecture Notes in Artificial Intelligence* **6954**, pp. 17–24.

[178] Peters, J. (2012a). How near are Zdzisław Pawlak's paintings? study of merotopic distances between digital picture regions-of-interest, in A. Skowron and Z. Suraj (eds.), *Rough Sets and Intelligent Systems* (Springer), pp. 89–114.

[179] Peters, J. (2012b). Local near sets. Pattern discovery in proximity spaces, *Math. in Comp. Sci. , to appear.*

[180] Peters, J. and Borkowski, M. (2011). ε-near collections, *Lecture Notes in Artificial Intelligence* **6954**, pp. 533–544.

[181] Peters, J. and Henry, C. (2006). Reinforcement learning with approximation spaces, *Fund. Informaticae* **71**, pp. 2,3.

[182] Peters, J. and Naimpally, S. (2011). Approach spaces for near families, *Gen. Math. Notes* **2**, 1, pp. 159–164.

[183] Peters, J. and Naimpally, S. (2012). Applications of near sets, *Notices of the Amer. Math. Soc.* **59**, 4, pp. 536–542.

[184] Peters, J. and Puzio, L. (2009). Image analysis with anisotropic wavelet-based nearness measures, *International Journal of Computational Intelligence Systems* **2**, 3, pp. 168–183, doi 10.1016/j.ins.2009.04.018.

[185] Peters, J. and Ramanna, S. (2012). Pattern discovery with local near sets, Jornadas Chilendas de Computacion (IEEE Press, Calparaiso, Chile).

[186] Peters, J. and Tiwari, S. (2011). Approach merotopies and near filters. Theory and application, *Gen. Math. Notes* **2**, 2, pp. 1–15.

[187] Peters, J. and Wasilewski, P. (2012). Tolerance spaces: Origins, theoretical aspects and applications, *Information Sciences* **195**, pp. 211–225.

[188] Peters, J. F. (2012c). Nearness of local admissible covers. Theory and application in micropalaeontology, *Fund. Informaticae*, pp. 139–161, *to appear.*

[189] Peters, J. F. and Wasilewski, P. (2009). Foundations of near sets, *Inf. Sci.* **179**, 18, pp. 3091–3109, http://dx.doi.org/10.1016/j.ins.2009.04.018.

[190] Pfeffer, C. (2010). Multimodal nonlinear optical imaging of collagen arrays, *J. of Structural Biol.* **164**, 2, pp. 0205011–02050113.

[191] Polkowski, L. (2002). *Rough Sets. Mathematical Foundations.* (Springer-Verlag, Heidelberg, Germany).

[192] Pompeiu, D. (1905). Sur la continuité des fonctions de variables complexes, *Annales de la Faculté des Sciences de Toulouse* **VII**.

[193] Pompeiu, D. (1907). Sur les fonctions dérivées, *Mathematische Annalen* **63**, pp. 326–332.

[194] Poppe, H. (1967). Über graphentopologien für abbildungsraume i, *Bull. Acad. Pol. Sci. Ser. Sci. Math.* **15**, pp. 71–80.

[195] Poppe, H. (1968). Über graphentopologien für abbildungsraume ii, *Math. Nachr.* **38**, pp. 89–96.

[196] Pottera, M. (1997). *The Design and Analysis of a Computational Model of Cooperative Coevolution*, Ph.D. thesis, Fairfax, Virginia, U.S.A., supervisor: K. DeJong.

[197] Raymond, K., Deugnier, M.-A., Faraldo, M. and Glukhova, M. (2009). Adhesion within the stem cell niches, *Curr. Op. Cell Biol.* **21**, 5, pp. 623–629.

[198] Riemann, G. (1857). Theorie der abel'schen functionen, *J. für die Reine und Angewandte Mathematik* **54**, pp. 101–155.

[199] Riesz, F. (1908). Stetigkeitsbegriff und abstrakte mengenlehre, *m IV Congresso Internazionale dei Matematici* **II**, pp. 18–24.

[200] Rocchi, N. (1936). *Parliamo di Insiemi* (Istituto Didattico Editoriale Felsineo, Bologna).

[201] Rosenfeld, A. (1979). Digital topology, *The Amer. Math. Monthly* **86**, 8, pp. 621–630.

[202] Rosenzweig, M. and MacArthur, R. (1963). Graphical representation and stability conditions of predator-prey interactions, *The Amer. Naturalist* **97**, 895, pp. 209–223.

[203] Rowe, C. (1926). Note on a pair of properties which characterize continuous functions, *Bull. Amer. Math. Soc.* **32**, pp. 285–287.

[204] Roy, B. and Mukherjee, M. (2007). On a type of compactness via grills, *Mat. Bech., UDK 515.122)* **59**, pp. 113–120.

[205] Russ, J. (2007). *The Image Processing Handbook, 5th Ed.* (Taylor & Francis, London).

[206] Sell, G. (1965). On the fundamental theory of ordinary differential equations, *J. Diff. Eqs* **1**, pp. 370–392.

[207] Smirnov, J. (1952). On proximity spaces, *Mat. Sb. (N.S.)* **31**, 73, pp. 543–574, English trans.: Amer. Math. Soc. Transl. Ser. 2, 38, 1964, 5-35.

[208] Smirnov, Y. (1951). On metrization of topological spaces, *Uspekhi. Matem. Nauk* **6**, pp. 100–111.

[209] Smith, D. (1966). Hyperspaces of a uniformizable space, *Proc. Camb. Phil. Soc.* **62**, pp. 25–28.

[210] Sobel, I. (1970). *Camera Models and Perception*, Ph.D. thesis, Stanford, CA.

[211] Sobel, I. (1990). *An Isotropic 3x3 Gradient Operator, Machine Vision for Three-Dimensional Scenes* (Freeman, H., Academic Press, NY), pp. 376-379.

[212] Solomon, C. and Breckon, T. (2011). *Fundamentals of Digital Image Processing. A Practical Approach with Examples in Matlab* (Wiley-Blackwell, Oxford, UK), xiv + 328 pp.

[213] Steen, L. and Seebach, J. (1970). *Counterexamples in Topology* (New York, Springer-Verlag), dover edition, 1995.

[214] Stone, A. (1948). Paracompactness and product spaces, *Bull. Amer. Math. Soc.* **54**, pp. 977–982.

[215] Stone, A. (1960). Sequences of coverings, *Pacific J. Math.* **10**, pp. 689–691.

[216] Strecker, G. and Wattel, E. (1966). On semi-regular and minimal hausdorff embeddings, *Mathematisch Centrum* **ZW1966-007**, pp. 1–8.

[217] Sutherland, W. (1974, 2009). *Introduction to Metric & Topological Spaces* (Oxford University Press, Oxford, UK), 2*nd* Ed., 2008.

[218] Taimanov, A. (1952). On the extension of continuous mappings of topological spaces, *Mat. Sb* **31**, pp. 451–463.

[219] Thomas, R. (2006). Isonemal prefabrics with only parallel axes of symmetry, *arXiv:0895.3791v1 [math.CO]* , pp. 1–26.

[220] Thomas, R. (2009). Isonemal prefabrics with only parallel axes of symmetry, *Discrete Math.* **309**, pp. 2696–2711.

[221] Thomas, R. (2010). Isonemal prefabrics with no axes of symmetry, *Discrete Math.* **310**, pp. 1307–1324.

[222] Thomas, R. (2011). Perfect colourings of isonemal fabrics by thin striping, *Bull. Aust. Math. Soc.* **83**, pp. 63–86.

[223] Thron, W. (1973). Proximity structures and grills, *Mathematische Annalen (Math. Ann.)* **206**, pp. 35–62.

[224] Tietze, H. (1915). Uber funktionen die auf einer abgeschlossenen menge stetig sind, *J. fur de reinund angew. Math.* **145**, pp. 9–14.

[225] Tinbergen, N. (1953). *Social Behaviour in Animals. With Special Reference to Vertebrates* (The Scientific Book Club, London).

[226] Tiwari, S. (2010). *Some Aspects of General Topology and Applications. Approach Merotopic Structures and Applications*, Ph.D. thesis, Allahabad, U.P., India, supervisor: M. Khare.

[227] Tukey, J. (1940). *Convergence and uniformity in Topology* (Princeton Univ. Press, Annals of Math. Studies AM-2, Princeton, NJ), 90pp.

[228] Tychonoff, A. (1935). Uber einen funktionenraum, *Mathematische Annalen (Math. Ann.)* **111**, pp. 726–766.

[229] Čech, E. (1966). *Topological Spaces* (John Wiley & Sons Ltd., London), fr seminar, Brno, 1936-1939; rev. ed. Z. Frolik, M. Katětov.

[230] Urysohn, P. (1925). Über die mächtigkeit der zusammenhängenden mengen, *Mathematische Annalen* **94**, pp. 262–295.
[231] Vaughan, J. (1984). Countably compact and sequentially compact spaces, in K. Kunen and J. Vaughan (eds.), *Handbook of Set-Theoretic Topology* (North Holland, Amsterdam), pp. 571–602.
[232] Čech, E. (1937). On bicompact spaces, *Annals of Math.* **38**, 4, pp. 823–844.
[233] Vietoris, L. (1921). Stetige mengen, *Monatshefte für Mathematik und Physik* **31**, pp. 173–204.
[234] Vietoris, L. (1922). Bereiche zweiter ordnungy, *Monatshefte für Mathematik und Physik* **32**, p. 258Ű280.
[235] Wallace, A. (1941). Separation spaces, *The Annals of Math.* **42**, 3, pp. 687–697.
[236] Wallace, A. (1951). Extensional invariance, *Trans. Amer. Math. Soc.* **70**, 1, pp. 97–102.
[237] Wallman, H. (1938). Lattices and topological spaces, *Annals of Math.* **39**, 1, pp. 112–126.
[238] Wang, L. and Jones, D. (2011). The effects of aging on stem cell behavior in *Drosophila*, *Experimental Gerontology* **46**, 5, pp. 623–629.
[239] Ward, A. (1966). A counter-example in uniformity theory, *Proc. Camb. Phil. Soc.* **62**, pp. 207–208.
[240] Ward, A. (1967). On H-equivalence of uniformities: the Isbell-Smith problem, *Pacific J. Math.* **22**, 1, pp. 189–196.
[241] Watson, R. (2003). Symbiotic architecture. Prehending digitality, *Biosystems* **69**, pp. 187–2009.
[242] Weierstrass, K. (1894-1903). *Mathematische Werke* (Mayer and Müller, Berlin).
[243] Weihrauch, K. (2010). Computable separation in topology, from t_0 to t_2, *J. Univ. Comp. Sci.* **16**, 18, pp. 2733–2753.
[244] Weil, A. (1938). *Sur les espaces à structure uniforme et sur la topologie générale* (Harmann & cie, Actualit¨es scientifique et industrielles, Paris).
[245] White, D. (1968). Functions preserving compactness and connectedness are continuous, *J. London Math. Soc.* **43**, pp. 767–768.
[246] Wijsman, R. (1966). Convergence of sequences of convex sets, cones, and functions, ii, *Trans. Amer. Math. Soc.* **123**, pp. 32–45.
[247] Zernike, F. (1934). Beugungstheorie des schneidenverfahrens und seiner verbesserten form, der pasenkontrastmethode, *Physica* **1**, pp. 689–701.
[248] Zorn, M. (1935). A remark on method in transfinite algebra, *Bull. Amer. Math. Soc.* **41**, 10, pp. 667–670, doi:10.1090/S0002-9904-1935-06166-X.

Author Index

Subject Index

www.ingramcontent.com/pod-product-compliance
Lightning Source LLC
Chambersburg PA
CBHW050546190326
41458CB00007B/1932

* 9 7 8 9 8 1 4 4 0 7 6 5 6 *